Centre for Appl[illegible] ...rary

INFRARED AND RAMAN SPECTROSCOPY

INFRARED AND RAMAN SPECTROSCOPY

PRINCIPLES AND SPECTRAL INTERPRETATION

PETER LARKIN

AMSTERDAM • BOSTON • HEIDELBERG • LONDON
NEW YORK • OXFORD • PARIS • SAN DIEGO
SAN FRANCISCO • SINGAPORE • SYDNEY • TOKYO

Elsevier
225 Wyman Street, Waltham, MA 02451, USA
525 B Street, Suite 1900, San Diego, CA 92101-4495, USA
The Boulevard, Langford Lane, Kidlington, Oxford, OX5 1GB, UK
Radarweg 29, PO Box 211, 1000 AE Amsterdam, The Netherlands

Copyright © 2011 Elsevier Inc. All rights reserved

No part of this publication may be reproduced, stored in a retrieval system or transmitted in any form or by any means electronic, mechanical, photocopying, recording or otherwise without the prior written permission of the publisher

Permissions may be sought directly from Elsevier's Science & Technology Rights Department in Oxford, UK: phone (+44) (0) 1865 843830; fax (+44) (0) 1865 853333; email: permissions@elsevier.com. Alternatively you can submit your request online by visiting the Elsevier web site at http://elsevier.com/locate/permissions, and selecting Obtaining permission to use Elsevier material

Notice
No responsibility is assumed by the publisher for any injury and/or damage to persons or property as a matter of products liability, negligence or otherwise, or from any use or operation of any methods, products, instructions or ideas contained in the material herein

Library of Congress Cataloging-in-Publication Data
Larkin, Peter (Peter J.)
Infrared and raman spectroscopy: principles and spectral interpretation/Peter Larkin.
p. cm.
ISBN: 978-0-12-386984-5 (hardback)
1. Infrared spectroscopy. 2. Raman Spectroscopy. I. Title.
QD96.I5L37 2011
535'.8'42—dc22

2011008524

British Library Cataloguing in Publication Data
A catalogue record for this book is available from the British Library

ISBN: 978-0-12-386984-5

For information on all **Elsevier** publications
visit our web site at elsevierdirect.com

Printed and bound by CPI Group (UK) Ltd, Croydon, CR0 4YY

Transferred to digital print 2012

Working together to grow
libraries in developing countries
www.elsevier.com | www.bookaid.org | www.sabre.org
ELSEVIER BOOK AID International Sabre Foundation

To my wife and family

Contents

Preface

IR and Raman spectroscopy have tremendous potential to solve a wide variety of complex problems. Both techniques are completely complementary providing characteristic fundamental vibrations that are extensively used for the determination and identification of molecular structure. The advent of new technologies has introduced a wide variety of options for implementing IR and Raman spectroscopy into the hands of both the specialist and the non-specialist alike. However, the successful application of both techniques has been limited since the acquisition of high level IR and Raman interpretation skills is not widespread among potential users. The full benefit of IR and Raman spectroscopy cannot be realized without an analyst with basic knowledge of spectral interpretation. This book is a response to the recent rapid growth of the field of vibrational spectroscopy. This has resulted in a corresponding need to educate new users on the value of both IR and Raman spectral interpretation skills.

To begin with, the end user must have a suitable knowledge base of the instrument and its capabilities. Furthermore, he must develop an understanding of the sampling options and limitations, available software tools, and a fundamental understanding of important characteristic group frequencies for both IR and Raman spectroscopy. A critical skill set an analyst may require to solve a wide variety of chemical questions and problems using vibrational spectroscopy is depicted in Figure 1 below.

Selecting the optimal spectroscopic technique to solve complex chemical problems encountered by the analyst requires the user to develop a skill set outlined in Figure 1. A knowledge of spectral interpretation enables the user to select the technique with the most favorable selection

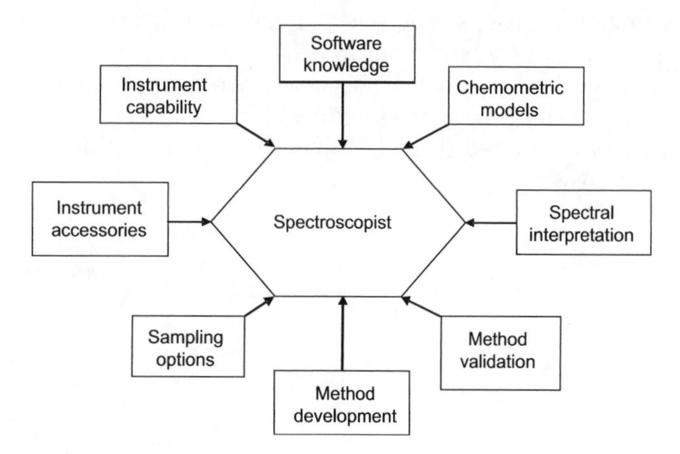

FIGURE 1 Skills required for a successful vibrational spectroscopist. *(Adapted from R.D. McDowall,* Spectroscopy Application Notebook, *February 2010).*

of characteristic group frequencies, optimize the sample options (including accessories if necessary), and use suitable software tools (both instrumental and chemometric), to provide a robust, sensitive analysis that is easily validated.

In this book we provide a suitable level of information to understand instrument capabilities, sample presentation, and selection of various accessories. The main thrust of this text is to develop high level of spectral interpretation skills. A broad understanding of the bands associated with functional groups for both IR and Raman spectroscopy is the basic spectroscopy necessary to make the most of the potential and set realistic expectations for vibrational spectroscopy applications in both academic and industrial settings.

A primary goal of this book has been to fully integrate the use of both IR and Raman spectroscopy as spectral interpretation tools. To this end we have integrated the discussion of IR and Raman group frequencies into different classes of organic groups. This is supplemented with paired generalized IR and Raman spectra, use of numerous tables that are discussed in text, and finally referenced to a selection of fully interpreted IR and Raman spectra. This fully integrated approach to IR and Raman interpretation enables the user to utilize the strengths of both techniques while also recognizing their weaknesses.

We have attempted to provide an integrated approach to the important group frequency of both infrared and Raman spectroscopy. Graphics is used extensively to describe the basic principles of vibrational spectroscopy and the origins of group frequencies. The book includes sections on basic principles in Chapters 1 and 2; instrumentation, sampling methods, and quantitative analysis in Chapter 3; a discussion of important environmental effects in Chapter 4; and a discussion of the origin of group frequencies in Chapter 5. Chapters 4 and 5 provide the essential background to understand the origin of group frequencies in order to assign them in a spectra and to explain why group frequencies may shift. Selected problems are included at the end of some of these chapters to help highlight important points. Chapters 6 and 7 provide a highly detailed description of important characteristic group frequencies and strategies for interpretation of IR and Raman spectra.

Chapter 8 is the culmination of the book and provides 110 fully interpreted paired IR and Raman spectra arranged in groups. The selected compounds are not intended to provide a comprehensive spectral library but rather to provide a significant selection of interpreted examples of functional group frequencies. This resource of interpreted IR and Raman spectra should be used to help verify proposed assignments that the user will encounter. The final chapter is comprised of the paired IR and Raman spectra of 44 different unknown spectra with a corresponding answer key.

Peter Larkin
Connecticut, August 2010

CHAPTER

1

Introduction: Infrared and Raman Spectroscopy

Vibrational spectroscopy includes several different techniques, the most important of which are mid-infrared (IR), near-IR, and Raman spectroscopy. Both mid-IR and Raman spectroscopy provide characteristic fundamental vibrations that are employed for the elucidation of molecular structure and are the topic of this chapter. Near-IR spectroscopy measures the broad overtone and combination bands of some of the fundamental vibrations (only the higher frequency modes) and is an excellent technique for rapid, accurate quantitation. All three techniques have various advantages and disadvantages with respect to instrumentation, sample handling, and applications.

Vibrational spectroscopy is used to study a very wide range of sample types and can be carried out from a simple identification test to an in-depth, full spectrum, qualitative and quantitative analysis. Samples may be examined either in bulk or in microscopic amounts over a wide range of temperatures and physical states (e.g., gases, liquids, latexes, powders, films, fibers, or as a surface or embedded layer). Vibrational spectroscopy has a very broad range of applications and provides solutions to a host of important and challenging analytical problems.

Raman and mid-IR spectroscopy are complementary techniques and usually both are required to completely measure the vibrational modes of a molecule. Although some vibrations may be active in both Raman and IR, these two forms of spectroscopy arise from different processes and different selection rules. In general, Raman spectroscopy is best at symmetric vibrations of non-polar groups while IR spectroscopy is best at the asymmetric vibrations of polar groups. Table 1.1 briefly summarizes some of the differences between the techniques.

Infrared and Raman spectroscopy involve the study of the interaction of radiation with molecular vibrations but differs in the manner in which photon energy is transferred to the molecule by changing its vibrational state. IR spectroscopy measures transitions between molecular vibrational energy levels as a result of the absorption of mid-IR radiation. This interaction between light and matter is a resonance condition involving the electric dipole-mediated transition between vibrational energy levels. Raman spectroscopy is a two-photon inelastic light-scattering event. Here, the incident photon is of much greater energy than the vibrational quantum energy, and loses part of its energy to the molecular vibration with the

TABLE 1.1 Comparison of Raman, Mid-IR and Near-IR Spectroscopy

	Raman	Infrared	Near-IR
Ease of sample preparation	Very simple	Variable	Simple
Liquids	Very simple	Very simple	Very simple
Powders	Very simple	Simple	Simple
Polymers	Very simple*	Simple	Simple
Gases	Simple	Very simple	Simple
Fingerprinting	Excellent	Excellent	Very good
Best vibrations	Symmetric	Asymmetric	Comb/overtone
Group Frequencies	Excellent	Excellent	Fair
Aqueous solutions	Very good	Very difficult	Fair
Quantitative analysis	Good	Good	Excellent
Low frequency modes	Excellent	Difficult	No

* True for FT-Raman at 1064 nm excitation.

remaining energy scattered as a photon with reduced frequency. In the case of Raman spectroscopy, the interaction between light and matter is an off-resonance condition involving the Raman polarizability of the molecule.

The IR and Raman vibrational bands are characterized by their frequency (energy), intensity (polar character or polarizability), and band shape (environment of bonds). Since the vibrational energy levels are unique to each molecule, the IR and Raman spectrum provide a "fingerprint" of a particular molecule. The frequencies of these molecular vibrations depend on the masses of the atoms, their geometric arrangement, and the strength of their chemical bonds. The spectra provide information on molecular structure, dynamics, and environment.

Two different approaches are used for the interpretation of vibrational spectroscopy and elucidation of molecular structure.

1) Use of group theory with mathematical calculations of the forms and frequencies of the molecular vibrations.
2) Use of empirical characteristic frequencies for chemical functional groups.

Many empirical group frequencies have been explained and refined using the mathematical theoretical approach (which also increases reliability).

In general, many identification problems are solved using the empirical approach. Certain functional groups show characteristic vibrations in which only the atoms in that particular group are displaced. Since these vibrations are mechanically independent from the rest of the molecule, these group vibrations will have a characteristic frequency, which remains relatively unchanged regardless of what molecule the group is in. Typically, group frequency

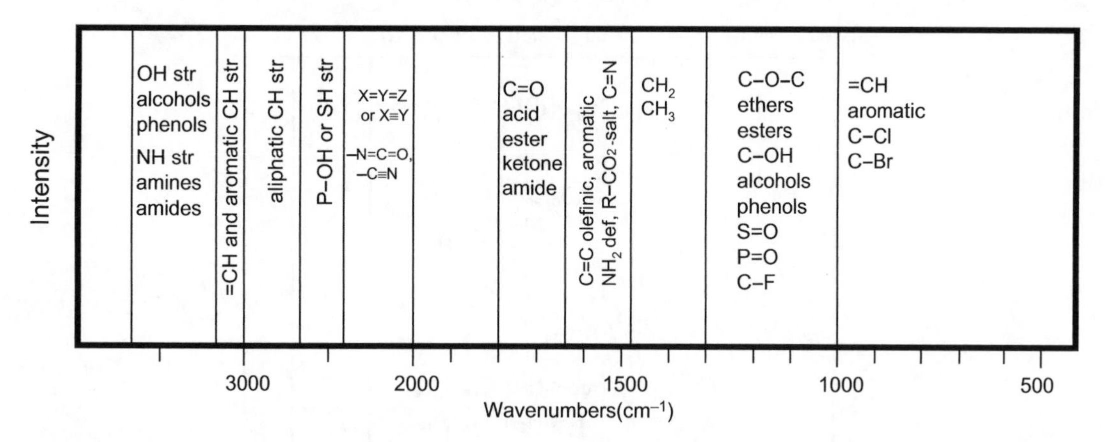

FIGURE 1.1 Regions of the fundamental vibrational spectrum with some characteristic group frequencies.

analysis is used to reveal the presence and absence of various functional groups in the molecule, thereby helping to elucidate the molecular structure.

The vibrational spectrum may be divided into typical regions shown in Fig. 1.1. These regions can be roughly divided as follows:

- X–H stretch (str) highest frequencies (3700–2500 cm^{-1})
- X≡Y stretch, and cumulated double bonds X=Y=Z asymmetric stretch (2500–2000 cm^{-1})
- X=Y stretch (2000–1500 cm^{-1})
- X–H deformation (def) (1500–1000 cm^{-1})
- X–Y stretch (1300–600 cm^{-1})

The above represents vibrations as simple, uncoupled oscillators (with the exception of the cumulated double bonds). The actual vibrations of molecules are often more complex and as we will see later, typically involve coupled vibrations.

1. HISTORICAL PERSPECTIVE: IR AND RAMAN SPECTROSCOPY

IR spectroscopy was the first structural spectroscopic technique widely used by organic chemists. In the 1930s and 1940s both IR and Raman techniques were experimentally challenging with only a few users. However, with conceptual and experimental advances, IR gradually became a more widely used technique. Important early work developing IR spectroscopy occurred in industry as well as academia. Early work using vibrating mechanical molecular models were employed to demonstrate the normal modes of vibration in various molecules.[1, 2] Here the nuclei were represented by steel balls and the interatomic bonds by helical springs. A ball and spring molecular model would be suspended by long threads attached to each ball enabling studies of planar vibrations. The source of oscillation for the ball and spring model was through coupling to an eccentric variable speed motor which enabled studies of the internal vibrations of molecules. When the oscillating frequency matched that of one of the natural frequencies of vibration for the mechanical model

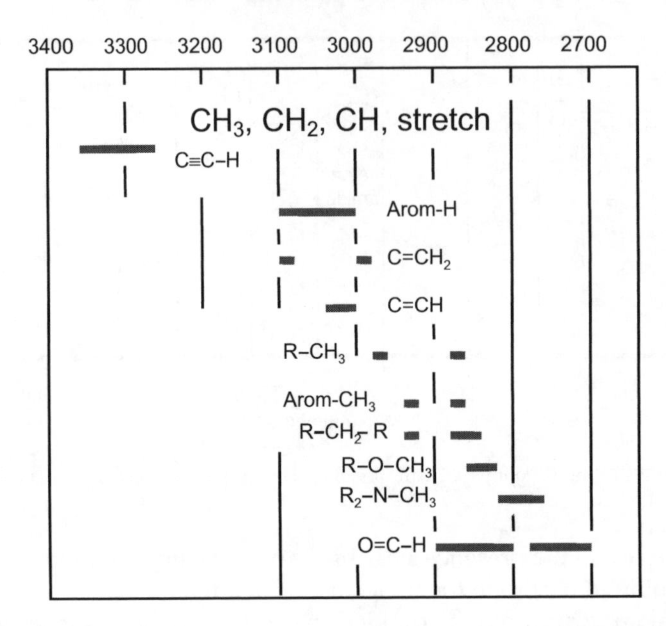

FIGURE 1.2 The correlation chart for CH_3, CH_2, and CH stretch IR bands.

a resonance occurred and the model responded by exhibiting one of the internal vibrations of the molecule (i.e. normal mode).

In the 1940s both Dow Chemical and American Cyanamid companies built their own NaCl prism-based, single beam, meter focal length instruments primarily to study hydrocarbons.

The development of commercially available IR instruments had its start in 1946 with American Cyanamid Stamford laboratories contracting with a small optical company called Perkin–Elmer (PE). The Stamford design produced by PE was a short focal length prism IR spectrometer. With the commercial availability of instrumentation, the technique then benefited from the conceptual idea of a correlation chart of important bands that concisely summarize where various functional groups can be expected to absorb. This introduction of the correlation chart enabled chemists to use the IR spectrum to determine the structure.[3, 4] The explosive growth of IR spectroscopy in the 1950s and 1960s were a result of the development of commercially available instrumentation as well as the conceptual breakthrough of a correlation chart. Appendix shows IR group frequency correlation charts for a variety of important functional groups. Shown in Fig. 1.2 is the correlation chart for CH_3, CH_2, and CH stretch IR bands.

The subsequent development of double beam IR instrumentation and IR correlation charts resulted in widespread use of IR spectroscopy as a structural technique. An extensive user base resulted in a great increase in available IR interpretation tools and the eventual development of FT-IR instrumentation. More recently, Raman spectroscopy has benefited from dramatic improvements in instrumentation and is becoming much more widely used than in the past.

References

1. Kettering, C. F.; Shultz, L. W.; Andrews, D. H. *Phys. Rev.* **1930**, *36*, 531.
2. Colthup, N. B. *J. Chem. Educ.* **1961**, *38* (8), 394–396.
3. Colthup, N. B. *J. Opt. Soc. Am.* **1950**, *40* (6), 397–400.
4. *Infrared Characteristic Group Frequencies* G. Socrates, 2nd ed.; John Wiley: New York, NY, 1994.

CHAPTER

2

Basic Principles

1. ELECTROMAGNETIC RADIATION

All light (including infrared) is classified as electromagnetic radiation and consists of alternating electric and magnetic fields and is described classically by a continuous sinusoidal wave like motion of the electric and magnetic fields. Typically, for IR and Raman spectroscopy we will only consider the electric field and neglect the magnetic field component. Figure 2.1 depicts the electric field amplitude of light as a function of time.

The important parameters are the wavelength (λ, length of 1 wave), frequency (v, number cycles per unit time), and wavenumbers ($\bar{v}$, number of waves per unit length) and are related to one another by the following expression:

$$\bar{\nu} = \frac{\nu}{(c/n)} = \frac{1}{\lambda}$$

where c is the speed of light and n the refractive index of the medium it is passing through. In quantum theory, radiation is emitted from a source in discrete units called photons where the photon frequency, v, and photon energy, E_p, are related by

$$E_p = h\nu$$

where h is Planck's constant (6.6256×10^{-27} erg sec). Photons of specific energy may be absorbed (or emitted) by a molecule resulting in a transfer of energy. In absorption spectroscopy this will result in raising the energy of molecule from ground to a specific excited state

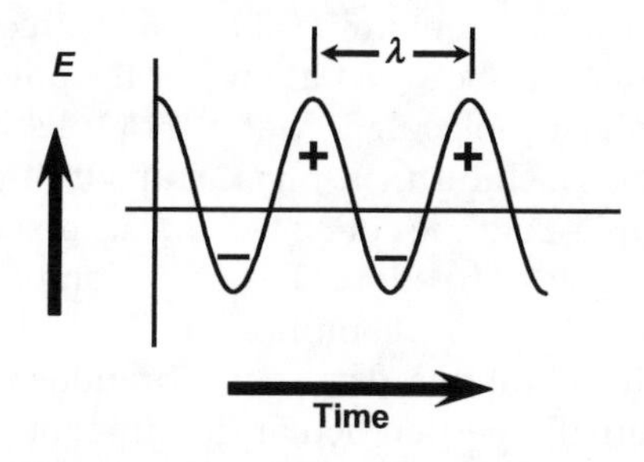

FIGURE 2.1 The amplitude of the electric vector of electromagnetic radiation as a function of time. The wavelength is the distance between two crests.

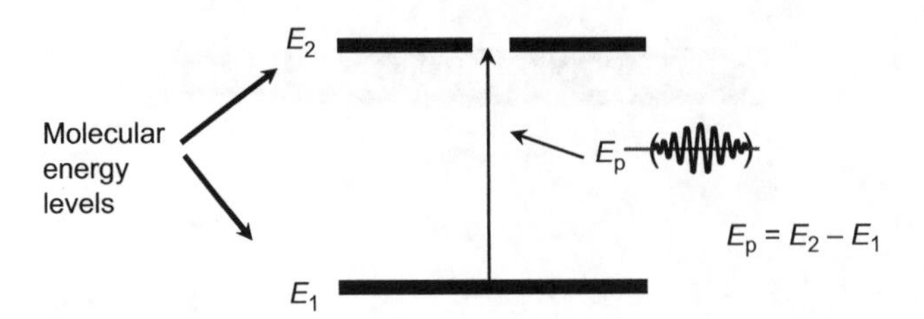

FIGURE 2.2 Absorption of electromagnetic radiation.

as shown in Fig. 2.2. Typically the rotational (E_{rot}), vibrational (E_{vib}), or electronic (E_{el}) energy of molecule is changed by ΔE:

$$\Delta E = E_p = h\nu = hc\bar{\nu}$$

In the absorption of a photon the energy of the molecule increases and ΔE is positive. To a first approximation, the rotational, vibrational, and electronic energies are additive:

$$E_T = E_{el} + E_{vib} + E_{rot}$$

We are concerned with photons of such energy that we consider E_{vib} alone and only for condensed phase measurements. Higher energy light results in electronic transitions (E_{el}) and lower energy light results in rotational transitions (E_{rot}). However, in the gas-state both IR and Raman measurements will include $E_{vib} + E_{rot}$.

2. MOLECULAR MOTION/DEGREES OF FREEDOM

2.1. Internal Degrees of Freedom

The molecular motion that results from characteristic vibrations of molecules is described by the internal degrees of freedom resulting in the well-known $3n - 6$ and $3n - 5$ rule-of-thumb for vibrations for non-linear and linear molecules, respectively. Figure 2.3 shows the fundamental vibrations for the simple water (non-linear) and carbon dioxide (linear) molecules.

The internal degrees of freedom for a molecule define n as the number of atoms in a molecule and define each atom with 3 degrees of freedom of motion in the X, Y, and Z directions resulting in $3n$ degrees of motional freedom. Here, three of these degrees are translation, while three describe rotations. The remaining $3n - 6$ degrees (non-linear molecule) are motions, which change the distance between atoms, or the angle between bonds. A simple example of the $3n - 6$ non-linear molecule is water (H_2O) which has $3(3) - 6 = 3$ degrees of freedom. The three vibrations include an in-phase and out-of-phase stretch and a deformation (bending) vibration. Simple examples of $3n - 5$ linear molecules include H_2, N_2, and O_2 which all have $3(2) - 5 = 1$ degree of freedom. The only vibration for these simple molecules is a simple stretching vibration. The more complicated CO_2 molecule has $3(3) - 5 = 4$ degrees of freedom and therefore four vibrations. The four vibrations include an in-phase and out-of-phase stretch and two mutually perpendicular deformation (bending) vibrations.

The molecular vibrations for water and carbon dioxide as shown in Fig. 2.3 are the normal mode of vibrations. For these vibrations, the Cartesian displacements of each atom in molecule

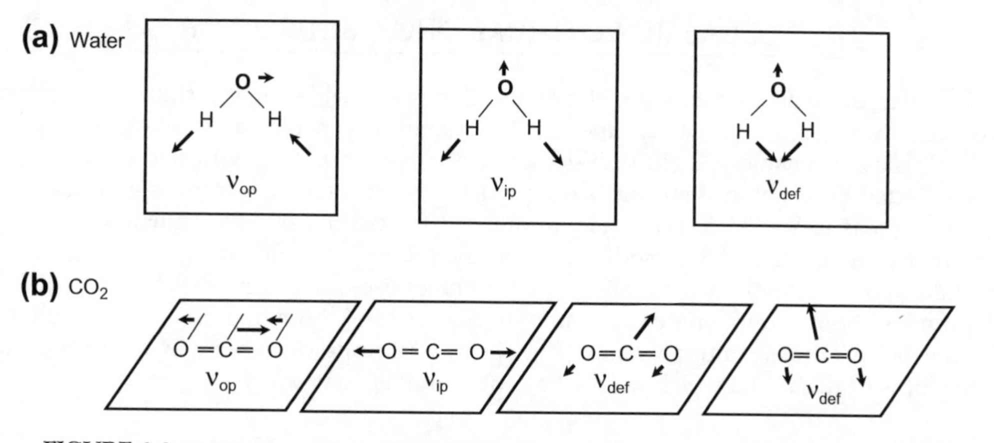

FIGURE 2.3 Molecular motions which change distance between atoms for water and CO_2.

change periodically with the same frequency and go through equilibrium positions simultaneously. The center of the mass does not move and the molecule does not rotate. Thus in the case of harmonic oscillator, the Cartesian coordinate displacements of each atom plotted as a function of time is a sinusoidal wave. The relative vibrational amplitudes may differ in either magnitude or direction. Figure 2.4 shows the normal mode of vibration for a simple diatomic such as HCl and a more complex totally symmetric CH stretch of benzene.

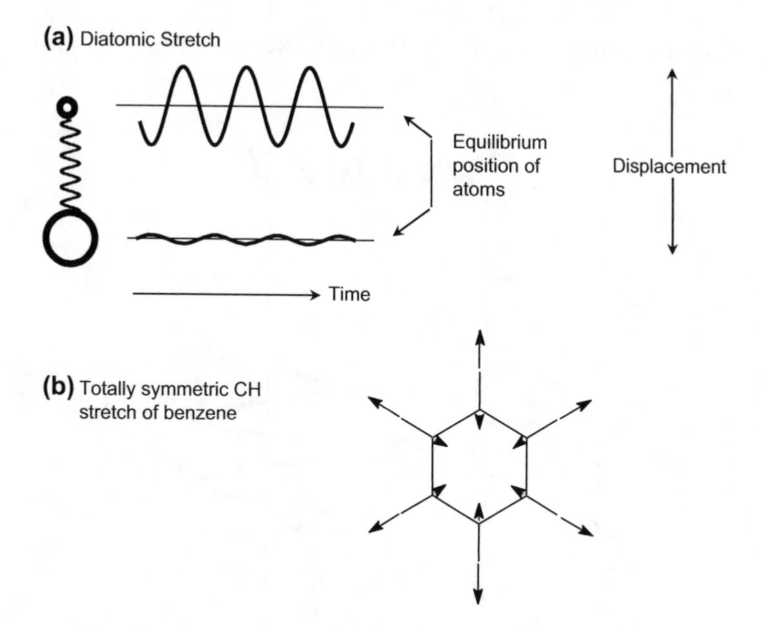

FIGURE 2.4 Normal mode of vibration for a simple diatomic such as HCl (a) and a more complex species such as benzene (b). The displacement versus time is sinusoidal, with equal frequency for all the atoms. The typical Cartesian displacement vectors are shown for the more complicated totally symmetric CH stretch of benzene.

3. CLASSICAL HARMONIC OSCILLATOR

To better understand the molecular vibrations responsible for the characteristic bands observed in infrared and Raman spectra it is useful to consider a simple model derived from classical mechanics.[1] Figure 2.5 depicts a diatomic molecule with two masses m_1 and m_2 connected by a massless spring. The displacement of each mass from equilibrium along the spring axis is X_1 and X_2. The displacement of the two masses as a function of time for a harmonic oscillator varies periodically as a sine (or cosine) function.

In the above diatomic system, although each mass oscillates along the axis with different amplitudes, both atoms share the same frequency and both masses go through their equilibrium positions simultaneously. The observed amplitudes are inversely proportional to the mass of the atoms which keeps the center of mass stationary

$$-\frac{X_1}{X_2} = \frac{m_2}{m_1}$$

The classical vibrational frequency for a diatomic molecule is:

$$\nu = \frac{1}{2\pi}\sqrt{K\left(\frac{1}{m_1} + \frac{1}{m_2}\right)}$$

where K is the force constant in dynes/cm and m_1 and m_2 are the masses in grams and ν is in cycles per second. This expression is also encountered using the reduced mass where

$$\frac{1}{\mu} = \frac{1}{m_1} + \frac{1}{m_2} \quad \text{or} \quad \mu = \frac{m_1 m_2}{m_1 + m_2}$$

In vibrational spectroscopy wavenumber units, $\bar{\nu}$ (waves per unit length) are more typically used

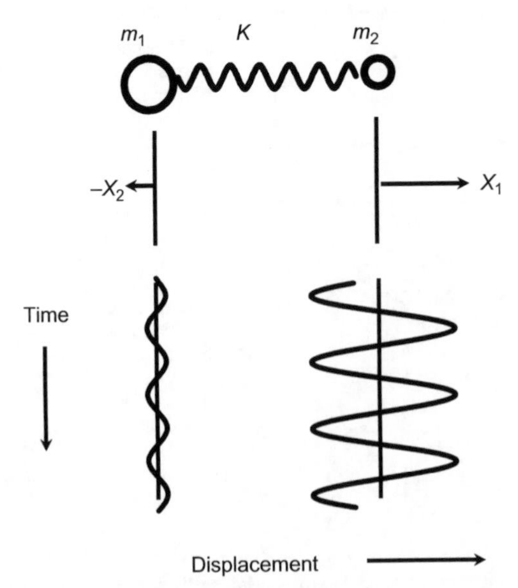

FIGURE 2.5 Motion of a simple diatomic molecule. The spring constant is K, the masses are m_1 and m_2, and X_1 and X_2 are the displacement vectors of each mass from equilibrium where the oscillator is assumed to be harmonic.

$$\bar{\nu} = \frac{1}{2\pi c}\sqrt{K\left(\frac{1}{m_1}+\frac{1}{m_2}\right)}$$

where $\bar{\nu}$ is in waves per centimeter and is sometimes called the frequency in cm^{-1} and c is the speed of light in cm/s.

If the masses are expressed in unified atomic mass units (u) and the force constant is expressed in millidynes/Ångström then:

$$\bar{\nu} = 1303\sqrt{K\left(\frac{1}{m_1}+\frac{1}{m_2}\right)}$$

where $1303 = [N_a \times 10^5)^{1/2}/2\pi c$ and N_a is Avogadro's number (6.023×10^{23} $mole^{-1}$)

This simple expression shows that the observed frequency of a diatomic oscillator is a function of

1. the force constant K, which is a function of the bond energy of a two atom bond (see Table 2.1)
2. the atomic masses of the two atoms involved in the vibration.

TABLE 2.1 Approximate Range of Force Constants for Single, Double, and Triple Bonds

Bond type	K (millidynes/Ångström)
Single	3–6
Double	10–12
Triple	15–18

Table 2.1 shows the approximate range of the force constants for single, double, and triple bonds.

Conversely, knowledge of the masses and frequency allows calculation of a diatomic force constant. For larger molecules the nature of the vibration can be quite complex and for more accurate calculations the harmonic oscillator assumption for a diatomic will not be appropriate.

The general wavenumber regions for various diatomic oscillator groups are shown in Table 2.2, where Z is an atom such as carbon, oxygen, nitrogen, sulfur, and phosphorus.

TABLE 2.2 General Wavenumber Regions for Various Simple Diatonic Oscillator Groups

Diatomic oscillator	Region (cm^{-1})
Z–H	4000–2000
C≡C, C≡N	2300–2000
C=O, C=N, C =C	1950–1550
C–O, C–N, C–C	1300–800
C–Cl	830–560

4. QUANTUM MECHANICAL HARMONIC OSCILLATOR

Vibrational spectroscopy relies heavily on the theoretical insight provided by quantum theory. However, given the numerous excellent texts discussing this topic only a very cursory review is presented here. For a more detailed review of the quantum mechanical principles relevant to vibrational spectroscopy the reader is referred elsewhere.[2-5]

For the classical harmonic oscillation of a diatomic the potential energy (PE) is given by

$$\mathrm{PE} = \frac{1}{2} KX^2$$

A plot of the potential energy of this diatomic system as a function of the distance, X between the masses, is thus a parabola that is symmetric about the equilibrium internuclear distance, X_e. Here X_e is at the energy minimum and the force constant, K is a measure of the curvature of the potential well near X_e.

From quantum mechanics we know that molecules can only exist in quantized energy states. Thus, vibrational energy is not continuously variable but rather can only have certain discrete values. Under certain conditions a molecule can transit from one energy state to another ($\Delta\upsilon = \pm 1$) which is what is probed by spectroscopy.

Figure 2.6 shows the vibrational levels in a potential energy diagram for the quantum mechanical harmonic oscillator. In the case of the harmonic potential these states are equidistant and have energy levels E given by

$$E_i = \left(\upsilon_i + \frac{1}{2}\right) h\upsilon \quad \upsilon_i = 0, 1, 2\ldots$$

Here, ν is the classical vibrational frequency of the oscillator and υ is a quantum number which can have only integer values. This can only change by $\Delta\upsilon = \pm 1$ in a harmonic oscillator model. The so-called zero point energy occurs when $\upsilon = 0$ where $E = ½\, h\nu$ and this vibrational energy cannot be removed from the molecule.

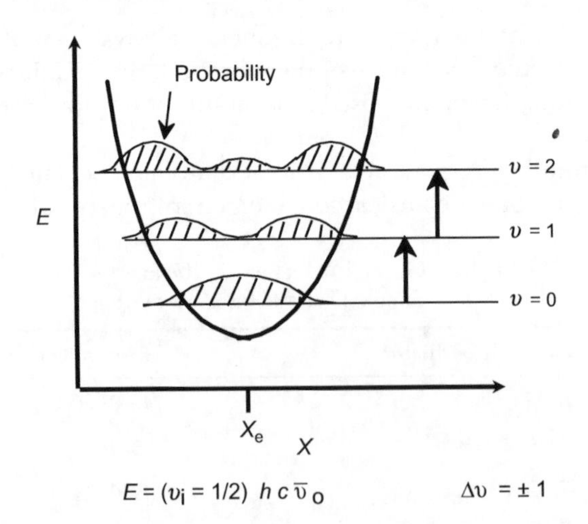

FIGURE 2.6 Potential energy, E, versus internuclear distance, X, for a diatomic harmonic oscillator.

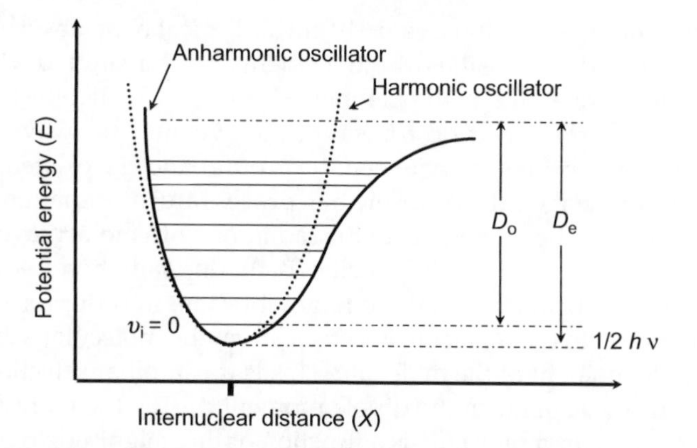

FIGURE 2.7 The potential energy diagram comparison of the anharmonic and the harmonic oscillator. Transitions originate from the $\upsilon = 0$ level, and D_o is the energy necessary to break the bond.

Figure 2.6 shows the curved potential wells for a harmonic oscillator with the probability functions for the internuclear distance X, within each energy level. These must be expressed as a probability of finding a particle at a given position since by quantum mechanics we cannot be certain of the position of the mass during the vibration (a consequence of Heisenberg's uncertainty principle).

Although we have only considered a harmonic oscillator, a more realistic approach is to introduce anharmonicity. Anharmonicity results if the change in the dipole moment is not linearly proportional to the nuclear displacement coordinate. Figure 2.7 shows the potential energy level diagram for a diatomic harmonic and anharmonic oscillator. Some of the features introduced by an anharmonic oscillator include the following.

The anharmonic oscillator provides a more realistic model where the deviation from harmonic oscillation becomes greater as the vibrational quantum number increases. The separation between adjacent levels becomes smaller at higher vibrational levels until finally the dissociation limit is reached. In the case of the harmonic oscillator only transitions to adjacent levels or so-called fundamental transitions are allowed (i.e., $\Delta \upsilon = \pm 1$) while for the anharmonic oscillator, overtones ($\Delta \upsilon = \pm 2$) and combination bands can also result. Transitions to higher vibrational states are far less probable than the fundamentals and are of much weaker intensity. The energy term corrected for anharmonicity is

$$E_\upsilon = h\nu_e\left(\upsilon + \frac{1}{2}\right) - h\chi_e\nu_e\left(\upsilon + \frac{1}{2}\right)^2$$

where $\chi_e\nu_e$ defines the magnitude of the anharmonicity.

5. IR ABSORPTION PROCESS

The typical IR spectrometer broad band source emits all IR frequencies of interest simultaneously where the near-IR region is 14,000–4000 cm^{-1}, the mid-IR region is

4000–400 cm^{-1}, and the far-IR region is 400–10 cm^{-1}. Typical of an absorption spectroscopy, the relationship between the intensities of the incident and transmitted IR radiation and the analyte concentration is governed by the Lambert–Beer law. The IR spectrum is obtained by plotting the intensity (absorbance or transmittance) versus the wavenumber, which is proportional to the energy difference between the ground and the excited vibrational states.

Two important components to the IR absorption process are the radiation frequency and the molecular dipole moment. The interaction of the radiation with molecules can be described in terms of a resonance condition where the specific oscillating radiation frequency matches the natural frequency of a particular normal mode of vibration. In order for energy to be transferred from the IR photon to the molecule via absorption, the molecular vibration must cause a change in the dipole moment of the molecule. This is the familiar selection rule for IR spectroscopy, which requires a change in the dipole moment during the vibration to be IR active.

The dipole moment, μ, for a molecule is a function of the magnitude of the atomic charges (e_i) and their positions (r_i)

$$\mu = \sum e_i r_i$$

The dipole moments of uncharged molecules derive from partial charges on the atoms, which can be determined from molecular orbital calculations. As a simple approximation, the partial charges can be estimated by comparison of the electronegativities of the atoms. Homonuclear diatomic molecules such as H_2, N_2, and O_2 have *no* dipole moment and are IR inactive (but Raman active) while heteronuclear diatomic molecules such as HCl, NO, and CO do have dipole moments and have IR active vibrations.

The IR absorption process involves absorption of energy by the molecule if the vibration causes a change in the dipole moment, resulting in a change in the vibrational energy level. Figure 2.8 shows the oscillating electric field of the IR radiation generates forces on the

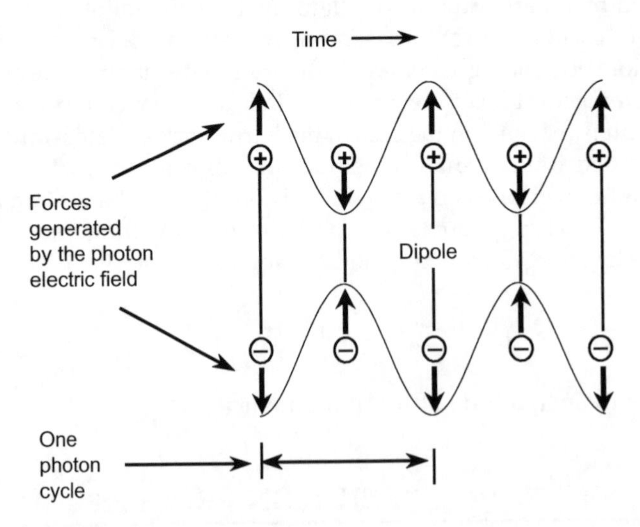

FIGURE 2.8 The oscillating electric field of the photon generates oscillating, oppositely directed forces on the positive and negative charges of the molecular dipole. The dipole spacing oscillates with the same frequency as the incident photon.

molecular dipole where the oscillating electric field drives the oscillation of the molecular dipole moment and alternately increases and decreases the dipole spacing.

Here, the electric field is considered to be uniform over the whole molecule since λ is much greater than the size of most molecules. In terms of quantum mechanics, the IR absorption is an electric dipole operator mediated transition where the change in the dipole moment, μ, with respect to a change in the vibrational amplitude, Q, is greater than zero.

$$\left(\frac{\partial \mu}{\partial Q}\right)_0 \neq 0$$

The measured IR band intensity is proportional to the square of the change in the dipole moment.

6. THE RAMAN SCATTERING PROCESS

Light scattering phenomena may be classically described in terms of electromagnetic (EM) radiation produced by oscillating dipoles induced in the molecule by the EM fields of the incident radiation. The light scattered photons include mostly the dominant Rayleigh and the very small amount of Raman scattered light.[6] The induced dipole moment occurs as a result of the molecular polarizability α, where the polarizability is the deformability of the electron cloud about the molecule by an external electric field.

Figure 2.9 shows the response of a non-polar diatomic placed in an oscillating electric field. Here we represent the static electric field by the plates of a charged capacitor. The negatively charged plate attracts the nuclei, while the positively charged plate attracts the least tightly

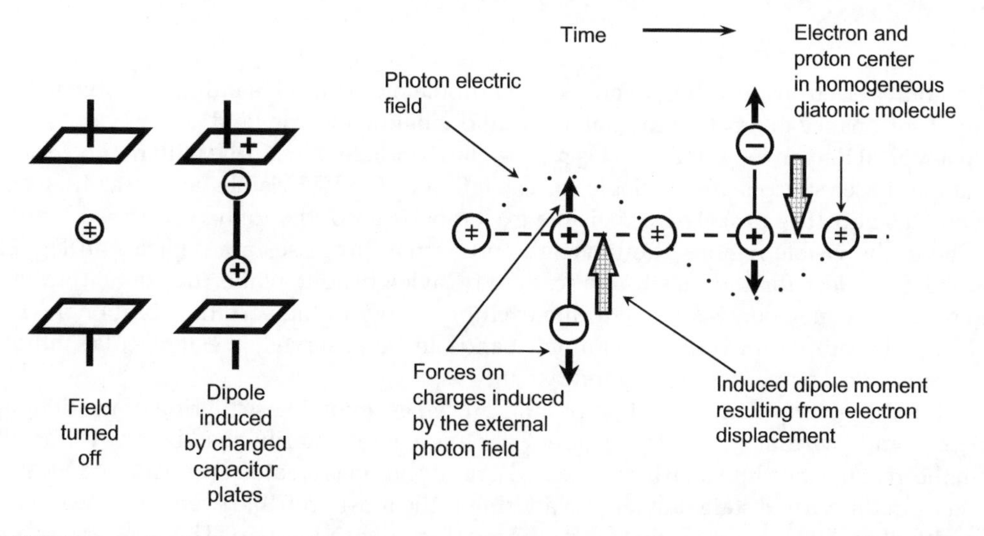

FIGURE 2.9 Induced dipole moment of a homonuclear diatomic originating from the oscillating electric field of the incident radiation. The field relative to the proton center displaces the electron center. The charged plates of a capacitor, which induces a dipole moment in the polarizable electron cloud, can represent the electric field.

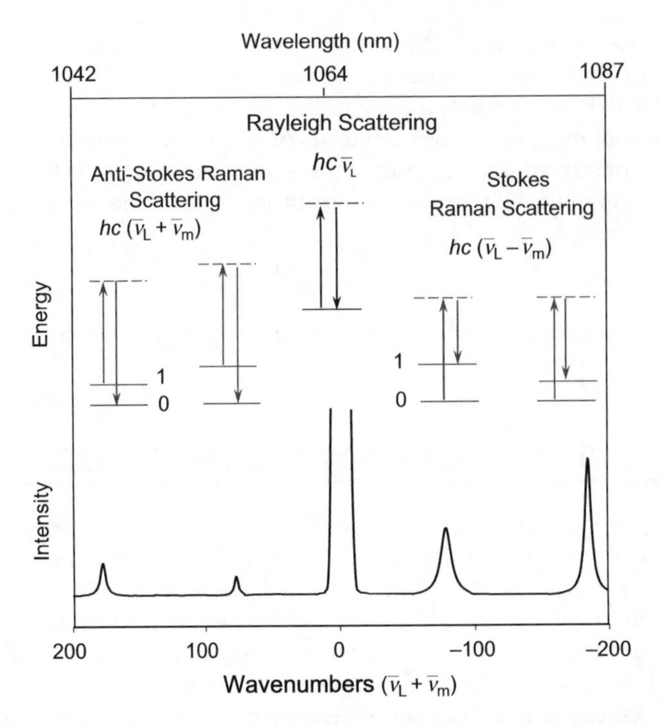

FIGURE 2.10 Schematic illustration of Rayleigh scattering as well as Stokes and anti-Stokes Raman scattering. The laser excitation frequency (ν_L) is represented by the upward arrows and is much higher in energy than the molecular vibrations. The frequency of the scattered photon (downward arrows) is unchanged in Rayleigh scattering but is of either lower or higher frequency in Raman scattering. The dashed lines indicate the "virtual state."

bound outer electrons resulting in an induced dipole moment. This induced dipole moment is an off-resonance interaction mediated by an oscillating electric field.

In a typical Raman experiment, a laser is used to irradiate the sample with monochromatic radiation. Laser sources are available for excitation in the UV, visible, and near-IR spectral region (785 and 1064 nm). Thus, if visible excitation is used, the Raman scattered light will also be in the visible region. The Rayleigh and Raman processes are depicted in Fig. 2.10. No energy is lost for the elastically scattered Rayleigh light while the Raman scattered photons lose some energy relative to the exciting energy to the specific vibrational coordinates of the sample. In order for Raman bands to be observed, the molecular vibration must cause a change in the polarizability.

Both Rayleigh and Raman are two photon processes involving scattering of incident light ($hc\bar{\nu}_L$), from a "virtual state." The incident photon is momentarily absorbed by a transition from the ground state into a virtual state and a new photon is created and scattered by a transition from this virtual state. Rayleigh scattering is the most probable event and the scattered intensity is ca. 10^{-3} less than that of the original incident radiation. This scattered photon results from a transition from the virtual state back to the ground state and is an elastic scattering of a photon resulting in no change in energy (i.e., occurs at the laser frequency).

Raman scattering is far less probable than Rayleigh scattering with an observed intensity that is ca. 10^{-6} that of the incident light for strong Raman scattering. This scattered photon results from a transition from the virtual state to the first excited state of the molecular vibration. This is described as an inelastic collision between photon and molecule, since the molecule acquires different vibrational energy ($\bar{\nu}_m$) and the scattered photon now has different energy and frequency.

As shown in Fig. 2.10 two types of Raman scattering exist: Stokes and anti-Stokes. Molecules initially in the ground vibrational state give rise to Stokes Raman scattering $hc(\bar{\nu}_L - \bar{\nu}_m)$while molecules initially in vibrational excited state give rise to anti-Stokes Raman scattering, $hc(\bar{\nu}_L + \bar{\nu}_m)$. The intensity ratio of the Stokes relative to the anti-Stokes Raman bands is governed by the absolute temperature of the sample, and the energy difference between the ground and excited vibrational states. At thermal equilibrium Boltzmann's law describes the ratio of Stokes relative to anti-Stokes Raman lines. The Stokes Raman lines are much more intense than anti-Stokes since at ambient temperature most molecules are found in the ground state.

The intensity of the Raman scattered radiation I_R is given by:

$$I_R \propto \nu^4 I_o N \left(\frac{\partial \alpha}{\partial Q} \right)^2$$

where I_o is the incident laser intensity, N is the number of scattering molecules in a given state, ν is the frequency of the exciting laser, α is the polarizability of the molecules, and Q is the vibrational amplitude.

The above expression indicates that the Raman signal has several important parameters for Raman spectroscopy. First, since the signal is concentration dependent, quantitation is possible. Secondly, using shorter wavelength excitation or increasing the laser flux power density can increase the Raman intensity. Lastly, only molecular vibrations which cause a change in polarizability are Raman active. Here the change in the polarizability with respect to a change in the vibrational amplitute, Q, is greater than zero.

$$\left(\frac{\partial \alpha}{\partial Q} \right) \neq 0$$

The Raman intensity is proportional to the square of the above quantity.

7. CLASSICAL DESCRIPTION OF THE RAMAN EFFECT

The most basic description of Raman spectroscopy describes the nature of the interaction of an oscillating electric field using classical arguments.[6] Figure 2.11 schematically represents this basic mathematical description of the Raman effect.

As discussed above, the electromagnetic field will perturb the charged particles of the molecule resulting in an induced dipole moment:

$$\mu = \alpha E$$

where α is the polarizability, E is the incident electric field, and μ is the induced dipole moment. Both E and α can vary with time. The electric field of the radiation is oscillating

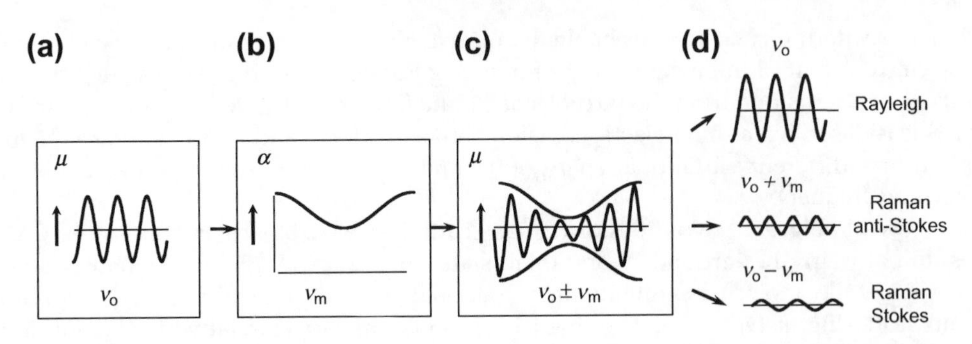

FIGURE 2.11 Schematic representing Rayleigh and Raman scattering. In (a) the incident radiation makes the induced dipole moment of the molecule oscillate at the photon frequency. In (b) the molecular vibration can change the polarizability, α, which changes the amplitude of the dipole moment oscillation. The result as shown in (c) is an amplitude modulated dipole moment oscillation. The image (d) shows the components with steady amplitudes which can emit electromagnetic radiation.

as a function of time at a frequency ν_0, which can induce an oscillation of the dipole moment μ of the molecule at this same frequency, as shown in Fig. 2.11a. The polarizability α of the molecule has a certain magnitude whose value can vary slightly with time at the much slower molecular vibrational frequency ν_m, as shown in Fig. 2.11b. The result is seen in Fig. 2.11c, which depicts an amplitude modulation of the dipole moment oscillation of the molecule. This type of modulated wave can be resolved mathematically into three steady amplitude components with frequencies ν_0, $\nu_0 + \nu_m$, and $\nu_0 - \nu_m$ as shown in Fig. 2.11d. These dipole moment oscillations of the molecule can emit scattered radiation with these same frequencies called Rayleigh, Raman anti-Stokes, and Raman Stokes frequencies. If a molecular vibration did not cause a variation in the polarizability, then there would be no amplitude modulation of the dipole moment oscillation and there would be no Raman Stokes or anti-Stokes emission.

8. SYMMETRY: IR AND RAMAN ACTIVE VIBRATIONS

The symmetry of a molecule, or the lack of it, will define what vibrations are Raman and IR active.[5] In general, symmetric or in-phase vibrations and non-polar groups are most easily studied by Raman while asymmetric or out-of-phase vibrations and polar groups are most easily studied by IR. The classification of a molecule by its symmetry enables understanding of the relationship between the molecular structure and the vibrational spectrum. Symmetry elements include planes of symmetry, axes of symmetry, and a center of symmetry.

Group Theory is the mathematical discipline, which applies symmetry concepts to vibrational spectroscopy and predicts which vibrations will be IR and Raman active.[1,5] The symmetry elements possessed by the molecule allow it to be classified by a point group and vibrational analysis can be applied to individual molecules. A thorough discussion of Group Theory is beyond the scope of this work and interested readers should examine texts dedicated to this topic.[7]

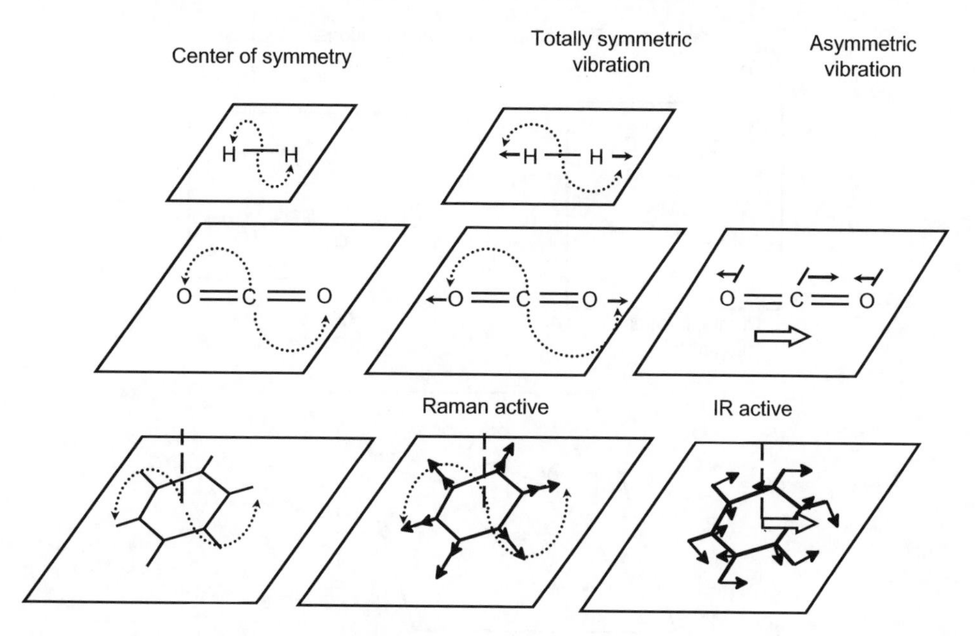

FIGURE 2.12 The center of symmetry for H_2, CO_2, and benzene. The Raman active symmetric stretching vibrations above are symmetric with respect to the center of symmetry. Some IR active asymmetric stretching vibrations are also shown.

For small molecules, the IR and Raman activities may often be determined by a simple inspection of the form of the vibrations. For molecules that have a center of symmetry, the rule of mutual exclusion states that no vibration can be active in both the IR and Raman spectra. For such highly symmetrical molecules vibrations which are Raman active are IR inactive and vice versa and some vibrations may be both IR and Raman inactive.

Figure 2.12 shows some examples of molecules with this important symmetry element, the center of symmetry. To define a center of symmetry simply start at any atom, go in a straight line through the center and an equal distance beyond to find another, identical atom. In such cases the molecule has no permanent dipole moment. Examples shown below include H_2, CO_2, and benzene and the rule of mutual exclusion holds.

In a molecule with a center of symmetry, vibrations that retain the center of symmetry are IR inactive and may be Raman active. Such vibrations, as shown in Fig. 2.12, generate a change in the polarizability during the vibration but no change in a dipole moment. Conversely, vibrations that do not retain the center of symmetry are Raman inactive, but may be IR active since a change in the dipole moment may occur.

For molecules without a center of symmetry, some vibrations can be active in both the IR and Raman spectra.

Molecules that do not have a center of symmetry may have other suitable symmetry elements so that some vibrations will be active only in Raman or only in the IR. Good examples of this are the in-phase stretches of inorganic nitrate and sulfate shown in Fig. 2.13. These are Raman active and IR inactive. Here, neither molecule has a center of symmetry but the negative oxygen atoms move radially simultaneously resulting in no dipole moment change.

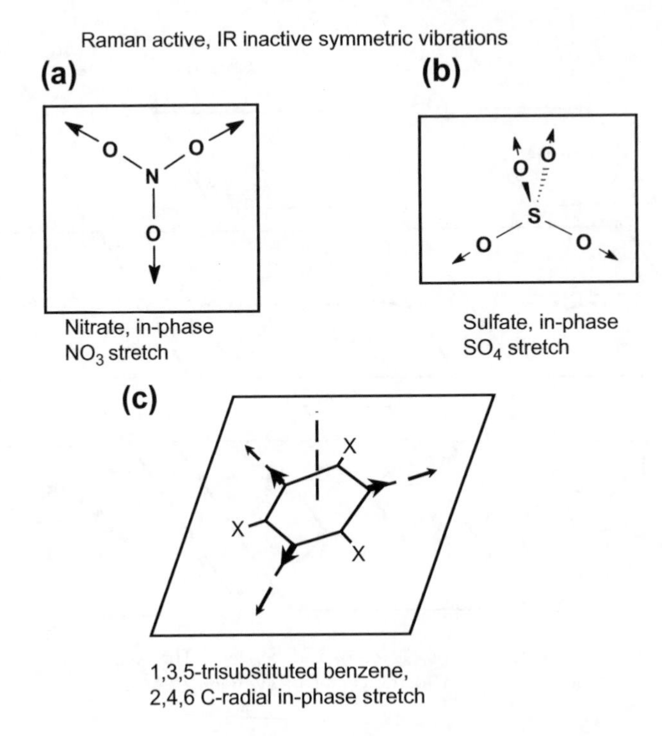

FIGURE 2.13 Three different molecules, nitrate, sulfate, and 1,3,5-trisubstituted benzene molecules that do not have a center of symmetry. The in-phase stretching vibrations of all three result in Raman active, but IR inactive vibrations.

Another example is the 1,3,5-trisubstituted benzene where the C-Radial in-phase stretch is Raman active and IR inactive.

In Figure 2.14 some additional symmetry operations are shown, other than that for a center of symmetry for an XY_2 molecule such as water. These include those for a plane of symmetry, a two-fold rotational axis of symmetry, and an identity operation (needed for group theory) which makes no change. If a molecule is symmetrical with respect to a given symmetry element, the symmetry operation will not make any discernible change from the original configuration. As shown in Fig. 2.14, such symmetry operations are equivalent to renumbering the symmetrically related hydrogen (Y) atoms.

Figure 2.15 shows the Cartesian displacement vectors (arrows) of the vibrational modes Q_1, Q_2, and Q_3 of the bent triatomic XY_2 molecule (such as water), and shows how they are modified by the symmetry operations C_2, σ_v, and σ'_v. For non-degenerate modes of vibration such as these, the displacement vectors in the first column (the identity column, *I*) are multiplied by either (+1) or (−1) as shown to give the forms in the other three columns. Multiplication by (+1) does not change the original form so the resulting form is said to be symmetrical with respect to that symmetry operation. Multiplication by (−1) reverses all the vectors of the original form and the resulting form is said to be anti-symmetrical with respect to that symmetry element. As seen in Fig. 2.15, Q_1 and Q_2 are both totally

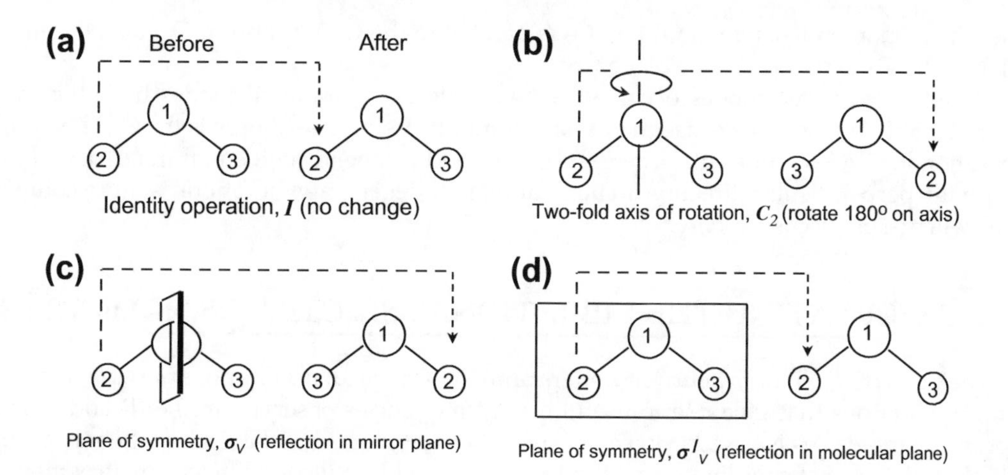

FIGURE 2.14 Symmetry operations for an XY_2 bent molecule such as water in the equilibrium configuration.

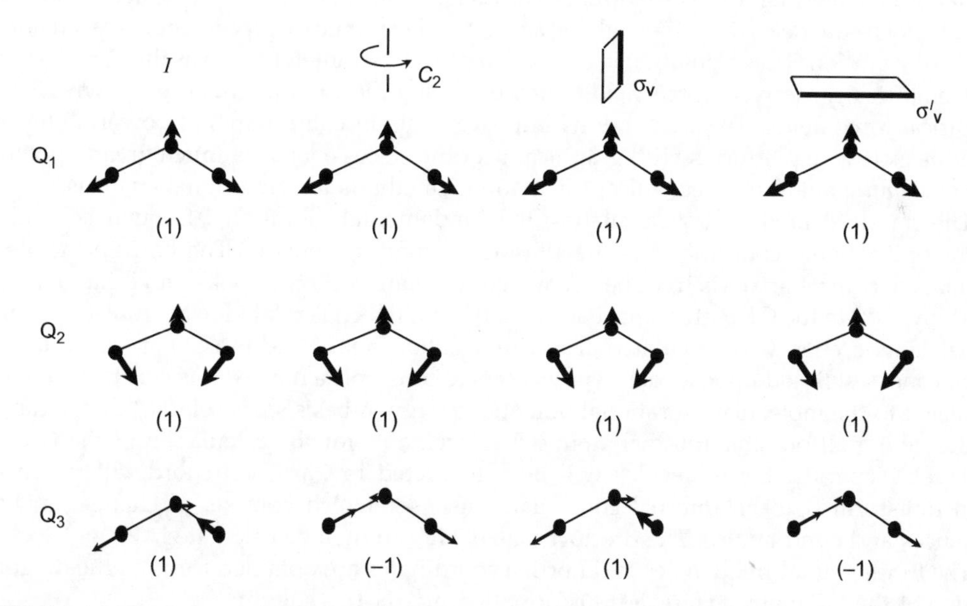

FIGURE 2.15 The bent symmetrical XY_2 molecule such as H_2O performing the three fundamental modes Q_1, Q_2, and Q_3. The vectors in column one (identity I) are transformed by the C_2, σ_v, and σ'_v operations into the forms in the remaining columns, where the vectors are like those in column one multiplied by (+1) symmetrical or (−1) anti-symmetrical.

symmetric modes (i.e., symmetric to all symmetry operations), whereas Q_3 is symmetric with respect to the σ'_v operation but anti-symmetric with respect to the C_2 and σ_v operations. The transformation numbers (+1 and −1) are used in group theory to characterize the symmetries of non-degenerate vibrational modes. From these symmetries one can deduce that Q_1, Q_2, and Q_3 are all active in both the IR and Raman spectra. In addition, the dipole moment change in Q_1 and Q_2 is parallel to the C_2 axis and in Q_3 it is perpendicular to the C_2 axis and the σ_v, plane.

Doubly degenerate modes occur when two different vibrational modes have the same vibrational frequency as a consequence of symmetry. A simple example is the C–H bending vibration in Cl_3C–H molecule where the C–H bond can bend with equal frequency in two mutually perpendicular directions. The treatment of degenerate vibrations is more complex and will not be discussed here.

9. CALCULATING THE VIBRATIONAL SPECTRA OF MOLECULES

The basis of much of the current understanding of molecular vibrations and the localized group vibrations that give rise to useful group frequencies observed in the IR and Raman spectra of molecules is based upon extensive historical work calculating vibrational spectra. Historically, normal coordinate analysis first developed by Wilson with a GF matrix method and using empirical molecular force fields has played a vital role in making precise assignments of observed bands. The normal coordinate computation involves calculation of the vibrational frequencies (i.e., eigenvalues) as well as the atomic displacements for each normal mode of vibration. The calculation itself uses structural parameters such as the atomic masses and empirically derived force fields. However, significant limitations exist when using empirical force fields. The tremendous improvements in computational power along with multiple software platforms with graphical user interfaces enables a much greater potential use of *ab initio* quantum mechanical calculational methods for vibrational analysis.

The standard method for calculating the fundamental vibrational frequencies and the normal vibrational coordinates is the Wilson GF matrix method.[8] The basic principles of normal coordinate analysis have been covered in detail in classic books on vibrational spectroscopy.[1,4,5,8] In the GF matrix approach a matrix, **G**, which is related to the molecular vibrational kinetic energy is calculated from information about the molecular geometry and atomic masses. Based upon a complete set of force constants a matrix, **F**, is constructed which is related to the molecular vibrational potential energy. A basis set is selected that is capable of describing all possible internal atomic displacements for the calculation of the **G** and **F** matrices. Typically, the molecules will be constructed in Cartesian coordinate space and then transformed to an internal coordinate basis set which consists of changes in bond distances and bond angles. The product matrix **GF** can then be calculated.

The fundamental frequencies and normal coordinates are obtained through the diagonalization of the **GF** matrix. Here, a transformation matrix **L** is sought:

$$L^{-1}\,GFL = \Lambda$$

Here, Λ is a diagonal matrix whose diagonal elements are λ_i's defined as:

$$\lambda_i = 4\pi^2 c^2\, v_i^{-2}$$

Where the frequency in cm^{-1} of the *i*th normal mode is $\bar{v}_i^2$. For the previous equation, it is the matrix $\mathbf{L}^{-1}$ which transforms the internal coordinates, **R**, into the normal coordinates, **Q**, as:

$$Q = L^{-1}R$$

In this equation, the column matrix of internal coordinates is **R** and **Q** is a column matrix that contains normal coordinates as a linear combination of internal coordinates.

In the case of vibrational spectroscopy, the polyatomic molecule is considered to oscillate with a small amplitude about the equilibrium position and the potential energy expression is expanded in a Taylor series and takes the form:

$$V = V_0 + \sum_{i=1}^{3N} \left(\frac{\partial V}{\partial_{q_i}} \right)_e dq_i + \frac{1}{2} \sum_{i=1}^{3N} \sum_{j=1}^{3N} \left(\frac{\partial^2 V}{\partial_{q_i} \partial_{q_j}} \right)_e dq_i dq_j + \cdots$$

expressed in internal coordinates, $q_{i,}$, which are directly connected to the internal bond lengths and angles. The above expression is simplified:

1. The first term $V_o = 0$ since the vibrational energy is chosen as vibrating atoms about the equilibrium position.
2. At the minimum energy configuration the first derivative is zero by definition.
3. Since the harmonic approximation is used all terms in the Taylor expansion greater than two can be neglected.

This leaves only the second order term in the potential energy expression for V. Using Newton's second law, the above is expressed as:

$$\frac{d^2 q_i}{dt^2} = -\left(\frac{\partial V}{\partial q_i} \right) = -\sum_{j=1}^{3N} \left(\frac{\partial^2 V}{\partial q_i \partial q_j} \right)_e q_j$$

The above equation of motion is solved to yield a determinant whose eigenvalues ($m\omega^2$) provides the vibrational frequencies (ω). The eigenvectors describe the atomic displacements for each of the vibrational modes characterized by the eigenvalues. These are the normal modes of vibration and the corresponding fundamental frequencies.

The force constant f_{ij} is defined as the second derivative of the potential energy with respect to the coordinates q_i and q_j in the equilibrium configuration as:

$$f_{ij} = \left(\frac{\partial^2 V}{\partial q_i \partial q_j} \right)$$

In order to obtain the molecular force field with the force constants given by the above equation, a variety of computational methods are available. In general, the calculations of the vibrational frequencies can be accomplished with either empirical force field method or quantum mechanical methods.[9,10] The quantum mechanical method is the most rigorous approach and is typically used for smaller to moderately sized molecules since it is computationally intensive. Since we are examining chemical systems with more than one electron, approximate methods known as *ab initio* methods utilizing a harmonic oscillator approximation are employed. Because actual molecular vibrations include both harmonic and anharmonic components, a difference is expected between the experimental and calculated vibrational frequencies. Other factors that contribute to the differences between the calculated and experimental frequencies include neglecting electron correlation and the limited size of the basis set. In order to obtain a better match with the experimental frequencies, scaling factors are typically introduced.

Quantum mechanical *ab initio* methods and hybrid methods are based upon force constants calculated by Hartree-Fock (HF) and density functional based methods. In general, these methods involve molecular orbital calculations of isolated molecules in a vacuum, such that environmental interactions typically encountered in the liquid and solid state are not taken into account. A full vibrational analysis of small to moderately sized molecules typically takes into account both the vibrational frequencies and intensities to insure reliable assignments of experimentally observed vibrational bands.

The *ab initio* Hartree-Fock (HF) method is an older quantum mechanical based approach.[9,10] The HF methodology neglects the mutual interaction (correlation) between electrons which affects the accuracy of the frequency calculations. In general, when using HF calculations with a moderate basis set there will be a difference of ca. 10–15% between the experimental and calculated frequencies and thus a scaling factor of 0.85–0.90. This issue can be resolved somewhat by use of post-HF methods such as configuration interaction (CI), multi-configuration self-consistent field (MCSF), and Møller Plessant perturbation (MP2) methods. Utilizing configuration interaction with a large basis set leads to a scaling factor between 0.92 and 0.96. However, use of these post-HF methods comes with a considerable computational cost that limits the size of the molecule since they scale with the number of electrons to the power of 5–7.

The *ab initio* density functional theory (DFT) based methods have arisen as highly effective computational techniques because they are computationally as efficient as the original HF calculations while taking into account a significant amount of the electron correlation.[9,10] The DFT has available a variety of gradient-corrected exchange functions to calculate the density functional force constants. Popular functions include the BLYP and B3LYP. The scaling factors encountered using a large basis set and BLYP or B3LYP often approach 1 (0.96–1.05).

Basis set selection is important in minimizing the energy state of the molecule and providing an accurate frequency calculation. Basis sets are Gaussian mathematical functions representative of the atomic orbitals which are linearly combined to describe the molecular orbitals. The simplest basis set is the STO-3G in which the Slater-type orbital (STO) is expanded with three Gaussian-type orbitals (GTO). The more complex split-valence basis sets, 3-21G and 6-31G are more typically used. Here, the 6-31G consists of a core of six GTO's that are not split and the valence orbitals are split into one basis function constructed from three GTO's and another that is a single GTO. Because the electron density of a nucleus can be polarized (by other nucleus), a polarization function can also be included. Such functions include the 6-31G* and the 6-31G**.

Accurate vibrational analysis requires optimizing the molecular structure and wavefunctions in order to obtain the minimum energy state of the molecule. In practice, this requires selection of a suitable basis set method for the electron correlation. The selection of the basis set and the HF or DFT parameters is important in acquiring acceptable calculated vibrational data necessary to assign experimental IR and Raman spectra.

References

1. Barrow, G. M. *Introduction to Molecular Spectroscopy*; McGraw-Hill: New York, NY, 1962.
2. Pauling, L.; Wilson, E. B. *Introduction to Quantum Mechanics with Applications to Chemistry*; McGraw-Hill: New York, NY, 1935.
3. Kauzmann, W. *Quantum Chemistry, An Introduction*; Academic: New York, NY, 1957.
4. Diem, M. *Introduction to Modern Vibrational Spectroscopy*; John Wiley: New York, NY, 1993.
5. Herzberg, G. *Infrared and Raman Spectra of Polyatomic Molecules*; D. Van Nostrand Company: New York, NY, 1945.
6. Long, D. A. *Raman Spectroscopy*; McGraw-Hill: New York, NY, 1977.
7. Cotton, F. A. *Chemical Applications of Group Theory*; Wiley (Interscience): New York, NY, 1963.
8. Wilson, E. B.; Decius, J. C.; Cross, P. C. *Molecular Vibrations: The Theory of Infrared and Raman Vibrational Spectra*; McGraw-Hill: New York, NY, 1955.
9. *Handbook of vibrational spectroscopy, vol. 3, Sample Characterization and Spectral Data Processing*; Chalmers, J. M., Griffiths, P. R., Eds.; John Wiley: 2002.
10. Meier, R. J. *Vib. Spectrosc.* **2007**, *43*, 26–37.

CHAPTER

3

Instrumentation and Sampling Methods

1. INSTRUMENTATION

Raman scattering and IR absorption are significantly different techniques and require very different instrumentation to measure their spectra. In IR spectroscopy, the image of the IR source through a sample is projected onto a detector whereas in Raman spectroscopy, it is the focused laser beam in the sample that is imaged. In both cases, the emitted light is collected and focused onto a wavelength-sorting device. The monochromators used in dispersive instruments and the interferometers used in Fourier transform instruments are the two basic devices. Historically, IR and Raman spectra were measured with a dispersive instrument. Today, almost all commercially available mid-IR instrumentation are (Fourier Transform) FT-IR spectrometers, which are based upon an interferometer (see below). Raman instruments include both grating-based instruments using multi-channel detectors and interferometer-based spectrometers.

1.1. Dispersive Systems

A monochromator consists of an entrance slit, followed by a mirror to insure the light is parallel, a diffraction grating, a focusing mirror, which directs the dispersed radiation to the exit slit, and onto a detector.[1, 2] In a scanning type monochromator, a scanning mechanism passes the dispersed radiation over a slit that isolates the frequency range falling on the detector. This type of instrument has limited sensitivity since at any one time, most of the light does not reach the detector.

Polychromatic radiation is sorted spatially into monochromatic components using a diffraction grating to bend the radiation by an angle that varies with wavelength. The diffraction grating contains many parallel lines (or grooves) on a reflective planar or on a concave support that are spaced a distance similar to the wavelength of light to be analyzed. Incident radiation approaching the adjacent grooves in-phase is reflected with a path length difference. The path length difference depends upon the groove spacing, the angle of incidence (α), and the angle of reflectance of the radiation (β).

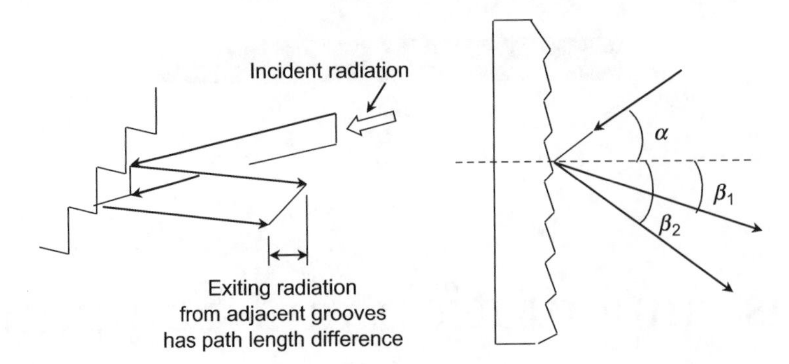

FIGURE 3.1 Schematic of a diffraction grating. Wavetrains from two adjacent grooves are displaced by the path length difference. Constructive interference can occur only when the path length difference is equal to the wavelength multiplied by an integer (first, second, third order). The polychromatic radiation will be diffracted at different angles resulting in spatial wavelength discrimination.

Figure 3.1 shows a schematic of a diffraction grating with the incident polychromatic radiation and the resultant diffracted light. When the in-phase incident radiation is reflected from the grating, radiation of suitable wavelength is focused onto the exit slit. At the exit slit, the focused radiation will be in-phase for only a selected wavelength and its whole number multiples, which will constructively interfere and pass through the exit slit. Other wavelengths will destructively interfere and will not exit the monochromator. Thus, each of the grooves acts as an individual slit-like source of radiation, diffracting it in various directions. Typically, a selective filter is used to remove the higher-order wavelengths. When the grating is slightly rotated a slightly different wavelength will reach the detector.

1.2. Dispersive Raman Instrumentation

Raman instrumentation must be capable of eliminating the overwhelmingly strong Rayleigh scattered radiation while analyzing the weak Raman scattered radiation. A Raman instrument typically consists of a laser excitation source (UV, visible, or near-IR), collection optics, a spectral analyzer (monochromator or interferometer), and a detector.[1, 2, 3] The choice of the optics material and the detector type will depend upon the laser excitation wavelength employed. Instrumental design considers how to maximize the two often-conflicting parameters: optical throughput and spectral resolution. The collection optics and the monochromator must be carefully designed to collect as much of the Raman scattered light from the sample and transfer it into the monochromator or interferometer.

Until recently, most Raman spectra were recorded using scanning instruments (typically double monochromators) with excitation in the visible region. An example of an array-based simple single grating-based Raman monochromator is depicted in Fig. 3.2. Use of highly sensitive array detectors and high throughput single monochromators with Rayleigh rejection filters have dramatically improved the performance of dispersive Raman systems.

The advent of highly efficient Rayleigh line filters to selectively reject the Rayleigh scattered radiation enables the instrument to use only one grating, thereby greatly improving

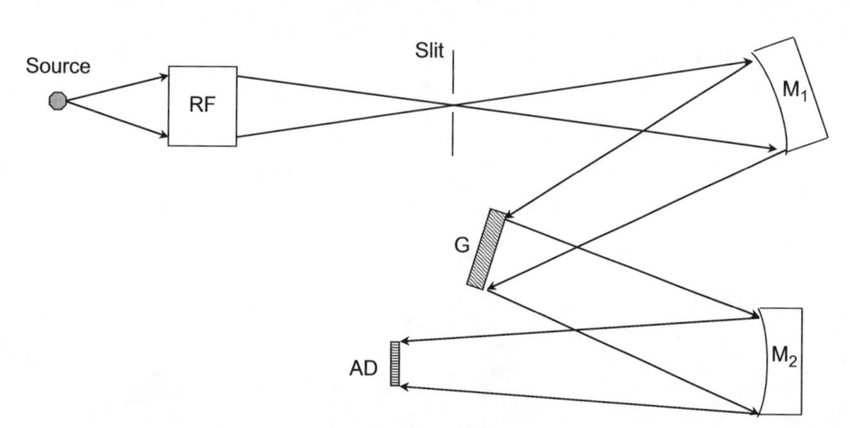

FIGURE 3.2 Schematic of a simple array-based high throughput, single monochromator-based Raman instrument. For the source, laser light is scattered at 90° or 180° and collection optics direct light to a Rayleigh filter, RF, to remove the Rayleigh scattered radiation. *G* is the grating, M_1 and M_2 are spherical mirrors, and AD is the array detector.

the optical throughput. Two commonly used filters include the "holographic notch" and dielectric band filters.[3]

Use of an array detector results in dramatically improved signal-to-noise ratio (SNR) as a result of the so-called multi-channel advantage. The array detectors (AD) are often photodiode arrays or CCD's (charge-coupled-devices), where each element (or pixel) records a different spectral band resulting in a multichannel advantage in the measured signal. However, only a limited number of detector elements (typically 256, 512, or 1064 pixels) are present in commercially available detector elements compared to the number of resolvable spectral elements. Thus, to cover the entire spectral range (4000–400 cm^{-1}) requires either low-resolution spectrum over the entire spectral range or high resolution over a limited spectral range. One solution to this is to scan the entire spectral range in sections using a multi-channel instrument and adding high resolution spectra together to give the full Raman spectrum.

1.3. Sample Arrangements for Raman Spectroscopy

Although the Raman scattered light occurs in all directions, the two most common experimental configurations for collecting Raman scattered radiation typically encountered are 90° and 180° backscattering geometry. Various collection systems have been used in Raman Spectroscopy based upon both reflective and refractive optics. [1, 2, 3]Figure 3.3 shows 90° and 180° collection geometries using refractive and reflective optics, respectively. The 180° collection optics are typically used in FT-Raman spectrometers and in Raman microscopes. The 180° collection geometry is the optimum sample arrangement for FT-Raman as a result of narrow band self-absorption in the near-IR region.

Raman spectroscopy is well known to be a technique requiring a minimum of sample handling and preparation. Typical Raman accessories include cuvette and tube holders, solids holders, and clamps for irregular solid objects. Often NMR or capillary tubes are used and many times the Raman spectra can be measured directly on the sample in their container.

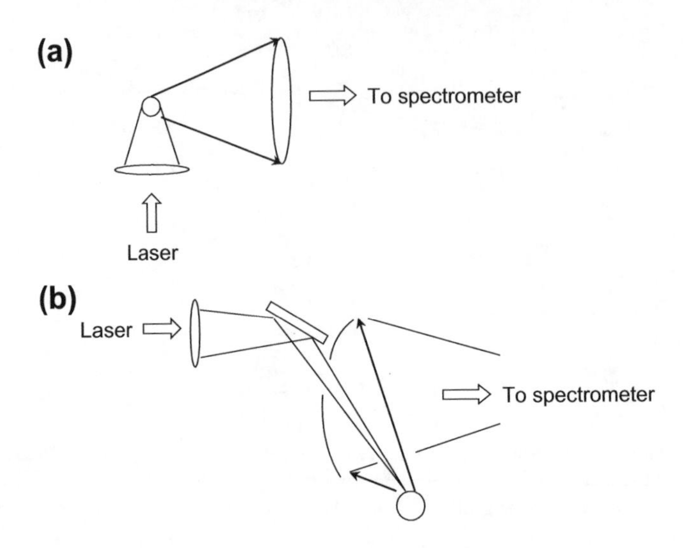

FIGURE 3.3 Two different collection optics for Raman spectroscopy. In (a) a simple 90° collection geometry using refractive optics found in older dispersive instruments. In (b) 180° collection geometry using reflective optics typical of FT-Raman instrumentation.

1.4. Interferometric Spectrometers

Both FT-IR and FT-Raman spectroscopy use an interferometer to separate light into individual components.[4, 5] The FT-Raman measurements use a Nd:YAG 1064 nm near-IR excitation laser and the Raman scattered light falls in the near-IR region. In the IR spectral region, significant improvements in instrumental performance are realized since the instrument measures all wavelengths of IR light simultaneously without use of an entrance slit. The simultaneous measurement of light results in a multiplex (or Felgett's) advantage while the latter results in a throughput (or Jacquinot) advantage. The relative weakness of the source (both IR and the Raman scattered light) along with the poor detector sensitivity make the Fourier transform measurements attractive in this spectral region.

Figure 3.4 shows the schematic of the Michelson interferometer. Use of an interferometer along with computation using the Fast Fourier Transform enables the generation of IR and Raman spectra from the time-dependent interferogram.

The components in a Michelson interferometer include a beam splitter, and a fixed and moving mirror. The collimated light from the source incident on an ideal beamsplitter will be divided into two equal intensity beams where 50% is transmitted to the moving mirror and the other 50% is reflected to the fixed mirror. The light is then reflected off both mirrors back to the beamsplitter where 50% is sent to the detector and the other 50% is lost to the source. As the moving mirror scans a defined distance (Δl), the path difference between the two beams is varied and is called the optical retardation and is two times the distance traveled by the moving mirror ($2\Delta l$). The interferometer records interferograms caused by phase-dependent interference of light with varying optical retardation.

The principle of operation can be easily described by first considering the source to contain only a single monochromatic wavelength, λ. When the position of the moving mirror with

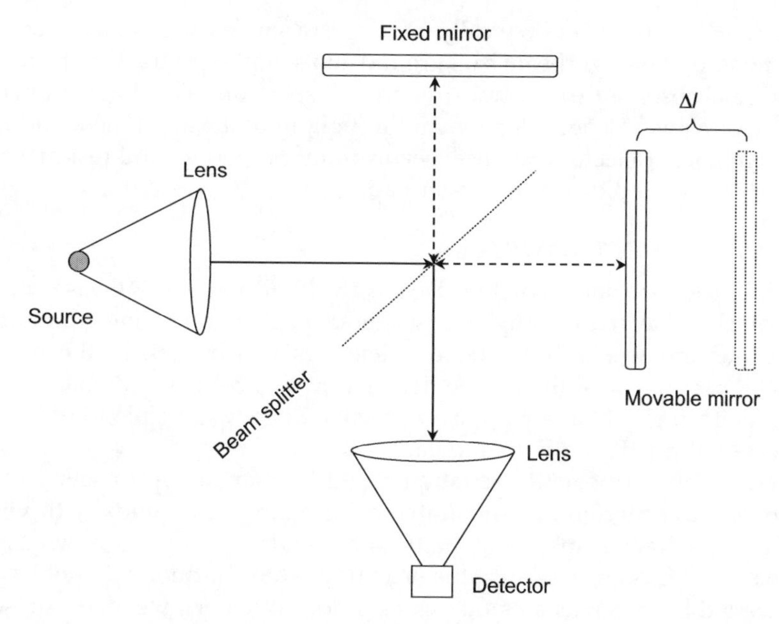

FIGURE 3.4 Schematic of a Michelson interferometer. The displacement of the moving mirror is indicated by Δl.

respect to the beamsplitter is identical to that of the fixed mirror, the optical retardation is zero (zero path difference) and the two beams will combine at the beamsplitter in-phase resulting in constructive interference for the beam passing through to the detector. The detector response will reach a maximum whenever the optical retardation is an integral number of the wavelengths ($0, \lambda, 2\lambda\ldots$). Similarly, a minimum detector response will result from destructive interference at optical retardation values with intervals of $\lambda/2$ ($\lambda/2, 3\lambda/2\ldots$). As the mirror scans at constant velocity, a simple sine wave will result as the two beams move in and out of phase. The Fourier transform of the sinusoidal interferogram will give rise to a single band with a characteristic frequency and intensity of the monochromatic source.

If a polychromatic source is used, the interferogram will consist of a summation of all the different cosine functions corresponding to all of the wavelengths and of the intensities in the source. Only at zero path difference will all the wavelengths be in-phase. Thus, the resulting interferogram in FT-IR and FT-Raman spectra have a very strong center-burst and rapidly damped intensity in the wings of the interferogram. It is necessary to precisely know the optical path differences in the interferometer and this is accomplished using a helium–neon laser. A computer is used to perform the fast Fourier transform to generate the spectrum, which can then be further processed. The computer interface in the FT-IR and FT-Raman spectrometers is used for setting measurement parameters, spectral processing, library searching, and for quantitation.

1.4.1. FT-IR SPECTROMETERS

In a FT-IR spectrometer, sampling occurs just prior to the detector and its collimating optics. Various sampling techniques employed with IR spectroscopy are discussed below.

Commercial FT-IR spectrometers typically operate in a single beam mode requiring sequential measurements of the single beam background and sample spectra, which are then ratioed to provide the final absorbance or %Transmittance IR spectrum. The single beam background spectrum provides a wavelength-dependent throughput of the instrument and is a function of the source emission, detector response, beamsplitter properties, and residual atmospheric absorptions from carbon dioxide and water vapor.

1.4.2. *FT-RAMAN SPECTROMETERS*

Commercial FT-Raman spectrometers use mostly a 1064 nm Nd:YAG laser, notch filters at the laser wavelength to reduce Rayleigh scattered light entrance into the interferometer, high quality interferometers, and sensitive detectors which peak in the near-IR region. For FT-Raman spectrometers, the source depicted in Fig. 3.4 is the Raman scattered light. Presently, by switching only a few optical elements it is now possible to utilize one instrument to collect both the IR and Raman data.

A major advantage of near-IR excitation based-FT-Raman spectroscopy is the greatly reduced fluorescence interference encountered for many compounds with visible excitation. This allows the Raman spectra of many compounds to be measured which was previously impossible. This occurs because the near-IR photons normally do not have sufficient energy to access the vibronic states that cause fluorescence. However, fluorescence problems are not completely eliminated for all samples and limitations due to sensitivity and absorption also exist. The υ^4 dependence shown by the Raman intensity is an intrinsic limitation to the sensitivity using near-IR excitation and can only be addressed by continued improvements in detector sensitivity. The inherent near-IR absorption profile of the sample itself can also pose significant problems. Many compounds have multiple absorption bands due to X–H type bonds in the near-IR spectral region which can attenuate the incident laser and/or the Raman lines. The unequal absorption throughout the Raman spectrum can affect the relative Raman band intensities and is termed as narrow band self-absorption. Thermal decomposition can sometimes occur due to absorption at the laser wavelength.

2. SAMPLING METHODS FOR IR SPECTROSCOPY

2.1. IR Transmitting Materials

Infrared spectroscopy can be used to study a wide range of sample types either in bulk or in microscopic amounts over a wide range of temperatures and physical states. Quite often, an IR transmitting material is needed to aid sampling. Since quartz strongly absorbs in the near-IR region, front-silvered or gold-coated reflective optics are often employed rather than refractive lenses. However, transmitting materials are needed for other materials including the beamsplitter. Table 3.1 summarizes some important infrared transmitting materials. For the majority of applications, NaCl or KBr windows are used and ZnSe is frequently used for aqueous or wet samples. The alkali metal halides fuse under pressure to give IR and optically transparent windows, but they are hygroscopic and fog if not treated properly.

TABLE 3.1 Selected Infrared Transmitting Material

Material	Wavenumber range (cm^{-1})	Refractive index	Comments
NaCl	5000–625	1.52	Common, low cost, hygroscopic
KBr	5000–400	1.54	Common, low cost, very hygroscopic
BaF_2	5000–870	1.45	Water insoluble, easily cracked
CaF_2	5000–1100	1.40	Water insoluble, good at high pressures
KRS-5	5000–275	2.38	Water insoluble, good ATR, deforms, poisonous
ZnSe	5000–550	2.41	Water insoluble, good ATR
Diamond	4500–2500, 1800–200	2.4	Very hard, inert, diamond Anvil cell also good for ATR

2.2. Sampling Techniques

The measured IR spectra are defined by the sample-preparation procedures and the optical spectroscopic techniques employed.[6, 7] However, the quality of the spectra often depends upon the spectroscopist's sample-preparation skills. A wide variety of sampling techniques exists; these are necessary to insure that the optimum spectrum will be measured for the varied physical sample states encountered by the analyst. For any given sample, there will usually be several different sampling techniques that can be used to obtain the spectrum. The available accessory equipment, personal preference of the analyst, or the nature of the particular sample will often dictate the selection of the technique.

In general, the IR radiation incident upon a sample can be reflected, absorbed, transmitted, or scattered.

$$I_o = I_r + I_a + I_t + I_s$$

Here I_o is the intensity of the incident radiation; I_r, I_a, I_t, and I_s are the intensities of the reflected, absorbed, transmitted, and scattered radiation, respectively. Experimentally, any of the above intensity components can be used to measure the IR spectrum of the sample. The extent that light is transmitted, reflected, scattered, or refracted is dependent upon the sample morphology, crystalline state, the angle of the incident light, and the difference in refractive indices of the sample and surrounding matrix.

Figure 3.5 depicts the angular dependence of the transmission, refraction, and reflected light. Transmitted light has a 0° angle of incidence while reflectance (which includes specular, external, internal, and diffuse) is observed at larger angles. The critical angle (Θ_c) beyond which refraction becomes imaginary and all light is reflected is also shown.

In general, measurement of acceptable quality FT-IR spectra requires development of requisite sample-preparation skills necessary to measure photometrically accurate spectra. General criteria for suitable IR sample preparations include the following:

1. Suitable band intensities in the spectrum. The strongest band in the spectrum should have intensity in the range of 5–15% transmittance. The resulting preparation must be uniformly thick and homogenously mixed without holes or voids to insure a good quality spectrum.

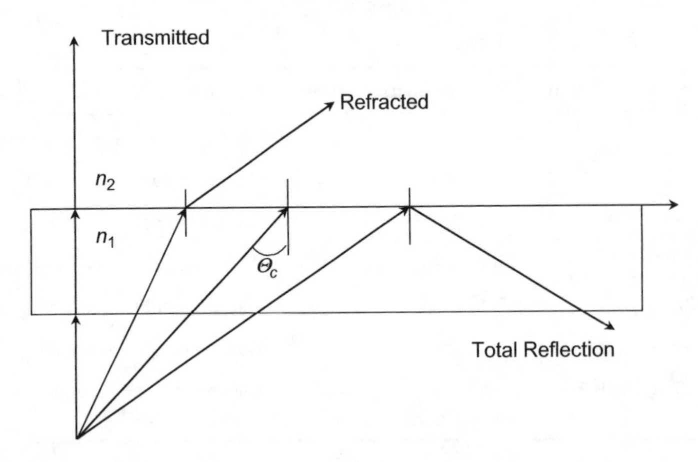

FIGURE 3.5 The angle of incidence of the light and the refractive index difference help define how much light is transmitted, refracted, and reflected.

2. Baseline. The baseline should be relatively flat. The highest point in the spectrum should lie between 95 and 100% transmittance.
3. Poor atmospheric compensation. Water vapor and carbon dioxide bands should be minimized. Since the sample scan is ratioed against a previously measured background spectral subtraction of water vapor/carbon dioxide, reference spectrum should be used.
4. *X*-axis scale. Plots should employ a 2*X*-axis expansion and not a simple linear *X*-axis. This provides an expansion of the important fingerprint region.

Figure 3.6 shows the FT-IR spectrum of starch with two different sample preparations. In the top spectrum (A), the starch was dissolved in water and prepared as a cast film on ZnSe resulting in a good quality, photometrically accurate spectrum. In the case of starch, this is one of the preferred sample-preparation techniques. The Nujol mull preparation of powdered starch provides an unacceptably poor quality FT-IR spectrum and a so-called false spectrum. A broadening of the most intense bands and a strengthening of the weak bands characterize false IR spectra. In general, this is caused by a non-uniform sample film or distribution of sample in a Nujol of KBr matrix. In the case of starch, good quality IR spectrum cannot be obtained by either Nujol mulls or KBr discs. Starch demonstrates that the nature of the particular sample dictates what sample-preparation technique is appropriate.

2.3. Transmission IR

Simple transmission IR measurements are commonly made of gas, liquid, and solid-state samples. The most accurate information is obtained if the path length or analyte concentration is adjusted such that the peaks of interest have an absorbance between 0.3 and 0.9. The two simplest types of samples that can be prepared are gas phase samples and neat (low volatility) liquids. Gas phase measurements require only the introduction of the sample into a gas

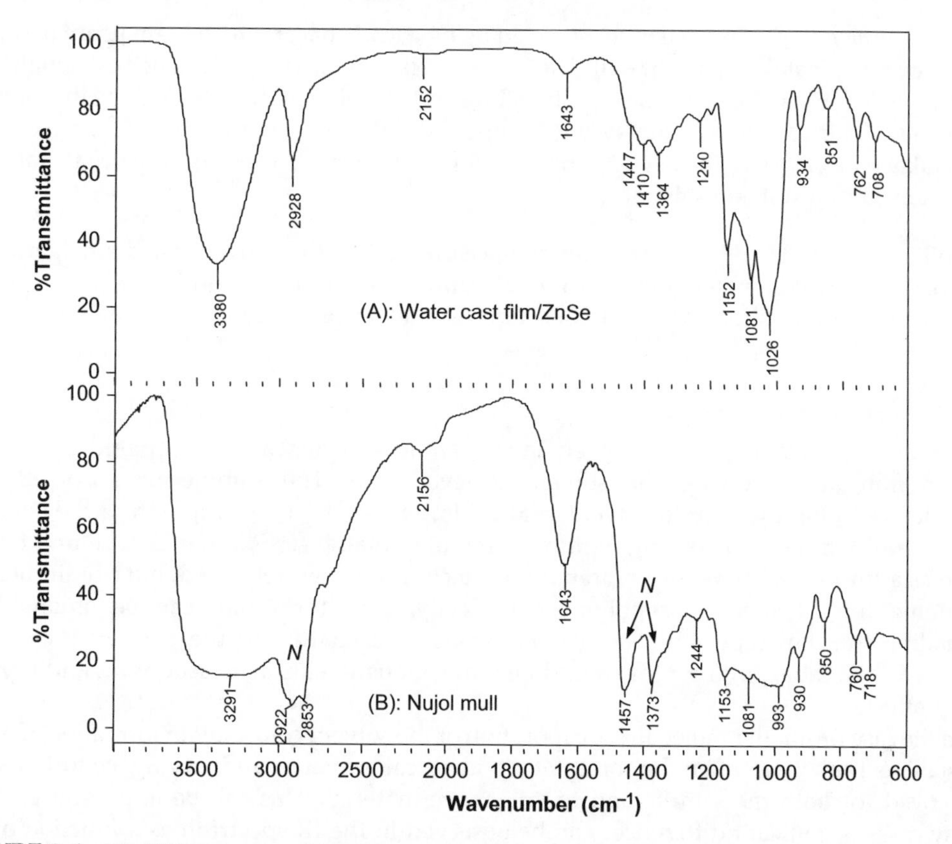

FIGURE 3.6 The FT-IR spectrum of starch prepared as (A) water cast film on a ZnSe plate and (B) Nujol Mull. The *N* marks the Nujol bands.

cell, which can vary in path length from 10 cm to 8 meters. Highly accurate quantitation of single and multiple components is possible because the path length is well defined and no solid state effects are present (see below).

2.3.1. Liquids and Solutions

Liquids and solutions are easily measured by FT-IR spectroscopy. Neat liquids can be easily prepared as a capillary film between two plates. The liquid film should be uniformly thick without holes or voids to insure a good, quantitative IR spectrum. Some common sample-preparation techniques for liquids include the following:

1. *Capillary film*. Neat liquids can be easily prepared as a capillary film. A drop of liquid sample is simply sandwiched between the cell windows.
2. *Demountable cell*. Here the technique is identical to the capillary film except a spacer (usually lead or teflon) is used to define the thickness of the liquid film. Disassembly allows access to the windows for easy cleaning or repolishing. The cell thickness is easily changed by selecting a different spacer. For a neat liquid sample, a thickness of about 0.01 mm is suitable.

3. *Sealed amalgam cell*. This cell is constructed using a lead spacer which is amalgamated with mercury to create a very tight seal. This type of cell provides very accurate path length and is suitable for very volatile liquids. The filling and cleaning of the cell are done through the use of a pair of hypodermic syringe fittings. Fixed path length cells are commercially available in path lengths from 15 μm to 500 μm and can be used to measure the analyte dissolved in a suitable solvent.

Both pure liquids and solutions can be measured using the above. Sealed amalgam cells are commonly used for solutions. Additional sample preparations include smears, cast films, and analytes in solution. More viscous liquids can be prepared as a smear on a single window.

2.3.2. *Cast Films*

Samples that have been dissolved in a volatile solvent can be prepared by casting a thin film and allowing the solvent to evaporate. The sample is dissolved into a moderate to highly volatile solvent, and a film is prepared by evaporating the solution directly onto an IR transmitting window. The non-volatile sample remains behind on the plate as a thin film. This sample-preparation technique is widely used, but not limited to, obtaining the IR spectra of polymers Typically, a uniform film can be obtained by spreading the film repeatedly on a window using a spatula or the edge of a capillary pipet. Ideally, the resulting film should be amorphous to eliminate scattering and crystallinity effects.

In the case of smaller molecules, evaporation of the solvent can result in formation of a thin crystalline film. It is often important that identical solvents and drying conditions are employed for both the sample and the reference material. This can be important since in many cases, significant differences can be observed in the IR spectrum as a function of the crystalline form.

When preparing a cast film, first select a suitable solvent and IR window. The solvent must not react with the sample, but should be volatile and dissolve the sample. For the majority of applications, NaCl or KBr windows are used and ZnSe is frequently used for aqueous solutions. Dissolve the sample in the selected solvent. The concentration is not critical, however, higher concentrations will result in faster sample preparation.

Now add a few drops of the solution on the central portion of the infrared window. Gentle heating using a hot plate and a flow of nitrogen can be used to hasten the solvent evaporation. Care should be taken since some materials will decompose even under gentle heat. With more volatile solvents, the solution can be allowed to spread out over the surface of the window and as the solvent evaporates a thin film will remain. However, some solutions will not spread out evenly by themselves and will require manipulation to obtain a thin uniform film (see below). With less volatile solvents (for example water) or solutions that will not spread out evenly on the surface of the window, a flexible spatula or glass pipet should be used to spread a thin film and facilitate the solvent evaporation. For a uniform film, the solution may require a thin film to be stroked on.

Examine the resultant film visually to confirm that a thin uniform film free of holes and voids has been prepared. The thickness of the film for FT-IR spectroscopy should be approximately 5 μm. There is only limited control over the thickness of a cast film. In general, the more concentrated the solution and the more drops placed on the window, the thicker the film. The process of obtaining the correct thickness is simply trial and error. Lastly, confirm from the IR spectrum that all of the solvent has evaporated leaving only the material of interest. The analyst should be familiar with the spectrum of the pure solvent.

2.3.3. SOLID-POWDERED SAMPLES: KBr DISCS AND NUJOL MULLS

Solid-powdered samples can be prepared as fine particles dispersed in NUJOL or as a disk in a transparent KBr matrix. A NUJOL mull preparation simply mixes mineral oil and finely ground sample to form a paste, which is then sandwiched, between two IR transmitting windows. A KBr disc uses a mixture of dried KBr powder and finely ground sample. The KBr/sample mixture forms a clear disc when pressed under high pressure using a hydraulic press.

Both preparations require that the sample must be sufficiently ground and mixed with the support matrix to insure a good, artifact free IR spectrum. Insufficiently mixed and ground sample particles result in undesirable spectral artifacts. Most obvious is a sloping baseline due to scattering. A distorted band shapes as a result of anomalous dispersion of the refractive index in an absorption band (Christiansen effect). False spectra due to non-uniform sample thickness are characterized by broadening of the bandshape of the strong bands and relative intensification of weak bands.

NUJOL MULL

A Nujol mull involves mixing of mineral oil and finely ground sample to form a paste, which is then sandwiched, between two IR transmitting windows (typically KBr or NaCl). In general, 30–50 mg of sample is needed for this sample preparation. The sample must be ground to an average particle size of at least 0.5 μm using an agate mortar and pestle. Larger particles approaching the wavelength of light being used will scatter light resulting in a sloping background.

A suitably ground sample will be evenly distributed over the mortar and takes on a smooth, glossy appearance. A very small amount of the mulling agent (mineral oil) is added to the mortar, and grinding is continued until the sample is suspended in the mulling agent to form a smooth, creamy paste. A small amount of the mull is transferred to the center of an infrared window. A second window is placed on top of the sample and the two are pressed together. The mull should be spread out into a thin translucent film, free of holes or voids.

The representative IR spectra of Nujol and Fluorolube is shown in Fig. 3.7. Nujol has bands characteristic of a long chain hydrocarbon. If detailed information is needed in the CH stretching region and the deformation region, a Fluorolube (perfluorinated hydrocarbon) mull should be used which has no appreciable absorption bands from 4000 to 1350 cm^{-1}. If necessary, a split-mull spectrum can be obtained by merging a Nujol and Fluorolube mull spectra.

A significant advantages of Nujol mull sample preparation is that no water is introduced into the sample. This can be particularly important when trying to confirm the presence of an OH or NH type species or when analyzing a particularly hydroscopic material. When faced with analyzing particularly toxic solid materials, the analyst can limit potential exposure by simply mulling the sample between the salt plates.

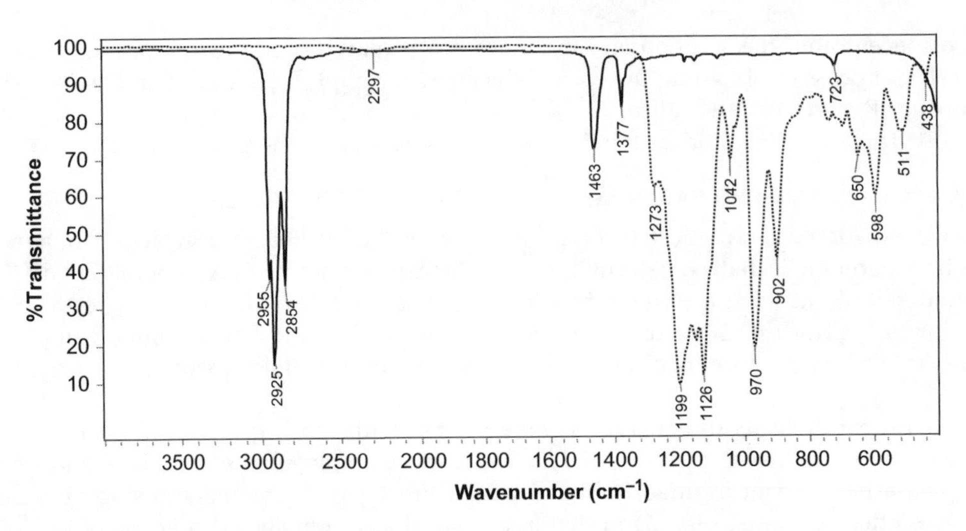

FIGURE 3.7 The FT-IR spectrum of Nujol and Fluorolube. The Nujol spectrum is shown as a solid line and the Fluorolube spectrum is shown as a dotted line.

KBr DISC SAMPLE PREPARATION

A KBr disc involves mixing of dried KBr powder and finely ground sample. The KBr/sample mixture forms a clear disc when pressed in a die under high pressure using a hydraulic press. Only 5 mg of sample is needed for this sample preparation. The sample must be ground to an average particle size of at least 0.5 μm using an agate mortar and pestle to minimize a scatter-induced sloping background. The sample concentration varies from 0.1% to 3% (by weight). In general, start with *ca.* 5 mg of sample and add 500 mg of KBr powder a little at a time and mix well. Transfer the thoroughly mixed KBr preparation into a 13 mm die. Press the disc in a bench top press. With a hydraulic press, anywhere from 2 to 8 tons is sufficient.

Solids are more frequently run as KBr discs than as mulls. This preference probably arises because KBr has no absorptions of its own over the entire transmission range and requires far less sample than a mull. However, the KBr disc method does have some disadvantages compared to a mull. KBr is very hygroscopic and some water is almost always introduced in the sample preparation. Figure 3.8 shows a KBr disc prepared without any organic sample. The water introduced from the sample grinding results in bands at *ca.* 3440, 1630, and 560 cm^{-1}, from the OH stretch, OH bend, and OH wag that can complicate interpretation of the spectrum. KBr discs of crystalline materials are also sometimes less reproducible than mulls. The large pressure along with the addition of water can cause changes in hydration state, crystallinity, or polymorphism. In some cases, the sample itself can ion exchange or react with KBr.

2.3.4. *Melts*

Melts are the last commonly used IR sample technique that is particularly useful for polymers. This method involves melting the sample on a salt plate and allowing it to resolidify. This method is not recommended for crystalline materials because of possible orientation

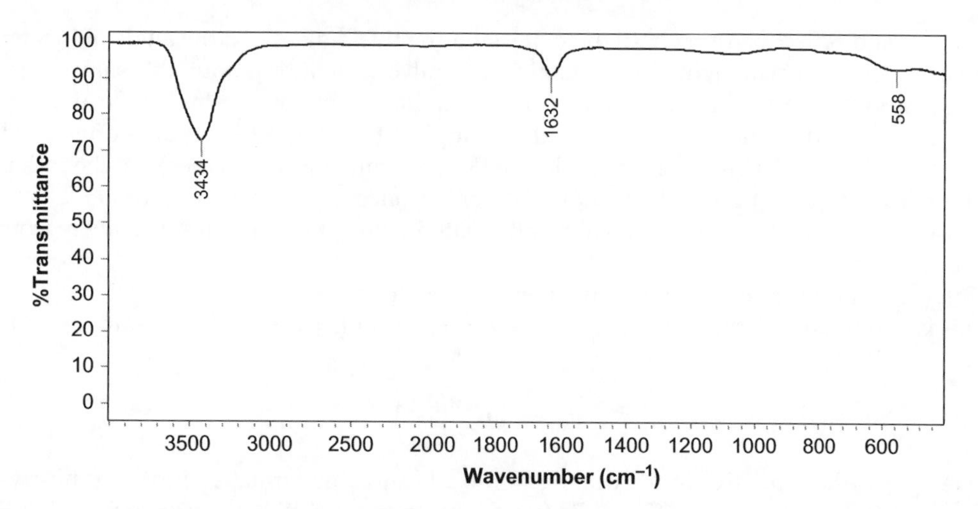

FIGURE 3.8 The FT-IR spectrum of a KBr disc prepared by mixing KBr powder using a mortar and pestle into a fine powder and preparing a clear disc using a hydraulic press. Characteristic bands from water are observed at 3434, 1632, and *ca.* 560 cm^{-1}.

effects. If a sample is known to decompose under heat, this method should not be employed. The technique involves using a hot plate, heat the sample above its melting point, and allowing the sample to become a liquid. Next, using a pipet take one drop of the melt and place it on an IR transmitting window such as KBr. Use the pipet to draw down the liquid into a uniform film and allow it to form a solid.

2.4. Reflection Techniques

Two of the more commonly used FT-IR sampling techniques that capitalize on the spectral information that can be obtained from reflection (and refraction) properties are ATR (attenuated total reflection or internal reflection) and DRIFTS (diffuse reflectance FT-IR spectroscopy).

2.4.1. Attenuated Total Reflectance (ATR)

ATR is a contact sampling method that involves a crystal with a high refractive index and excellent IR transmitting properties. ATR is one of the more popular sampling techniques used by FT-IR Spectroscopists because it is quick, non-destructive and requires no sample preparation. It is a contact sampling method that involves an IRE (Internal Reflection Element) with a high refractive index and excellent IR transmitting properties. The sampling technique capitalizes on the spectral information that can be obtained from reflection phenomena. The technique is used to measure the IR spectra of surfaces or of material that is too thick or strongly absorbing to be analyzed by more traditional transmission methods. A variety of ATR accessory designs exist and are available from almost all manufacturers of accessories for infrared spectrometry.

In the case of ATR, the angle of incidence must be greater than the critical angle so that total internal reflectance occurs. Figure 3.5 depicted the dependence of light transmission, refraction and reflection upon both the material refractive indices and the angle of incidence. However, the reflected light does contain spectral information about the sample at the sample—crystal interface and the ATR technique capitalizes on this. A single reflection crystal is commonly used in micro-ATR work and a multiple reflectance ATR crystal is often used for bulk samples. An ATR accessory provides an IR spectrum of a sample because the radiation at the reflection point probes the sample with an "evanescent wave." At a frequency within an absorption band, the reflection will be attenuated while at frequencies well away from an absorption band, all light is reflected.

Internal reflectance occurs when the angle of incidence exceeds the "critical" angle. This angle is a function of the real parts of the refractive indices of both the sample and the ATR crystal

$$\theta_C = \sin^{-1}\left(\frac{\eta_2}{\eta_1}\right)$$

Here, η_2 is the refractive index of the sample and η_1 is the refractive index of the crystal. Typical crystals such as Ge, ZnSe, or diamond have high refractive indices much higher than that of organics resulting in internal reflections at moderate angles of incidence. Figure 3.9 depicts some of the basic principals of ATR measurements.

The refractive index of the sample is a complex with both a real and imaginary component:

$$\eta = n + ik$$

The real part of this expression n in the regular refractive index holds when there is no absorption. The imaginary component of the refractive index, ik, holds within an absorption

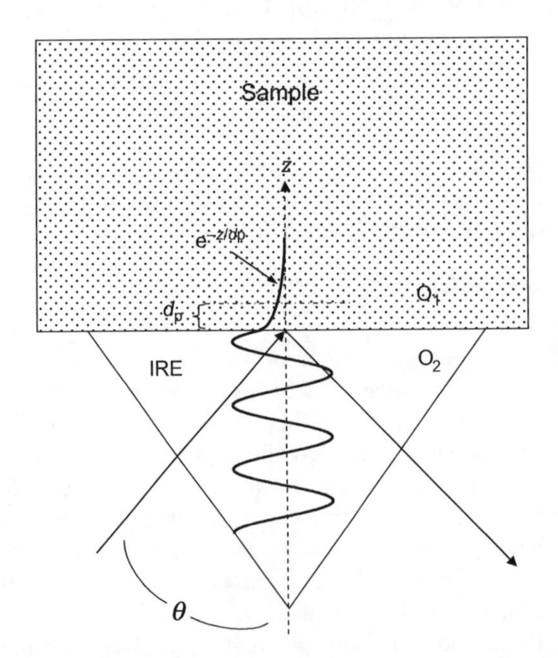

FIGURE 3.9 Diagram of a single reflection ATR crystal depicting the basic principles of the technique.

band resulting in an effect known as anomalous dispersion. The parameter k is directly related to the extinction coefficient found in the Lambert–Beer law. The total absorption intensity, A, as a function of θ can be expressed as:

$$A(\theta) = \int_0^\infty \alpha(z) e^{-z/\mathrm{dp}} \mathrm{d}z$$

where z is the depth into the sample and $\alpha(z)$ is the absorption coefficient of the sample as a function of depth. As depicted in Fig. 3.9, the reflected radiation penetrates the sample boundary as a so-called evanescent wave. The electric field amplitude of this evanescent wave exhibits an exponential dependence. The parameter, d_p, the effective penetration depth is expressed as:

$$d_\mathrm{p} = \frac{\lambda}{2\pi n_\mathrm{p}\left(\sin^2\theta - n_\mathrm{sp}^2\right)^{1/2}}$$

Where, λ equals the wavelength of the radiation, θ is the angle of incidence, n_p is the refractive index of the IRE, and $n_\mathrm{sp} = n_2/n_1$ the ratio of the refractive indices of the sample and IRE. This depth of penetration is defined as the distance from the IRE-sample boundary where the intensity of the evanescent wave decays to $1/e$ (37%) of its original value.

There are a number of important consequences to the above expression. The sampling depth of the ATR method is approximately 2–15 μm, is wavelength dependent, and increases with decreasing wavenumber. Experimentally, varying the angle of incidence of the light and the refractive index of the crystal used can control the sampling depth. Increasing either results in a decreased sampling depth. Careful comparison with classic transmission spectrum also indicates that ATR measurements result in small differences in the peak frequency and bandshape as a result of refractive index effects.

ATR requires excellent contact between the sample and the crystal and is therefore an excellent method for liquids or soft, easily deformed solids. Use of micro ATR crystals with a protective ATR coating enable much greater pressure to be used to insure good contact between the ATR crystal and harder materials. Table 3.2 summarizes some of the commonly used ATR crystal materials.

2.4.2. *Diffuse Reflectance*

Diffuse reflectance (DRIFTS, Diffuse Reflectance Infrared Fourier Transform Spectroscopy) is applied to analyze powders and rough surface solids. Typically, the technique is applied to powders since the technique relies upon scattering of radiation within the sample. The incident light on a sample may result in a single reflection from the surface (specular reflectance) or be multiply reflected giving rise to diffusely scattered light over a wide area which is used in DRIFTS measurements. Collection optics in the DRIFTS accessory are designed to reject the specularly reflected radiation and collect as much of the diffuse reflected light as feasible.

In general, theories to describe the diffuse reflectance phenomena consider plane-parallel layers within the sample with light scattering in all directions. The observed reflectance

TABLE 3.2 Commonly Used ATR*Crystal Materials and Characteristics

Material	Wavenumber range (cm^{-1})	Refractive index	Comments
Germanium	5000–850	4.0	Hard and brittle, good ATR material but temperature sensitive.
KRS-5	5000–275	2.38	Water insoluble, conventional ATR material. Relatively soft, deforms, poisonous. Reacts with complexing agents.
ZnSe	5000–550	2.41	Water insoluble, hard and brittle, good ATR material. Attacked by acids and strong alkalines.
Diamond	4500–2500 1800–200	2.4	Very hard, inert, used in micro-ATR crystals. Often used as a protective film for ATR elements such as ZnSe
Silicon	8300–1500	3.4	Limited wavelength range for ATR. Material is inert, hard and brittle.

* *ATR is attenuated total reflectance.*

spectrum using the Kubelka–Munk equation is a ratio of the single beam spectra of the sample diluted by KBr powder against pure powder KBr and is defined as:

$$f(R_\infty) = \frac{(1 - R_\infty)^2}{2R_\infty} = \frac{K}{s} = \frac{C}{k}$$

where K is the absorption coefficient and s is the scattering coefficient of the sample material. The reflectance, R_∞, indicates the sample is thick enough that no radiation reaches the back surface. The above predicts a linear relationship between K and the maximum value of $f(R_\infty)$ for each peak provided s is constant. The scattering coefficient is dependent on the particle size and packing and must remain constant for reproducible results. Lastly, C is the concentration of the sample and k is related to the particle size and molar absorptivity of the sample, $k = s/2.303e$.

The DRIFTS spectrum can show a strong dependence on the refractive index of the sample, the particle size and size distribution, packing density, and sample homogeneity. For quantitative analysis, the particle size, sample packing, and dilution must be carefully controlled. To obtain the highest quality DRIFTS spectrum, the following precautions are taken.

1. Particle size should be small and uniform. This will result in narrower bandwidths and more accurate relative intensities. The sample/matrix particulates size should be 50 μm or less.
2. Dilute sample in a non-absorbing matrix powder such as KBr or KCl. The finely powdered sample is ~5% relative to the matrix powder. This insures a deeper penetration of the incident light into the sample, which will increase the diffuse scattered light.
3. Homogeneity. The samples should be well mixed.
4. Packing. The sample should be loosely packed in the cup to maximize the IR beam penetration.

For quantitative analysis, the particle size, sample packing, and dilution must be carefully controlled. Despite compliance with the precautions listed above, the ratioed diffuse reflectance spectrum will appear different from that obtained with a classic transmission IR measurement. In general, low intensity bands will be increased relative to intense bands and the strong intensity bands will have broader, rounder peak shapes. A Kubelka–Munk conversion available in most FT-IR software packages can compensate for some of these effects.

2.5. Microscopy

Both IR and Raman spectroscopy have been very successfully coupled to microscopy with widespread applications in heterogeneous samples in industrial, forensic, and biological sciences. Vibrational microscopy is used for particle and contaminant analysis, studies of laminar structures, characterization of fibers, and domain structure and chemical compositional mapping.

IR microspectroscopy involves coupling a microscope to an infrared (typically FT-IR) spectrometer and provides the capability to obtain spectra of extremely small samples (as small as 10 μm in diameter in certain specialized applications). The technique has widespread applications in the study of heterogeneous samples and is used for particle and contaminant analysis, studies of laminar structures, domain structures, and chemical composition mapping. The microscope system allows one to focus IR radiation onto the sample, to collect and image transmitted or reflected IR radiation from the sample onto a sensitive detector and to view and position the sample. The system includes an optical microscope system for viewing the sample and reflective optics for imaging the IR radiation. The system also includes remote apertures for defining the spatial area of the sample to be examined.

Commercial FT-IR microscopes are capable of a variety of sampling techniques including: specular reflectance, diffuse reflectance, micro-ATR, grazing angle, and conventional transmission measurements. Just as when studying materials in bulk, it is the sample itself that determines the most appropriate sampling technique.

Several commercial Raman microspectrometers are available with excitation sources that span the UV to the near-IR. The ease of sampling makes Raman spectroscopy an excellent choice for microscopy. However, fluorescence remains a considerable obstacle to application of Raman microscopy to many samples. More recently, IR and Raman imaging have been successfully applied to a variety of heterogeneous systems.

2.5.1. *Transmission IR Microscopy*

The ideal sample for transmission IR microscopy is a thin (5–10 μm), smooth, flat sample that does little to alter the optical path of the IR radiation. This requires the sample to also be large enough (*ca.* 25 μm) to minimize diffraction of light. Sample-preparation techniques can include:

a) slicing and microtoming
b) flattening sample with a roller
c) squeezing sample between salt plates or a diamond compression cell

As with macroscopic IR measurements, sample preparation plays an important role in obtaining quality IR spectra when using an IR microscope. Important experimental parameters include the diffraction light limit, light scattering, and focal shifts due to high refractive index samples.

When the sample or aperture size is on the order of the wavelength of light being used, diffraction plays a major role in the measured IR spectrum. Significant distortion of the measured IR spectrum occurs. Diffraction effects can be minimized by the:

a) use of widest possible apertures. Keep aperture closed within the sample image but not closed enough to reduce energy throughput significantly.
b) use of the same aperture setting in the background and sample scan.
c) flattening of sample which will enlarge the sample area.

Surface irregularities and fine particles can scatter light. Remedies include adding KBr or KI and applying pressure to obtain a clear plate or adding NUJOL oil to the a roughened surface. Samples or substrates with very high refractive index can give rise to focal shifts. This can be corrected by adjusting the focal system of the microscope by observing either the sharpness of the aperture edge or the energy throughput. At times unwanted internal reflections can also interfere. An example of this is interference fringes that can occur when using a diamond anvil cell.

2.5.2. *Reflection IR Microscopy*

Infrared reflectance measurements are desirable because of the reduced sample preparation required. The IR microscope will be set to reflectance rather than transmittance mode and if necessary, to an additional attachment placed on the IR objective. Commonly used reflectance techniques include specular reflectance and ATR measurements.

The most commonly used reflectance technique is ATR. A micro-ATR accessory is employed that attaches to the IR microscope objective. Once contact is established between the sample and the ATR element, the IR spectrum is measured. The resultant spectrum is typical of that obtained using any ATR accessory.

Specular reflectance is a less commonly used technique for IR microscopy and involves a front surface mirror-like reflection from the exterior of a material where the angle of incidence is equal to the angle of reflection. The measured spectrum depends on the angle of incidence, the sample index of refraction, absorption properties, and sample surface morphology. The technique requires no sample preparation and is best suited for smooth, highly reflective surfaces and can include samples such as single crystals and thin films on a reflective metal surfaces.

Not surprisingly, the resultant specular reflectance IR spectra are very different from typical transmission IR data. The spectra show derivative-like shape within the absorption bands. A Kramers–Kronig transformation which is often available in FT-IR software is often used to deconvolute the specular reflectance data and provide more absorbance-like spectra.

3. QUANTITATIVE ANALYSIS

Any quantitative analyses involve measurement of a test sample and comparison with standards of known concentration. The resulting calibration plot reflects the relationship between the measured quantity (such as infrared absorbance or the Raman intensity) and the analyte concentration. Provided appropriate standards for calibration of an analytical method selected, vibrational spectroscopy is a valuable potential technique for quantitative analysis.[1] Quantitative analysis methods can include both classical and chemometric PLS (principal least squares) analysis. Only classical analysis techniques will be discussed here. For an understanding of chemometric applications to spectroscopy the reader should consult a reference dedicated to this subject.[8, 9]

Advantages of using vibrational spectroscopy include the following:

1. Wide variety of sample types (gases, liquids, solids)
2. Clearly separated peaks are often found in the IR and Raman spectra
3. Unique bands frequently allow multicomponent analyses of mixtures
4. Typically not much effort is needed to acquire or prepare sample for analysis
5. Sample quantity is in milligram quantity

Both the analyte and non-analyte components will be observed by both IR and Raman spectroscopy. Consequently, the main components of the sample must be confirmed prior attempting quantitation. In general, Raman and IR spectroscopy of solids can be performed at concentrations of 1–100%. When strong isolated bands are available, this can be extended to lower concentrations (~0.1 %). IR spectroscopy can be employed to quantitate much lower concentrations (down to ppm levels) by greatly increasing the sample path length for solutions and for gases.

3.1. Relationship of IR and Raman Signal to Analyte Concentration

The basis of quantitative analyses in absorption spectroscopy is the Lambert–Beer law.

$$\log \frac{I_0}{I} = \log \frac{1}{T} = A = abC$$

where I_0 is the incident light, I is light after it has passed through the sample, T is the transmittance, A is the absorbance, a is the absorption coefficient, b is the sample thickness, and C is the sample concentration. The absorption coefficient is a constant specific for the material at a particular wavelength. Where the Lambert–Beer law holds, the measured absorbance is a linear function of the concentration (and path length). Both single component and multicomponent analysis is possible.

Analytical methods based upon IR and Raman spectroscopy often use either peak heights or peak areas. Commercial instrument software presently allows both peak height and areas to be used when developing analytical methods. The integrated peak areas may provide a more robust calibration if the analyte band changes with increasing concentration or is influenced by molecular interactions.

Deviations from the Lambert–Beer law resulting in a non-linear relationship between the measured absorbance and the analyte concentration often occur. At the high concentrations typically encountered in IR measurements of solids, the molecular interactions of the sample itself will often cause non-linearity. However, if the calibration takes such non-linearity into account, it will not affect the quantitative accuracy of the analytical method.

As discussed in the previous chapter, the intensity of the Raman scattered radiation is given by:

$$I_R \propto \nu^4 I_o N \left(\frac{\partial \alpha}{\partial Q}\right)^2$$

where I_o is the incident laser intensity, N is the number of scattering molecules in a given state, ν is the frequency of the exciting laser, α is the polarizability of the molecules, and Q is the vibrational amplitude, and the parameter $(\partial\alpha/\partial Q)$is the Raman cross section. Analogous to infrared spectroscopy, the Raman cross section is a constant specific for the material at a particular wavelength. Thus, if the Raman cross section and the incident laser intensity (I_0) remain constant, the intensity of the Raman band (I_R) is directly proportional to the sample concentration. Once again, at the high concentrations often encountered in Raman measurements interactions of the components in the sample itself often result in non-linearity.

To develop a quantitative analysis method for either IR or Raman spectroscopy, well-characterized standards covering a broad range of compositions are required. In the case of the IR spectra of solid-state samples, neither the sample thickness nor the concentration of the sample in the matrix is known. With Raman spectroscopy, the incident laser intensity can be very difficult to precisely control. For this reason, a ratio of an analyte band is often normalized against another component to eliminate these sources of variation.[6] The selection of appropriate peaks in the spectrum are chosen as the inputs for the quantitative analysis based on the spectral separation and knowledge of the underlying vibrational modes.

3.2. Quantitative Analysis: Ratio Method

A calibration curve defines the relationship between an analytical signal produced by the analyte and its concentration. In the common case of linear calibration, a linear regression will be used to fit the analytical signal y to the concentration x of the analyte in calibration samples.

$$y = a + bx$$

For determination of an unknown, the calibration equation is inverted.

$$x = -\frac{a}{b} + \frac{y}{b}$$

A different form of calibration equation is often applicable in spectroscopy. The simplest case involves only two components ($C_1 + C_2 = 100\%$) with two isolated analytical bands with absorbances A_1 and A_2. Here, we assume a linear relationship between the concentration of the analytes and the observed signal with a slope that differs depending upon the analytes response function. For IR spectroscopy, this response will depend upon the molar absorptivities of the bands and for the Raman cross section. This simple case is commonly used when employing the ratio method of IR analysis that relies upon Beer's law and enables quantitation of components without knowledge of the sample path length.

$$\frac{A_1}{A_2} = \frac{b_1}{b_2}\frac{a_1 C_1}{a_2 C_2} = k\frac{C_1}{C_2}$$

With simple rearrangement, we then obtain the following:

$$\%C_1 = \frac{100}{1 + kA_2/A_1}$$

In general, the analysis is more complex since the calibration curves are often affected by overlapping bands, additional interfering components, molecular interactions effecting the absorption coefficient and Raman cross sections and in the case of FT-Raman the self-absorption phenomena discussed below. Thus, the expressions for both IR and Raman are more complex than that suggested by the simple two component case. For IR, we could simply expand the original expression as:

$$A_1 = a_1 b_1 C_1 + a_2 b_2 C_2$$

For FT-Raman spectroscopy, the expression is more complex and below we add an additional term $(1-e^{-ax})$ which takes into account near-IR self-absorption of the Raman scattered light.

$$I_R \sim N I_o \left(\frac{\partial \sigma}{\partial \Omega}\right)(1 - e^{-ax}) + K$$

The resultant calibration curves will often be non-linear. Most instrumental software will enable the data to be fit with empirical equations such as quadratic or higher-order polynomial fits. An alternative is the use of a class of semi-empirical equations and linear fractional transformations. The equation form is suitable for the peak ratio method. Unlike quadratic equations, linear fractional transformations are easily inverted, which facilitates the development of calibration equations in a manner similar to the usual practice encountered in the linear calibration case.

3.2.1. *Fractional Linear Calibration Equations*

Fractional linear transformation can be applied to account for typically encountered variation from the simple two-component case discussed above.[10] For a mixture of two components A and B, a spectral response (absorbance or Raman intensity) in which there may be interference with the two analytical bands is

$$R_1 = k_{1A} C_A + k_{1B} C_B$$
$$R_2 = k_{2A} C_A + k_{2B} C_B$$

The ratio of concentrations $C_A/(C_A + C_B)$ (mole fraction) as a function of the ratio of the spectral response $R_1/(R_1 + R_2)$ has the form of a fractional linear transformation

$$\frac{C_A}{(C_A + C_B)} = \frac{aR_1 + bR_2}{R_1 + dR_2}$$

where coefficients a, b, and d are constants which can be determined empirically. In our case, b will often equal zero and the above simplifies to an expression with only two variables:

$$\frac{C_A}{(C_A + C_B)} = \frac{aR_1}{R_1 + dR_2}$$

The above equation can be employed for the non-linear calibration curves typically encountered in many IR and Raman quantitation procedures.

$$\frac{C_A}{(C_A + C_B)} = \frac{R_1}{R_1 + dR_2} = \frac{1}{1 + d^{R_2/R_1}}$$

In a specialized case where $a = 1$ and $b = 0$, the above expression simplifies to the standard expression for the ratio method of a linear system that we started with.

3.3. Example of Quantitation Using Ratio Method of Analysis

Below is an example of quantitation of the components in PAM–AETAC water-soluble copolymers using both FT-IR and FT-Raman spectroscopy. In this example, a broad selection of nine well-characterized PAM–AETAC standards are used to develop calibration curves for both IR and FT-Raman spectroscopy. Since the calibration curves are non-linear, a linear fractional transformation equation is used to model the non-linear calibration curves.

The structure of poly AMD–AETAC is shown below.

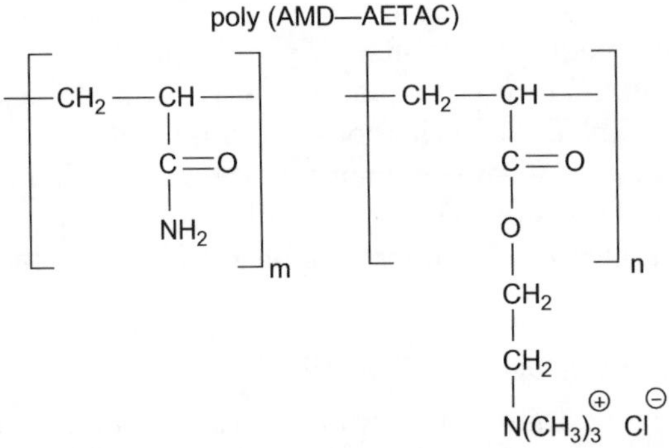

poly (AMD—AETAC)

Copolymers of acrylamide (AMD) and 2-(acryloyloxy) ethyltrimethylammonium chloride (AETAC)

Figure 3.10 shows the FT-IR and FT-Raman spectra of pure PAM, and two PAM–AETAC copolymers with 53 and 90% AETAC(Q9) respectively. These spectra illustrate the many strong characteristic bands that exist for the polymer backbone, primary acrylamide, the ester carbonyl, and the positively charged trimethylammonium chloride groups. Table 3.3 lists assignments of some of the important IR and Raman bands for PAM and AETAC.

Figure 3.11 shows the FT-IR and FT-Raman spectra of 47% AMD: 53% AETAC copolymers along with the selected baselines and peak heights used for this analysis. Bands characteristic of both groups are isolated and intense enough to provide quantitative bands for both components. In the IR spectrum, the quantitation method uses the AETAC ester carbonyl and the AMD amide carbonyl at 1734 and 1671 cm^{-1}, respectively. In the FT-Raman spectrum, the quantitation method uses the AETAC trimethylammonium chloride band at 719 cm^{-1} and the AMD primary amide NH_2 rock band at *ca.* 1105 cm^{-1}.

Figure 3.12 shows the FT-IR calibration plot for AMD–AETEC copolymers using a peak height ratio and plotting the calculated fit using nine calibration standards. The FT-IR calibration plot uses the peak height ratio of the AETAC ester carbonyl relative to the total carbonyl (ester + amide). The non-linearity observed in the above IR calibration curve is a result of deviations in Beer's law and overlapping contributions of the two carbonyl bands. The two variable functions chosen to model the calibration curve provides an excellent semi-empirical method to take into account these sources of non-linearity. The mole fraction AETAC with can be calculated from the FT-IR spectrum using the following expression:

$$\text{Mole fraction AETAC} = \frac{1.088(\text{Abs.}1734\ \text{cm}^{-1})}{(\text{Abs. }1734\ \text{cm}^{-1}) + 0.874(\text{Abs.}1670\ \text{cm}^{-1})}$$

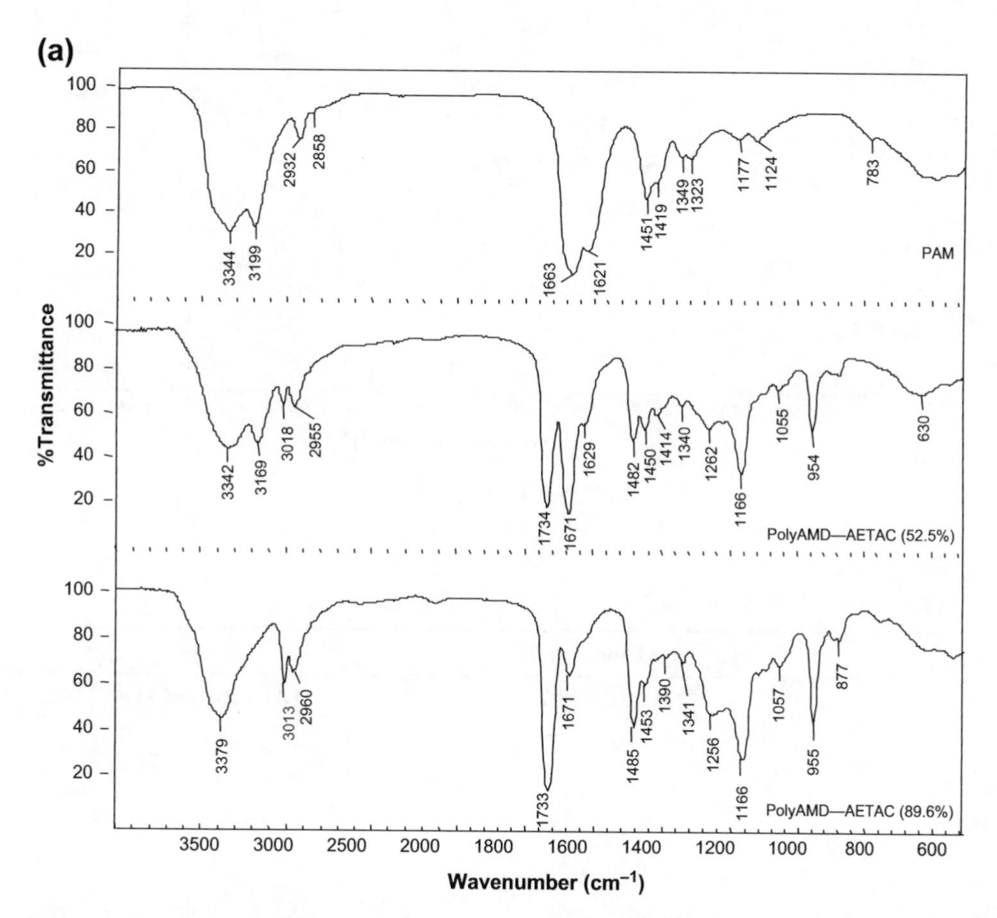

FIGURE 3.10 The FT-IR and FT-Raman spectra of the following: pure PAM and two PAM–AETAC copolymers (47/53 and 10/90). The FT-IR spectra are shown on the left and the FT-Raman spectra on the right. A cast film sample preparation on ZnSe of all three samples after dissolving the powders in water was used for the FT-IR data. For the FT-Raman data, the samples were in the supplied crystalline form using an NMR tube for sampling.

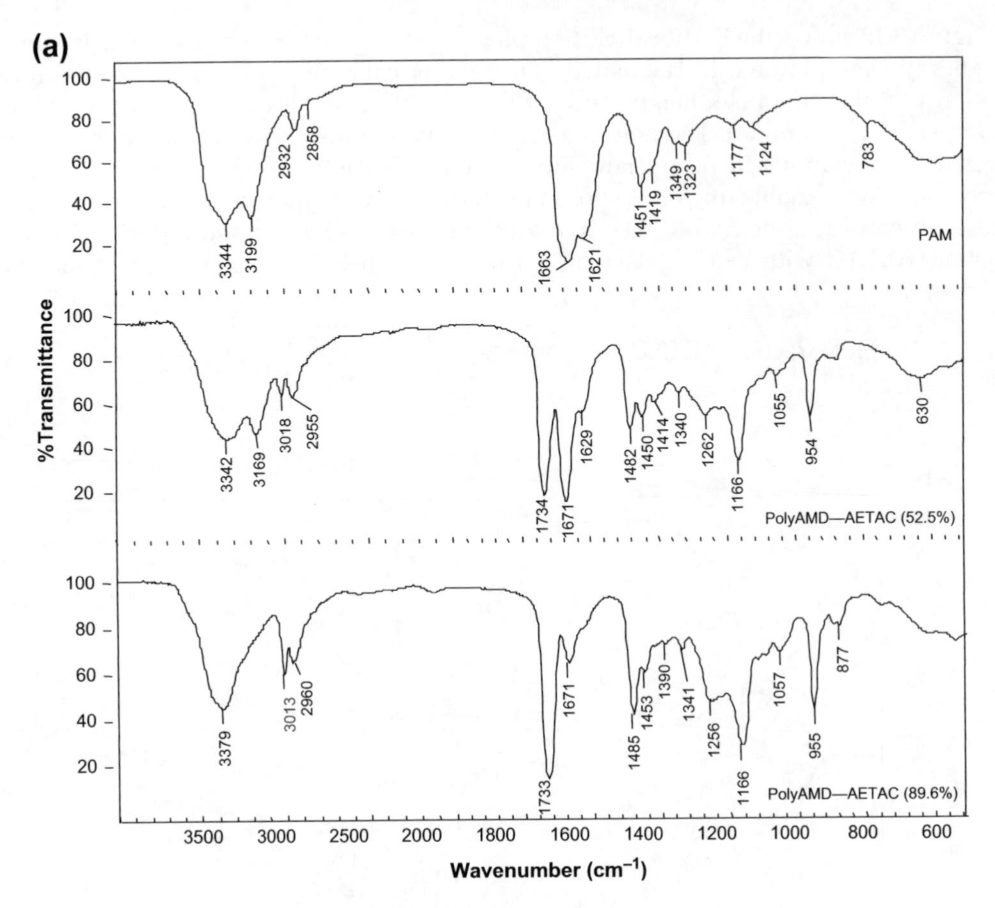

FIGURE 3.10 *(continued)*.

TABLE 3.3 Some Band Assignments for the Poly AMD-AETEC Samples

Group	PAM Bands (cm^{-1})	Assignments
NH_2	3350, 3200 (IR,R)	NH_2 i.ph. str. and o. ph. str.
NH_2	1608–1621 (IR,R)	NH_2 bend
NH_2	1107 (R)	NH_2 rock
C=O	1663–1656 (IR,R)	C=O str.
CH_2–CH	2936 (IR)	CH_2 o. ph. str.
CH_2–CH	2925 (R)	CH(C=O) str.
CH_2–CH	1455 (IR,R) 1324 (IR,R)	CH_2 bend
C–N	1424 (IR,R)	Involves C–N str.

TABLE 3.3 Some Band Assignments for the Poly AMD-AETEC Samples—cont'd

Group	PAM Bands (cm^{-1})	Assignments
C=O	1608–1621 (IR,R)	NH_2 bend
AETEC		
Ester C=O	1734 (IR,R)	C=O str.
Ester C=O	1166 (IR)	Involves C–O str.
$CH_2–N(CH_3)_3$	3010–3022 (IR,R)	CH_3 o. ph. str.
$CH_2–N(CH_3)_3$	955 (IR,R)	Involves NC_4 o. ph. str. + CH_3 rock
$CH_2–N(CH_3)_3$	720 (R)	Involves NC_4 i. ph. str.
Ester C=O	1734 (IR,R)	C=O str.

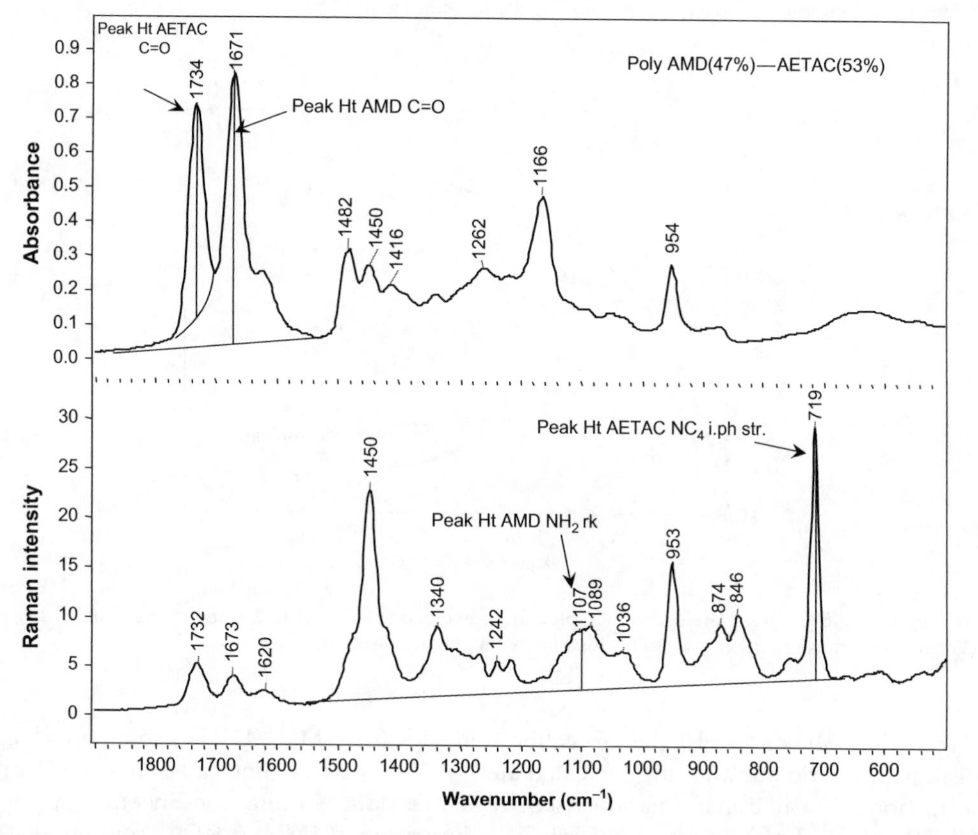

FIGURE 3.11 The baseline and peak height measurements used for quantitation. The top FT-IR spectrum shows the peak heights used to quantitate the ester and amide carbonyl bands. The bottom FT-Raman spectrum shows the peak heights used to quantitate the amide NH_2 rock and the cationic ammonium NC_4 in-phase stretch bands.

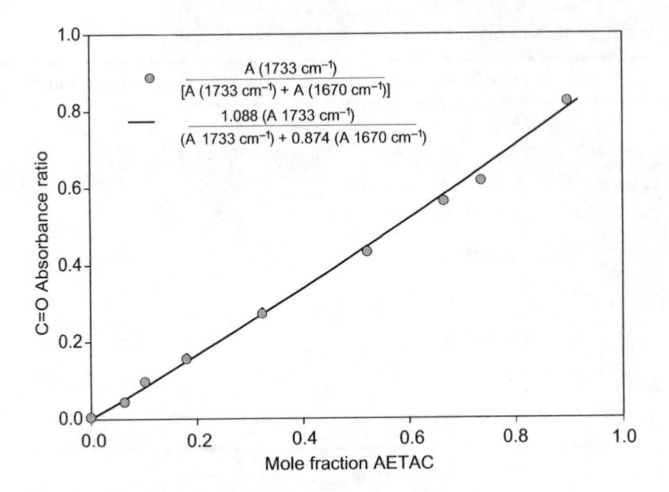

FIGURE 3.12 FT-IR calibration plot for AMD–AETAC copolymers. The form of the fractional linear transformation used to fit this data is only very slightly non-linear.

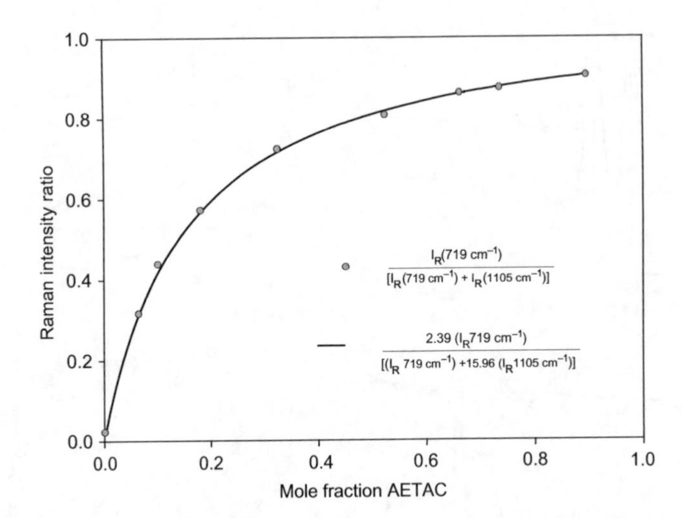

FIGURE 3.13 The FT-Raman calibration plot for powdered AMD–AETAC copolymers. The form of the fractional linear transformation used to fit this data is quite non-linear.

Figure 3.13 shows the FT-Raman calibration plot for AMD–AETEC copolymers using a peak height ratio and plotting the calculated fit for the same nine standards. The form of the fractional linear transformation used to fit this data is quite non-linear. Despite this, the FT-Raman technique provides excellent quantitation of PAM–AETAC copolymers. The mole fraction AETAC with can be calculated from the FT-Raman spectrum using the following expression:

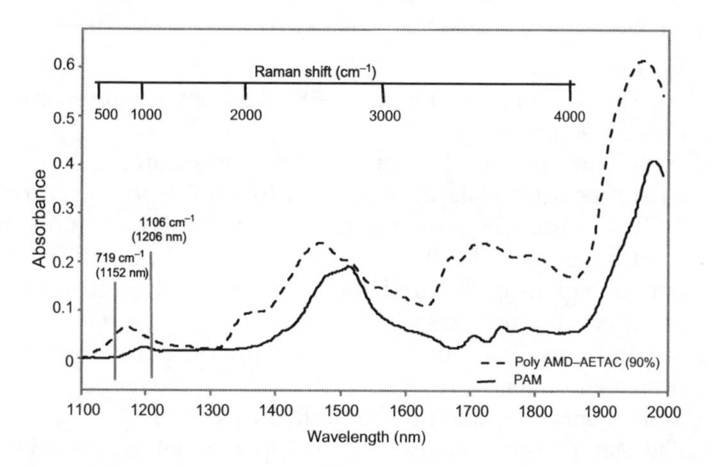

FIGURE 3.14 The diffuse reflectance near-IR spectra of pure PAM and poly AMD–AETAC (90%) copolymer. Note the different absorbances at 1152 and 1206 nm where the analytical Raman bands are located.

$$\text{Mole fraction AETAC} = \frac{2.39(I_R 719\ \text{cm}^{-1})}{[(I_R 719\ \text{cm}^{-1}) + 15.96(I_R 1105\ \text{cm}^{-1})]}$$

The significant non-linearity in the calibration curve above is due largely to narrow band self-absorption phenomena that are typical of 1064 nm excitation-based Raman spectra.[1] Such self-absorption is defined as the absorption of the Raman scattered light by the sample near-IR absorption bands themselves before the light can escape the sample in the NMR tube used in this experiment. Where there is a near-IR absorption band, the Raman band intensity will be decreased. The magnitude of this effect will vary significantly in powdered samples because of the variation in the path lengths the scattered light will follow. In a liquid sample, the magnitude of the Raman band suppression will follow a simple the Lambert–Beer law.

Figure 3.14 shows the diffuse reflectance near-IR spectra of pure PAM and poly AMD–AETAC (90%) copolymer. The 719 cm^{-1} AETAC band and the 1105 cm^{-1} PAM band both have different near-IR absorbances for these two components. Much of the observed curvature in the correlation plot is a consequence of this self-absorption process and not molecular environmental induced changes in the Raman cross sections.

Chapter 3: Questions

1. Outline the dispersive unit used in an FT-IR spectrometer to separate light into individual components.
2. Summarize the four guidelines for quality FT-IR spectrum.
3. Good sample-preparation skills are necessary to acquire photometrically accurate, high quality IR spectra. How can one differentiate between a good and a false IR spectrum? What causes a false IR spectrum?
4. Describe general regions of the fundamental vibrational spectrum with some characteristic group frequencies.
5. Outline one type of dispersive unit used in a Raman spectrometer to separate light into individual components.
6. Describe components of a Raman Fiber optic Probe.
7. List advantages and disadvantages of Raman spectrometers. Consider wavelength selection.
8. What is the difference between Rayleigh and Raman scattered light?
9. Summarize the advantages and disadvantages to KBr disc and Nujol mull sample preparations.
10. Does Raman or IR spectroscopy differentiate between amorphous and different crystalline materials?
11. Illustrate how light can be transmitted, refracted, and reflected through a crystal. What are the important variables?
12. What samples will provide the best quality IR spectrum using an ATR accessory and why? The worst?
13. How does an ATR measured spectrum differ from a classically measured transmission spectrum? What important parameters can the experimenter control?
14. Define the evanescent wave described by ATR theory. How does it define the sampling depth?
15. Identify three important criteria for selecting suitable ATR crystals.
16. What are the ideal sample characteristics for transmission IR microscopy?
17. List unwanted optical effects encountered in IR microscopy.
18. What is the diffraction light limit and what limits does it impose on IR microscopy?

References

1. *Infrared and Raman Spectroscopy, Methods and Applications*; Schrader, B., Ed.; VCH: New York, NY, 1995.
2. Diem, M. *Introduction to Modern Vibrational Spectroscopy*; John Wiley: New York, NY, 1993.
3. *Analytical Applications of Raman Spectroscopy*; Pelletier, M. J., Ed.; Blackwell Science: London, 1999.
4. Griffiths, P. R.; de Haseth, J. A. *Fourier Transform Infrared Spectroscopy*; John Wiley: New York, NY, 1986.
5. Chase, D. B.; Rabolt, J. F. *Fourier Transform Raman Spectroscopy, From Concept to Experiment*; Academic: 1994.
6. Colthup, N. B.; Daly, L. H.; Wiberley, S. E. *Introduction to Infrared and Raman Spectroscopy*, 3rd ed.; Academic: New York, NY, 1990.
7. Coleman, P. B. In *Practical Sampling Techniques for Infrared Analysis*; CRC: Boca Raton, 1993.
8. Beebe, K. R.; Pell, R. J.; Seasholtz, M. B. *Chemometrics A Practical Guide*; John Wiley: New York, NY, 1998.
9. Martens, H.; Naes, T. *Multivariate Calibration*; John Wiley: New York, NY, 1989.
10. Peter Fortini, personal communication.

CHAPTER

4

Environmental Dependence of Vibrational Spectra

1. SOLID, LIQUID, GASEOUS STATES

Vibrational spectroscopy shows a strong dependence on the environment of the molecule being examined. The IR and Raman spectra are significantly different for molecules in the gaseous, liquid, or solid state.

In the gaseous state, the molecules can be considered to be non-interacting and can rotate freely with a characteristic angular momentum. (We are ignoring pressure broadening.) Coupling of the vibrational transition with the rotational degrees of freedom of the molecule occurs resulting in band structure that is highly characteristic of the molecular structure and rotational–vibrational interactions.[1,2] This band structure is highly dependent upon the molecular symmetry and falls into four basic types: linear (e.g., diatomics, CO_2) spherical top (e.g., CH_4, SF_6), symmetric top (oblate e.g. $CHCl_3$ prolate e.g., CH_3Cl), and asymmetric top (e.g., H_2O). Numerous qualitative and quantitative gas phase applications exist using both IR and Raman techniques but will not be covered here.

Rotational fine structure can be observed in the vibrational–rotational band cluster if the molecular moment of inertias are sufficiently low as it is typical for simpler molecules. Fine structure occurs because angular momentum is also quantized. For larger molecules, the fine structure is typically unresolved resulting in broad bands. However, the band contour reveals the molecule symmetry and the direction of the dipole moment change during the vibration. Figure 4.1 shows the classical description of the IR gas phase contours of carbon dioxide, a linear type rotator. In Fig. 4.1a, CO_2 is bending and rotating about an axis parallel to the change in the dipole moment which is not affected by the rotation. As seen in Fig. 4.1b, this results in a single frequency depicted in Fig. 4.1c. In Fig. 4.1d, this rotation changes the orientation of the dipole moment vector, the vertical component which shown in Fig. 4.1e interacts with the radiation oscillating in the vertical plane. This complex wave function can be resolved trigonometrically into two steady amplitude waves which give rise to two lines in the spectrum at $V - R$ and $V + R$ (see Fig. 4.1f). Although the molecule vibrates at only one frequency, the rotational frequency varies, giving rise to two broad wings on the central band shown without fine structure. The total band cluster is shown in Fig. 4.1g as the sum of Fig. 4.1c and Fig. 4.1f. As shown in Fig. 4.1h and Fig. 4.1j, the CO_2 molecule is

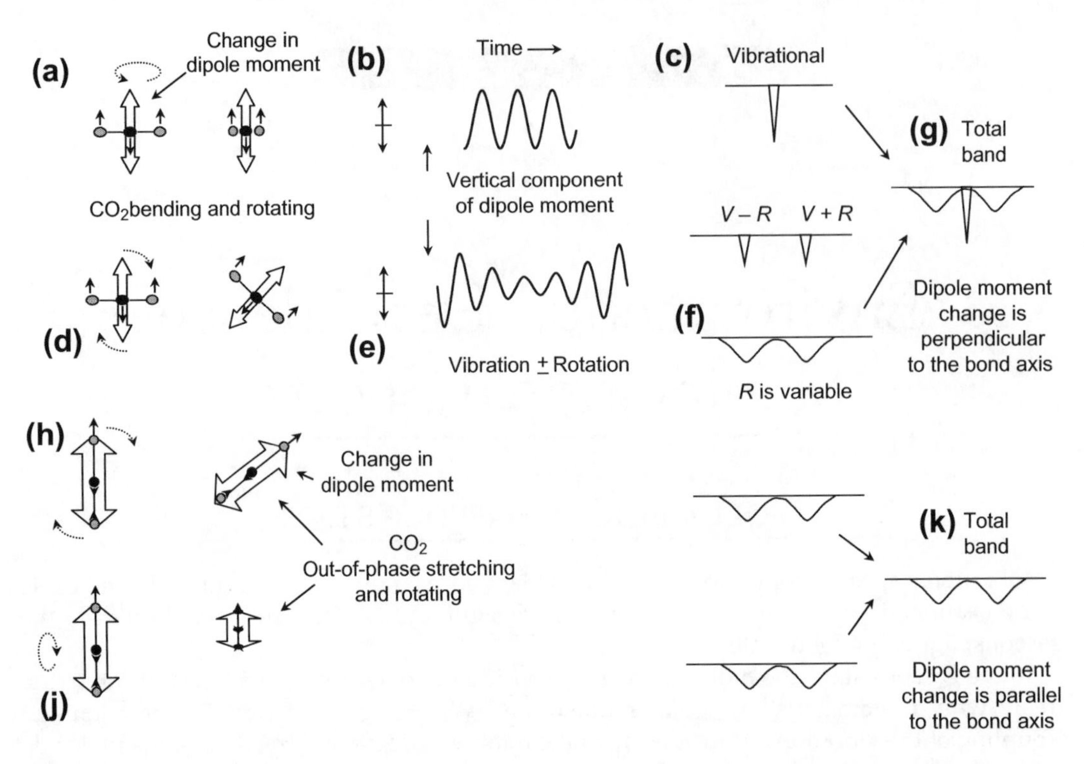

FIGURE 4.1 Classical treatment describing the IR gas phase contours of CO_2. Although the molecule vibrates at only one frequency, the rotational frequency varies, giving rise to two broad wings on the central band shown without fine structure. In this case the CO_2 molecule cannot rotate without rotating the dipole moment vector and the total band contour shown in (k) is a broad doublet with no central peak.

stretching out-of-phase. In this case, the CO_2 molecule cannot rotate without rotating the dipole moment vector and the total band contour shown in Fig. 4.1k is a broad doublet with no central peak.

Figure 4.2 shows the characteristic band clusters for three different vibrational modes of a para-substituted aromatic, an asymmetric top rotator. In larger molecules such as these, the rotational fine structure is typically not observed. The white arrows in the first two in-plane vibrations and a white plus sign for the last out-of-plane vibration indicate the dipole moment changes. These vectors are each parallel to one of the axes of rotational moments of inertia as shown in the figure. The rotational moments in the condensed phase state molecules can and do interact and these interactions strongly effect the vibrational spectrum.

In the liquid state, molecules may have any orientation relative to the spectrometer. Further, a molecule may be found in several conformations, all of which may have different spectra (e.g., rotational isomers). No rotational fine structure is observed and the resulting overall spectrum may consist of a large number of fairly broad bands. Here, hydrogen bonding can have a dramatic effect on the IR and Raman spectra. Typically, if the molecule is associated, the bands will also be broader than for the non-associated state.

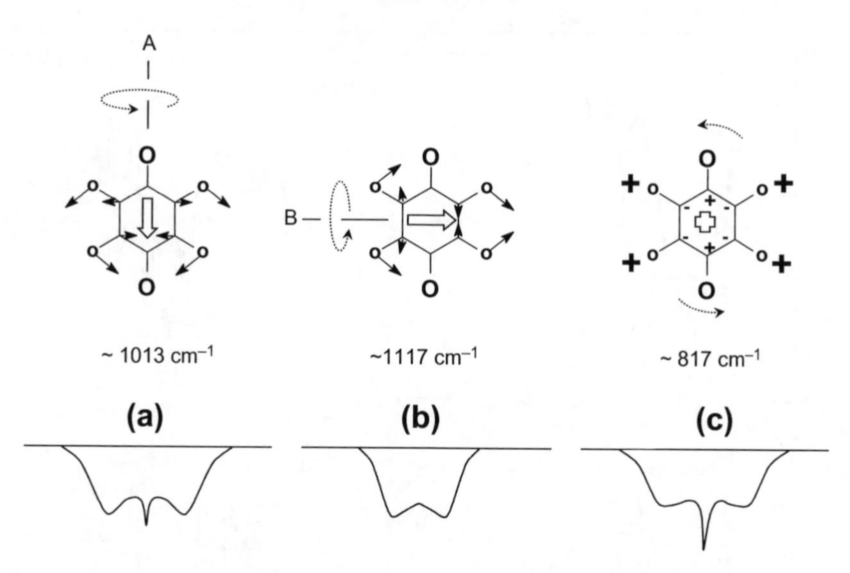

FIGURE 4.2 Asymmetric top gas phase band contour (para-substituted benzenes). inertia are at a minimum at (a), intermediate at (b), and at a maximum for (c). In the bottom row are typical gas phase IR spectra showing the gas phase contour for a, b, and c+ type bands. The b type bands have no central peak. In a and c type bands, the relative intensities of the central peaks to the band wings vary with changes in the relative a, b, and c moments of inertia.

Lastly, in the solid state, the crystallinity (or lack of it) of the molecule can be quite important and the nature of the sample preparation techniques employed can also be important. Figure 4.3 shows the IR spectra of common table sugar (sucrose) prepared as a KBr disc and as a cast film from water and clearly illustrate the differences observed in the IR spectrum of a crystalline and amorphous material. The large number of sharp bands evident in the upper spectrum is characteristic of a crystalline solid. When sugar is prepared as a water cast film, it becomes an amorphous material resulting in broader less distinct bands seen in the lower spectrum. Both spectra have some water present.

In general, crystalline solids can be frozen into one or more configurations fixed in a lattice resulting in sharper bands. Often, crystalline splitting of degenerate vibrations can occur and forbidden vibrations may result in weak bands. Vibrational spectroscopy can distinguish between different crystalline forms and quantitatively assay mixtures. This ability has been exploited for various applications.

2. HYDROGEN BONDING

Both IR and Raman spectroscopies are very sensitive to the molecular interactions involved in hydrogen bonding.[3] The hydrogen bond itself involves an association of a hydrogen atom from an X–H group of a molecule with the Y atom of another group (X–H–Y).[4,5] The X type atoms involved often have high electronegativities resulting in an

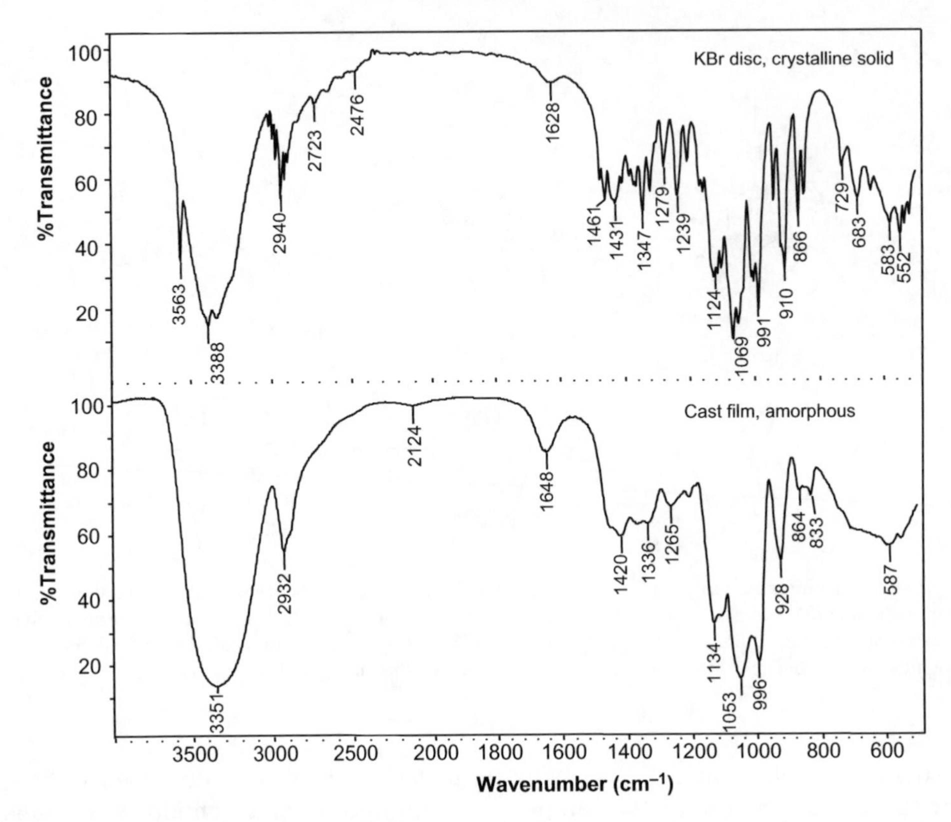

FIGURE 4.3 The FT-IR spectra of common table sugar. The upper spectrum was prepared as a KBr disc while the lower spectrum was prepared as a water cast film on ZnSe.

X—H bond which is partially ionic while the H—Y bond is suggested to be mostly electrostatic with some weak covalent characteristics. In general, hydrogen bonding to an X—H group results in a decrease in the X—H stretching frequency accompanied by a broadening and intensification of this band. Both OH and NH groups are found to show highly characteristic changes with hydrogen bonding. Figure 4.4 shows the OH stretches for two different phenolic OH stretches. The steric hindrance provided in 2,6-di-tert-butylphenol results in a strong sharp band at 3645 cm^{-1} characteristic of a non-hydrogen bonded OH group. Conversely, the easily accessible OH group of p-creosol results in a broad strong band at 3330 cm^{-1} characteristic of intermolecular hydrogen bonding between different phenolic OH groups. Intermolecular hydrogen bonding in chemical systems are very common. Examples of the condensed phase hydrogen bonded OH and NH stretching frequencies for carboxylic acid dimer and isocyanuric acid are shown below.

Carboxylic Acid Dimer

O---H—O
R—C C—R OH Str. ~ 3000 cm^{-1}
O—H---O

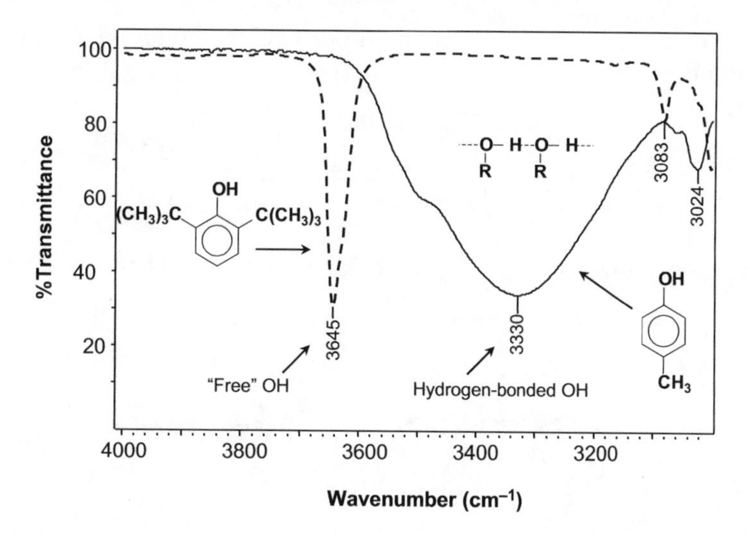

FIGURE 4.4 The solid state IR spectrum of para-creosol and 2,6-di-tert-butylphenol illustrating the effect of hydrogen bonding on the OH stretching frequency.

Isocyanuric acid: Note that all of the NH and C=O groups are involved in hydrogen bonding but only one set is shown for simplicity.

NH Str. ~ 3209, 3080 cm^{-1}

The carbonyl C=O stretching frequency for the carboxylic acid dimer and the isocyanuric acid molecules also have a lower frequency and broader bandwidth as a result of hydrogen bonding. Additional changes in the vibrational spectrum of molecules involved in hydrogen bonding also occur. Bands involving X–H bending will usually increase in wavenumber upon formation of the hydrogen bond. The resultant changes in frequency and bandwidth are not as pronounced as observed in the X–H stretching region. Lastly, new low frequency bands can often be seen involving the stretching and bending of the H–Y bond. The OH–O out-of plane hydrogen bend of the carboxylic acid dimer (960–875 cm^{-1}) is a classic well-known example of this.

Intramolecular hydrogen bonding can have a dramatic effect on the IR and Raman spectra and typically favors formation of five- and six-membered rings. Conjugated intramolecular hydrogen bonding systems in particular can result in very strong hydrogen bonds. Examples of this include the keto-enol conjugated system which involves conjugated ketone and hydroxy groups. As shown below, a resonance structure can be drawn that places the negative charge on the carbonyl oxygen and a positive charge on the oxygen carrying the hydrogen bonding proton.

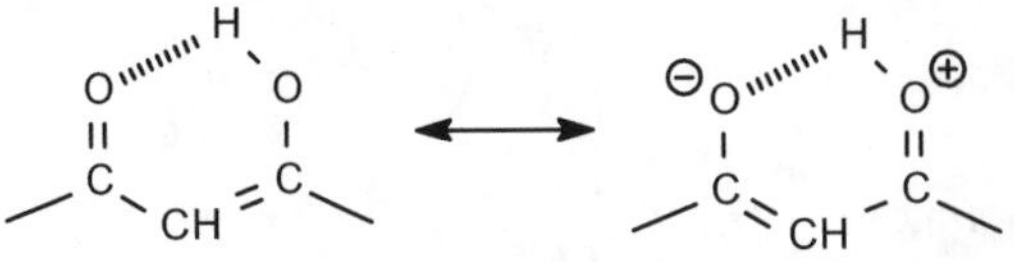

Some examples of intramolecularly hydrogen bonded systems are shown below.

2-Hydroxy-4-(methoxy)-benzophenone

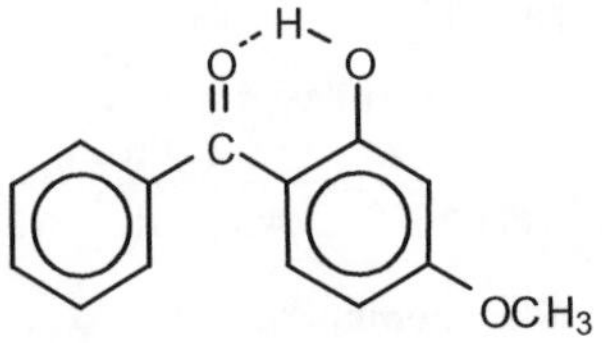

OH Str. ~2900 cm^{-1}

2-(4,6–Diphenyl–1,3,5–triazine–2–yl)–5 alkyloxy phenol

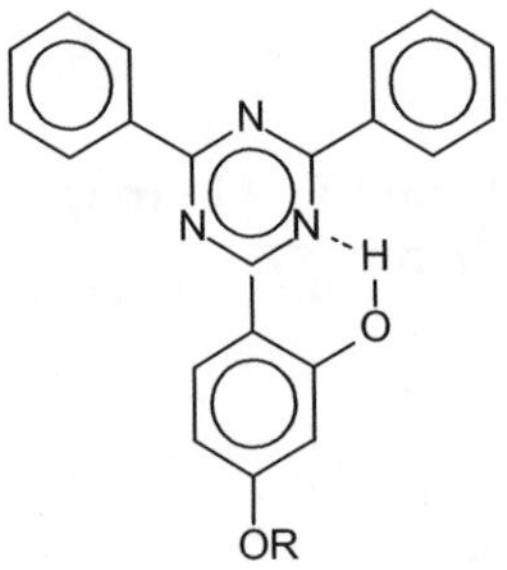

OH Str. ~ 2900 cm^{-1}

3. FERMI RESONANCE

The vibrational spectrum of a molecule typically is considered to have isolated vibrational coordinates resulting fundamental vibrations. For example, carbon dioxide which has four total vibrations ($3n - 5$) has two IR active vibrations, an out-of-phase stretch and a doubly degenerate deformation and a Raman active in-phase stretch. However, under certain conditions the vibrational levels are not well separated and can interact leading to phenomena such as Fermi resonance.[6] Such interaction requires suitable symmetry for the two vibrations, an anharmonic nature of the vibration, and a similar energy for the two states (typically within 30 cm^{-1}). The vibrations cannot be significantly separated within distinctly different parts of the molecule and they must mechanically interact in order for the two vibrations to mix.

In most cases, Fermi resonance involves a fundamental that interacts with a combination or overtone band resulting in two relatively strong bands where only one nearly coincident fundamental is expected. Figure 4.5 shows the idealized vibrational spectrum for a stretching and bending vibration before (Fig. 4.5a) and after Fermi resonance along with the idealized energy level diagram. The idealized Fermi resonance shown involves the overtone of

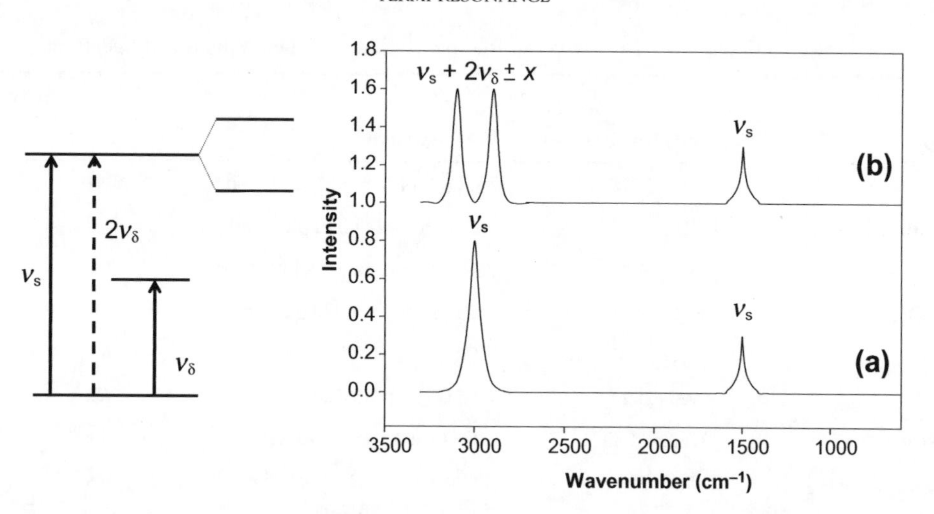

FIGURE 4.5 Schematic diagram illustrating Fermi resonance. The idealized vibrational spectrum for a stretching and bending vibration are shown before (a) and after Fermi resonance (b). To the left of the idealized spectra is the energy level diagram.

a bending vibration (ν_δ) with a symmetric stretching vibration (ν_S) that results in a classic Fermi resonance doublet that straddles the original symmetric band frequency. The Raman spectrum of carbon dioxide shown in Fig. 4.6 provides a classic example of Fermi resonance.[6] Only one band for the CO_2 in-phase stretch fundamental is expected (1332 cm^{-1}) but a classic Fermi resonance doublet is observed at 1385 and 1278 cm^{-1} instead. The two Raman bands are a linear combination of the fundamental in-phase stretch and the CO_2 bend overtone.

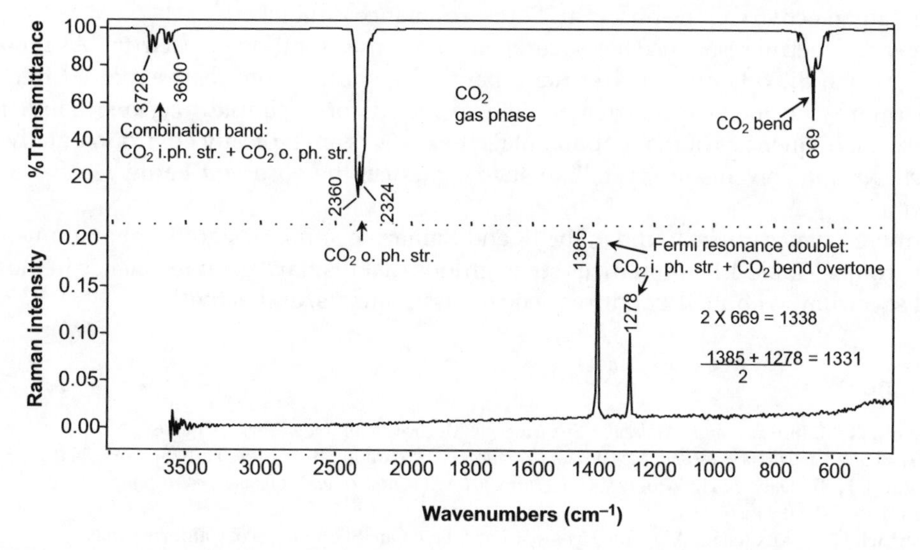

FIGURE 4.6 The IR and Raman spectra of carbon dioxide. Note the Fermi resonance doublet at 1385 and 1278 cm^{-1} resulting from interaction of the $\upsilon = 1$ CO_2 i.ph. str. and the $\upsilon = 2$ (overtone) of the 669 cm^{-1} CO_2 bend.

TABLE 4.1 Selected Groups that Display Fermi Resonance Bands in their Vibrational Spectrum.

Group	Bands (cm^{-1})	Assignment	Active	
			IR	Raman
CO_2	1385, 1278	CO_2 i. ph. str + 2x CO_2 bend	No	Yes
R–C(=0)–H	~2820, ~2710	CH str. + 2x CH i.ph. bend	Yes	Yes
$N{=}C{=}O^-$	~1300, ~1200	NCO i.ph. str. + 2x NCO bend	Yes	Yes
R–COOH	~3000–2650	OH str. + 2x OH i. pl bend	Yes	Yes
	~3000–2550	OH str. + 2x C-O str.	Yes	Yes
CH_2	~2920–2890	CH_2 str. + 2 x CH_2 bend	Yes	Yes
Aryl–C(=O)–Cl	~1780, ~1730	C=O str. + 2x 880 aryl-C str.	Yes	Yes
$N{\equiv}C{-}N{=}C(NH_2)_2$	2200, 2170	C≡N str. + (1250 +925 cm^{-1}) Combination band	Yes	–
R–C≡C–R	2300, 2235	C≡C str + C–C≡C bend	No	Yes
C=O	1810, 1780	C=O str. + 2x ~900 ring bend	Yes	–
Aryl–NH_2	3355, 3205	NH_2 i. ph. str. + 2x NH_2 bend	Yes	Yes
NH_4Cl	2825	NH_4 i. ph. str. + 2x NH_4 bend	Yes	Yes
– C(=O)–NH–C	3300 (s), 3100(m)	MH str. + 2x CNH str.-bend	Yes	Yes

Also noteworthy is that the combination band deriving from the bands involving the CO_2 in-phase str. and o. ph. str. are observed as two doublets in the 3728–3600 cm^{-1} region because of the Raman active CO_2 in-phase str. Fermi resonance doublet.

In general, Fermi resonance has several effects on the vibrational spectrum. As illustrated by the carbon dioxide case, the resultant bands are shifted from their expected frequency. Furthermore, the intensity is greater than expected for a simple overtone band. Lastly, because the frequencies of the fundamentals themselves are typically environmentally sensitive, changes in solvents or crystalline state can affect the observed Fermi resonance band intensities.

Fermi resonance can be found in the IR and Raman spectra of a significant amount of functional groups.[3] Table 4.1 summarizes some groups that display Fermi resonance in the vibrational spectrum, with their band frequencies, assignments, and activities.

References

1. Barrow, G. M. *Introduction to Molecular Spectroscopy*; McGraw-Hill: New York, NY, 1962.
2. *Infrared and Raman Spectroscopy, Methods and Applications*; Schrader, B., Ed.; VCH: New York, NY, 1995.
3. Colthup, N. B.; Daly, L. H.; Wiberley, S. E. *Introduction to Infrared and Raman Spectroscopy*, 3rd ed.; Academic: New York, NY, 1990.
4. Pimentel, G. C.; McClellan, A. L. *The Hydrogen Bond*; Freeman: San Francisco, California, 1960.
5. Bene, J. E.; Jordan, M. J. T. *Int. Rev. Phys. Chem.* **1990**, *18* (1), 119–162.
6. Diem, M. *Introduction to Modern Vibrational Spectroscopy*; John Wiley: New York, NY, 1993.

CHAPTER

5

Origin of Group Frequencies

When a molecule contains a bond or group with a vibrational frequency significantly different from adjacent groups, then the vibrations are localized primarily at this group. Such characteristic IR and Raman bands for particular bond or group are called "group frequencies" and are extremely useful for both qualitative and quantitative analysis. In order to understand the origin of these group frequencies, coupled oscillator systems are presented here and related to actual chemical groups. These model systems are based upon a classical description of different general types of mechanical coupling of bond vibrations.[1,2,3] This overview enables us to provide a framework to understand the sensitivity of vibrational spectra to molecular structure. Lastly, some general rules concerning mechanical coupling of vibrations are presented.

1. COUPLED OSCILLATORS

The simplest coupled oscillator system to consider is the stretch–stretch interaction of a linear triatomic molecule.[1,2] Here, only two bonds with two stretching vibrations are present which will mechanically couple. The two-coupled oscillators exert forces on each other when they oscillate, resulting in both in-phase and out-of-phase stretching vibrations (see Fig. 5.1). The out-of-phase stretch is of higher frequency than the in-phase stretch. The key difference between the two vibrations is the displacement of the central atom. Here we label the central atom as Mc and the two equivalent external atoms as M.

The out-of-phase (asymmetric) stretch involves movement of all three atoms (see Fig. 5.1a) and the two vectors associated with the central mass reinforce one another (i.e., cooperate), resulting in larger movement of the central atom during the vibration. Characteristic phasing of the out-of-phase stretch is that one bond stretches while the other contracts. We can regard the middle atom as split into two halves moving with the same frequency and phase. Hence we consider the central atom, Mc, as 2 × (M/2). Thus, the two oscillators of $M - (M/2)$ vibrate at 180° out-of-phase with each other. The result of this is that the frequency of this type of vibration is strongly dependent on the mass of the central atom. We can extend the classical vibrational frequency definition presented earlier to the out-of-phase stretching vibration by splitting the central mass in half and the simple expression is:

$$\bar{\nu}_{\mathrm{op}} = 1303\sqrt{K\left(\frac{1}{m_1}+\frac{2}{m_2}\right)} = 1303\sqrt{\frac{K}{m}}\sqrt{3}$$

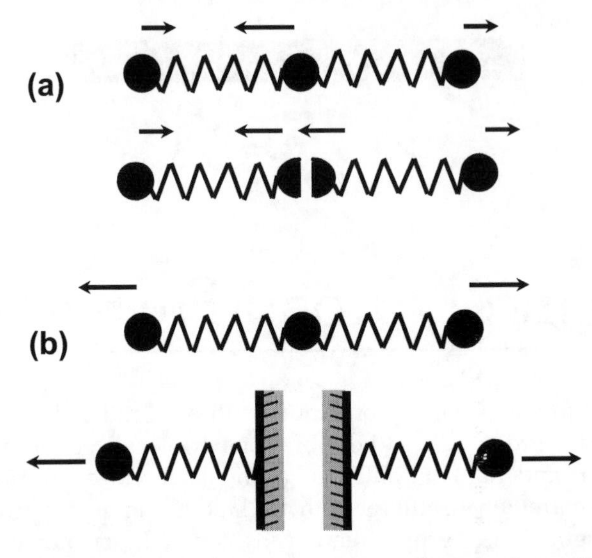

FIGURE 5.1 The stretching vibrations of a linear triatomic. The in-phase and out-of-phase stretches have been resolved into diatomic M–(∞) and M–(M/2) components. The out-of-phase stretch is shown in diagram a and the in-phase stretch in diagram b.

where K is the force constant and gives the restoring force for unit displacement from the equilibrium position, and m_1 and m_2 are the masses for the linear triatomic in Fig. 5.1. For three equivalent masses $\bar{\nu}_{op}$ simplifies to the last term in the above equation.

In the case of the in-phase (symmetric) stretch, the central atom does not move (see Fig. 5.1b). This is equivalent to the two exterior masses being connected to an infinite mass (∞) where both m—∞ systems oscillate with same frequency and phase. In this case, the two spring force vectors exerted on the central mass oppose one another, resulting in no movement of the central atom during the vibration. Here, the frequency of this vibration is independent of the mass of the central atom in the symmetrical molecule and can be expressed classically as:

$$\bar{\nu}_{ip} = 1303\sqrt{K\left(\frac{1}{m_1}+\frac{1}{\infty}\right)} = 1303\sqrt{\frac{K}{m}}\sqrt{1}$$

Comparison of the above simple expressions predicts that the out-of-phase vibrations will be higher in frequency than the in-phase vibration.

For a bent triatomic molecule, we find that the frequencies of the in-phase and out-of-phase stretching vibrations are a function of the bond angle, α, between the three atoms.[1,2] We again consider the force vectors for the central atom m_2 relative to the two exterior atoms, m_1 and m_3 for the in-phase and out-of-phase stretches.

Figure 5.2 (lower left) shows the total restoring force tending to move the central atom along the right-hand bond. The restoring force vectors will by definition be opposite in sign from the Cartesian displacement vectors. For the in-phase mode, the forces act in opposition so there is less restoring force, while for the out-of-phase mode the forces cooperate resulting in more restoring force. The same analysis holds for the left-hand bond. The

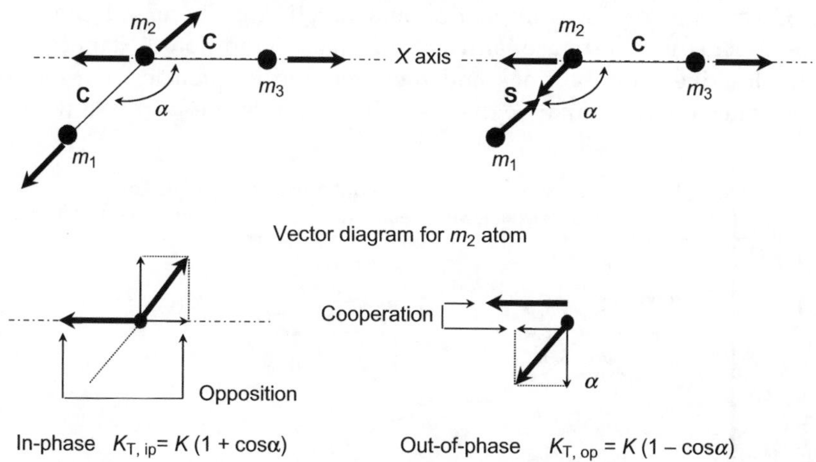

FIGURE 5.2 The force diagrams for the central atom in a bent triatomic. The nature of the interaction of the two bond oscillators is a function of the bond angle, α. S and C indicate stretching and contraction of a bond and the bold arrows are the restoring forces. Both the in-phase (contract-contract) and the out-of-phase (stretch-contract) modes are shown.

resultant mechanical coupling tends to lower and raise the in-phase and out-of-phase frequencies, respectively relative to that of the m_2–m_3 diatomic model.

Taking into account the bond angle dependence, we can calculate the frequency for the in-phase and out-of-phase stretching vibrations for bent triatomics. The out-of-phase stretch can be approximately calculated by:

$$\bar{\nu}_{\mathrm{op}} = 1303\sqrt{K\left(\frac{1}{m_1} + \frac{1 - \cos\alpha}{m_2}\right)}$$

where the bending force constant can be assumed to be zero. The in-phase stretch can be approximately calculated by:

$$\bar{\nu}_{\mathrm{ip}} = 1303\sqrt{K\left(\frac{1}{m_1} + \frac{1 + \cos\alpha}{m_2}\right)}$$

Here, we again neglect the bending force constant, but in this case it does contribute slightly.

Figure 5.3 demonstrates the frequency dependence of triatomic in-phase and out-of-phase stretches on the bond angle, α. When α is greater than 90°, then the in-phase stretch is lower in frequency than the out-of-phase stretch, while α is less than 90°, then the in-phase stretch is higher in frequency than the out-of-phase stretch. Lastly, when $\alpha = 90°$ the in-phase and out-of-phase stretches are equal because there is no mechanical interaction (vectors are orthogonal).

An important extension of the above discussion is to consider the effects of mechanical coupling as a function of chain length.[1,2] This is important since it applies to linear aliphatic systems. Shown in Fig. 5.4 is a plot of vibrational frequencies for coupled C–C stretches of linear aliphatics as a function of increased chain length (i.e., 2,3,4... atoms) showing in-phase and out-of-phase vibrations. Standing waves are used to help describe the vibrations and their relative frequencies. For clarity, a plane (or node) separates the molecular segments, which vibrate out-of-phase with each other.

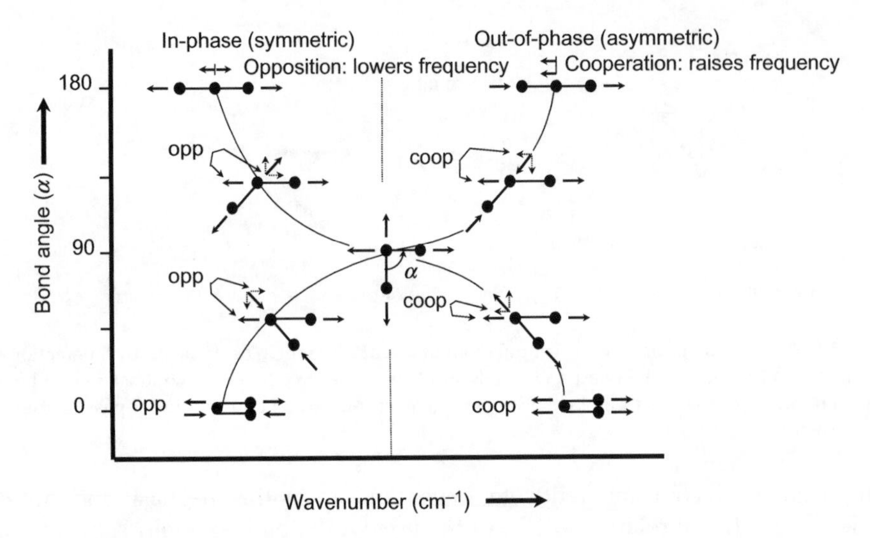

FIGURE 5.3 The frequency dependence of the in-phase and out-of-phase stretches of a triatomic molecule as a function of the bond angle, α. The arrows shown are displacement vectors.

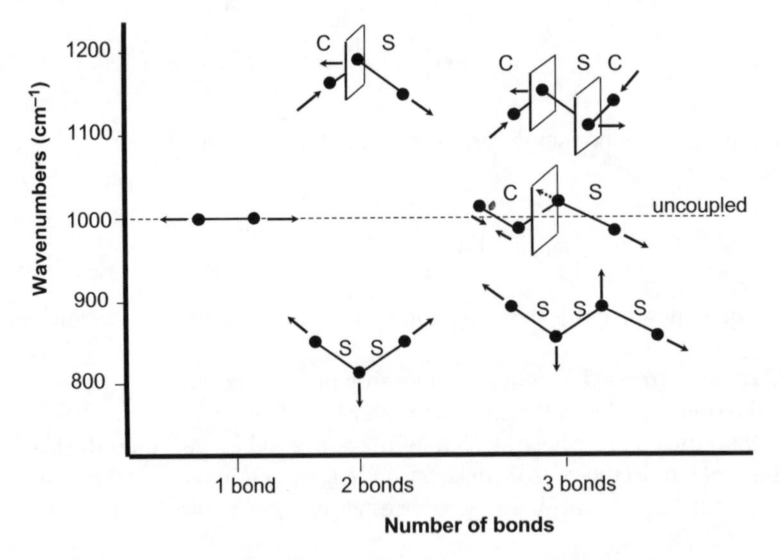

FIGURE 5.4 Plot of vibrations for coupled C–C stretch of linear aliphatics, as a function of increased chain length.

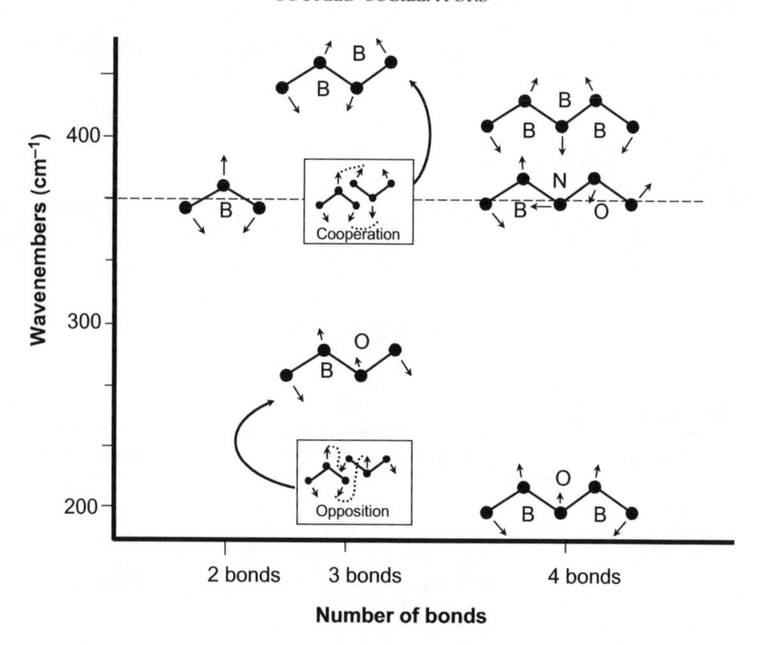

FIGURE 5.5 Plot of vibrations for coupled C–C bending vibrations of linear aliphatics, as a function of increased chain length. Note the out-of-phase vibration (opposition) is lower in frequency than the in-phase vibration (cooperation). Bending and open vibrations are indicated with B and O. No change in the bond angle is indicated by N.

Usually, the greater the number of nodes for stretching vibrations, the higher is the frequency.

Figure 5.4 shows that a connection of several similar oscillators in linear chains results in as many chain frequencies as there are bonds. Thus, similar C–C, C–O, and C–N bonds will interact resulting in prominent IR and Raman bands. The relatively constant mechanical interaction gives results in group frequencies. Fortunately, for longer chains only a few of these vibrations provide important bands.

We must also consider bending vibrations. The skeletal single-bond bending vibrations in aliphatic molecules vibrate at much lower frequencies than C–C stretches, CH stretch, or CH bending vibrations and generally fall in the far-IR region. The vibration involves mostly a deformation of the C–C–C bonds and the out-of-phase vibration is lower in frequency than the in-phase vibration. Figure 5.5 plots the vibrations for coupled C–C bending vibrations of linear aliphatics, as a function of increased chain length.

2. Rules of Thumb for Various Oscillator Combinations

For oscillator combinations of equal or nearly equal frequency both in-phase and out-of-phase oscillator combinations are involved.[1,2] This simplest case considers equal oscillator combinations and includes stretch–stretch and bend–stretch vibrations. Some examples are shown in Fig. 5.6 and Table 5.1.

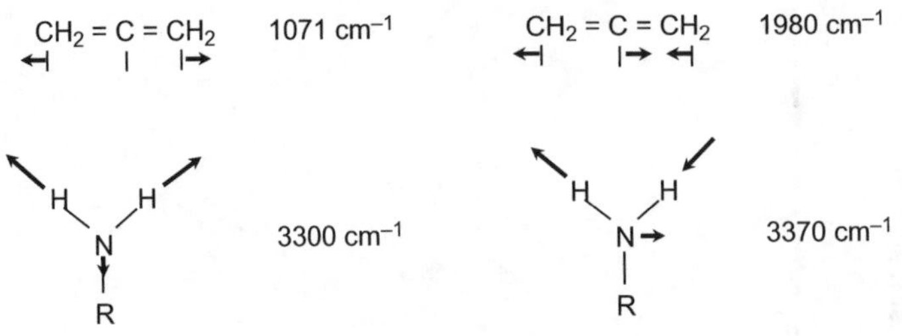

$CH_2 = C = CH_2$ 1071 cm^{-1}

$CH_2 = C = CH_2$ 1980 cm^{-1}

3300 cm^{-1}

3370 cm^{-1}

FIGURE 5.6 Examples of equal oscillator combinations.

TABLE 5.1 Examples of XY_2 Type Identical Oscillators

XY_2 Group	Out-of-phase stretch	In-phase stretch
$R–NH_2$	3370 cm^{-1}	3300 cm^{-1}
$R–CH_2–R$	2925 cm^{-1}	2850 cm^{-1}
$R–CO_2^-$	1600 cm^{-1}	1420 cm^{-1}
$R–NO_2$	1550 cm^{-1}	1375 cm^{-1}
$R–SO_2–R$	1310 cm^{-1}	1130 cm^{-1}

Examples of some two identical oscillator groups (non-linear $M_1–M_2–M_1$) shown in Table 5.1 include: two hydrogen atoms on groups such as a methylene or primary amine and two oxygen atoms on groups such as carboxylate, nitro or sulfones.

The XY_2 group in the above can be considered approximately as an isolated oscillator since it has stretching frequencies significantly higher than the frequencies of the C–C bond oscillators connected to it. As a result only the XY_2 group atoms move appreciably during the stretching vibration. Because of this the above groups result in good group frequencies.

The out-of-phase stretch is also an important IR group frequency for linear alkane alcohols, ethers, and primary and secondary amines.[2] The vibration involves stretches of the C–C, C–O, or C–N bonds, which have very similar force constants. The frequencies shown below support this.

$$CH_3 - CH_3 - 992\ cm^{-1}$$
$$CH_3 - OH - 1033\ cm^{-1}$$
$$CH_3 - NH_2 - 1036\ cm^{-1}$$

Consequently, stretches involving C–C, C–O, and C–N will strongly couple giving rise to skeletal out-of-phase stretching vibrations in the following general spectral regions:

$$C - C - O \sim 1050\ cm^{-1}$$
$$C - O - C \sim 1125\ cm^{-1}$$
$$C - C - N \sim 1080\ cm^{-1}$$
$$C - N - C \sim 1140\ cm^{-1}$$

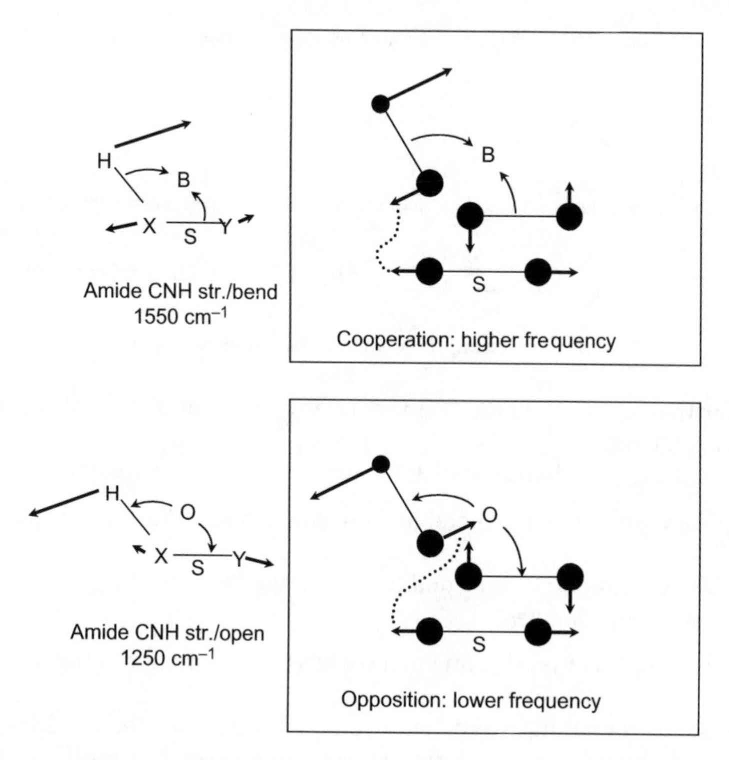

FIGURE 5.7 The stretching–bending interaction for the H–X–Y group. S stands for stretch, B for bend, and O for open. The approximate band position is also included for the amide group CNH stretch/bend and stretch open vibrations.

The C–C–O and C–O–C species in particular give good strong IR group frequencies as a result of a relatively constant interaction with other C–C bonds.

Figure 5.7 shows the second type of important equal oscillator vibration, the bend-stretch vibration.[2] Mechanical coupling of bending and stretching vibrations can occur if both are of nearly equal frequency. In this case, we consider a simple bent H–X–Y oscillator where both X and Y are heavier than the hydrogen. The H–X bending frequency may be nearly identical to that of the X–Y stretching frequency, allowing the two oscillators to couple. Two vibrations result involving the X–Y stretch and the H–X–Y bend which differ only in the relative phases. An example of this is the amide group where X = nitrogen and Y = carbon. The H–X stretching oscillator has much higher frequency than the X–Y stretching oscillator and is therefore an isolated vibration (see below for rules for unequal oscillators).

As shown above, there are three components for this molecular vibration: the rotation of the X–Y bond, the stretching of the X–Y bond, and the rotation of the H–X bond. Here, the rotation of the X–Y bond has little effect since the vectors involved are orthogonal (i.e., 90°) with respect to both the X–Y stretch and the H–X rotation. However, rotation of the H–X bond results in movement of the X atom either in the same or in opposite direction that it moves during the X–Y stretch. The bend/stretch involves cooperation of the X atom vectors

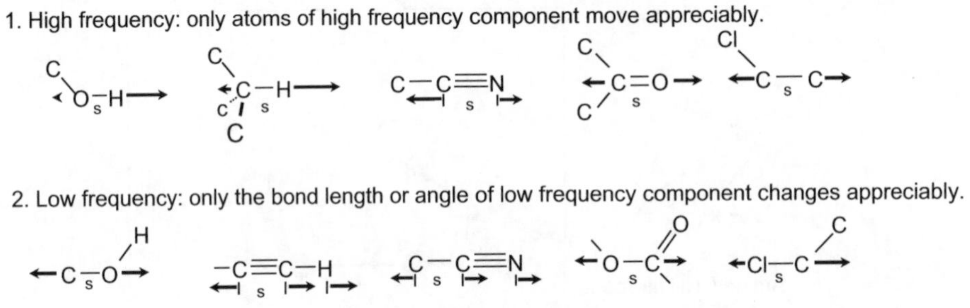

1. High frequency: only atoms of high frequency component move appreciably.

2. Low frequency: only the bond length or angle of low frequency component changes appreciably.

FIGURE 5.8 Examples of unequal oscillator combinations.

and thus higher frequency while the bend/open involves opposition of the X atom vectors and thus lower frequency.

Oscillators with different frequencies have two important approximate rules of thumb:

1. In high frequency vibration of the combination, only atoms of the high frequency component move.
2. In low frequency vibration of the combination, only the bond length or angle of the low frequency component changes.

Some simple examples of typical unequal oscillator combinations encountered are shown in Fig. 5.8.

Lastly, for systems containing more than two equal oscillators, the resulting vibrations can become much more complex and standing waves can be used to simplify the description of the vibrations.[2] An excellent example of this is the stretching vibrations for a benzene ring. Figure 5.9 shows the standing vibrational waves employed to describe the stretching vibrations of six-membered rings.

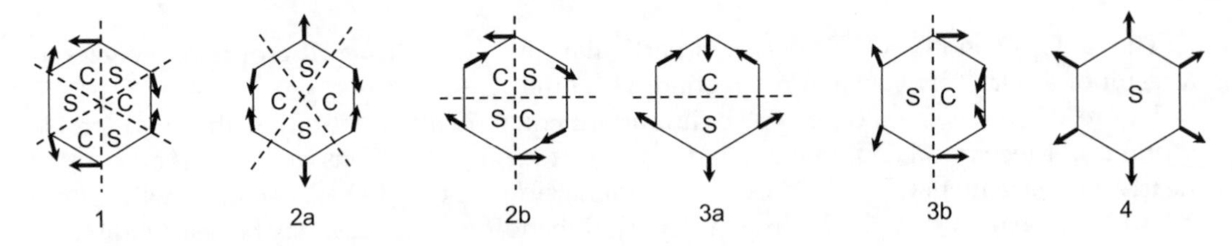

FIGURE 5.9 Standing vibrational waves for stretching vibrations of six-membered rings.

Chapter 5: Questions

1. Using the mathematical expression for a harmonic oscillator, calculate the force constant, *K*, for the following species: CH 3000 cm^{-1}, NH 3400 cm^{-1}, OH 3600 cm^{-1}. Note that H = 1 amu, C = 12 amu, O = 16 amu, and N = 14 amu. What are the possible sources of the observed variation in the frequencies for the above?
2. The following ethers have bands associated with the in-phase and out-of-phase C–O stretching vibrations: 1100, 810 cm^{-1} 1080, 890 cm^{-1} 1020, 990 cm^{-1} 1220, 890 cm^{-1}. Using a scatter plot of frequency versus bond angle, explain the origin of the observed trends.

1100, 810 cm^{-1} 1080, 890 cm^{-1} 1020, 990 cm^{-1} 1220, 890 cm^{-1}

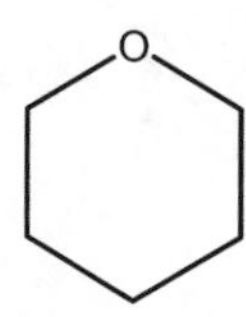

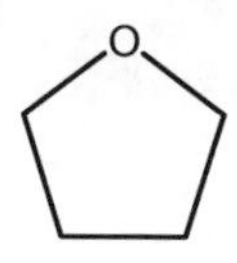

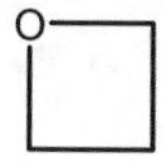

3. Calculate the in-phase and out-of-phase stretching frequencies for six and four membered cyclohexane, cyclopropane, and cyclobutane rings. Here $k = 4.2$ mdynes/A and $M = 14$ amu for the methylene group.

The in-phase ring stretch frequency equation is:

$$\upsilon_{ip} = 1303[k(1+\cos\alpha)(M_1 + 1/M_2)]^{1/2} = 1303[k/M]^{1/2} 2\cos\alpha/2$$

Ring Size	α (deg)	In-phase stretch (exp)	Calculated
N = 6	109.5	802	
N = 5	104	886	
N = 4	90	1003	

4. Calculate the in-phase and out-of-phase stretching frequencies for the following XY_2 groups: R–NH_2 (6.4 mdynes/A), R–CH_2–R (4.9 mdynes/A).
5. Provide a summary of the general types of vibrations encountered. In general, where in the vibrational spectrum will these vibrations be found?
6. The bending vibration for propane is observed at 375 cm^{-1} while for butane it is observed at 427 and 271 cm^{-1}. Draw the vibrations and explain the frequency differences.
7. Suggest other possible functional groups beyond those listed in table 5.1 that would fit the XY_2 type identical oscillators.

References

1. *Infrared and Raman Spectroscopy, Methods and Applications*; Schrader, B., Ed.; VCH: New York, NY, 1995.
2. Colthup, N. B.; Daly, L. H.; Wiberley, S. E. *Introduction to Infrared and Raman Spectroscopy*, 3rd ed.; Academic: New York, NY, 1990.
3. Barrow, G. M. *Introduction to Molecular Spectroscopy*; McGraw-Hill: New York, NY, 1962.

CHAPTER

6

IR and Raman Spectra-Structure Correlations: Characteristic Group Frequencies

In the following chapter we include generalized Infrared and Raman spectra of selected regions which illustrate the generalized appearance of bands for selected characteristic group frequencies. In each case, the upper spectrum shows the transmission infrared spectra with bands pointing downward and the lower spectrum shows the Raman spectrum with bands pointing up. The vertical axes are expressed in percent infrared transmission and Raman intensity. We also describe the approximate forms of selected vibrations and include tables summarizing important bands and their assignments.

1. X–H STRETCHING GROUP (X=O, S, P, N, SI, B)

The set of spectra in Fig. 6.1, illustrates some useful X–H stretching bands. The large differences in X–H stretching frequencies are mostly due to differences in bonding and therefore the force constants. Quite often these bands are more intense in the IR than in the Raman spectra.[1-4]

As shown in the top two spectra in Fig. 6.1, the OH stretch is particularly weak in the Raman but strong in the IR. The hydrogen-bonded alcoholic OH stretch vibration gives rise to a broad band near 3350 cm^{-1} in the IR spectrum. A more strongly hydrogen-bonded OH in a carboxylic acid dimer has a much broader band centering about 3000 cm^{-1}. An exception to this is the SH group of a mercaptan which gives rise to a sharp weak IR band and a strong Raman band at about 2560 cm^{-1}.

Aliphatic primary amines have a weak NH_2 doublet in the IR near 3380 and 3300 cm^{-1} which derive from the out-of-phase and in-phase NH_2 stretching vibrations, respectively. The weak shoulder observed near 3200 cm^{-1} is the Fermi-resonance enhanced overtone of the 1620 cm^{-1} NH_2 bend. Typically, the in-phase NH_2 stretch is stronger in the Raman spectra than the out-of-phase stretch. In primary amides, O=C–NH_2, bands involving the NH_2 stretches are found at 3370 and 3200 cm^{-1}, and are typically stronger in the IR than in the Raman spectra. The PH_2 group gives rise to a medium–weak single band in both the IR and Raman spectra near 2280 cm^{-1}.

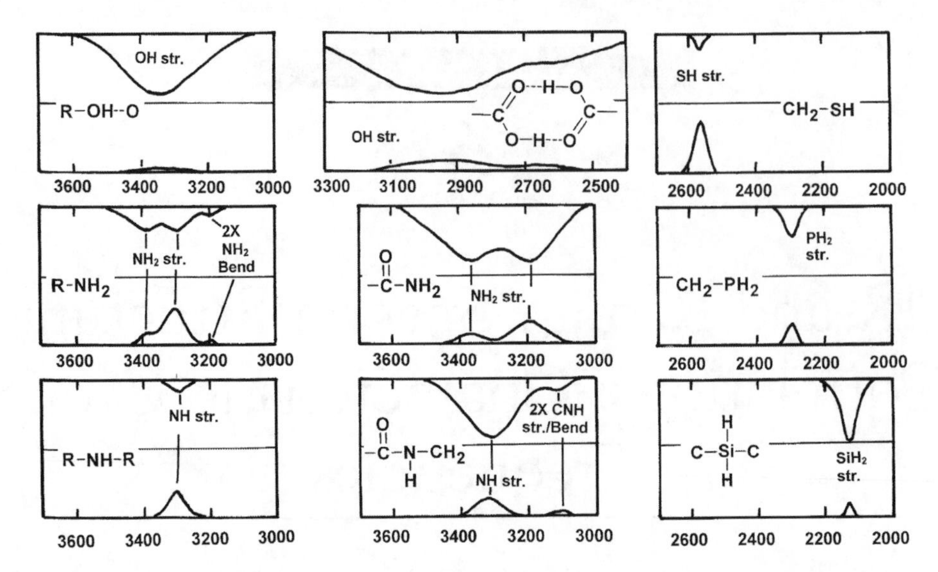

FIGURE 6.1 The generalized IR and Raman spectra of selected X–H stretching bands, where X = oxygen, sulfur, phosphorus, nitrogen, or silicon.

Secondary aliphatic amines have a NH stretch band near 3300 cm^{-1} which is quite weak in the IR but stronger in the Raman spectrum. Monosubstituted amides, O=C–NH–, have a stronger IR band near 3300 cm^{-1} which derives from the NH stretch. An additional weak band near 3100 cm^{-1} is the overtone of the 1550 cm^{-1} CNH stretch/bend. The last group shown is SiH_2 species that has a single band at ca. 2130 cm^{-1} which is quite strong in the IR but weak in the Raman.

Although not shown in Fig. 6.1, B–OH and BH/BH_2 are characterized by moderate to strong IR OH and BH stretching bands but weak Raman bands. The B–OH group has a strong broad band between 3300 and 3200 cm^{-1} from the OH stretching vibration. The BH and BH_2 groups have moderate to strong IR bands in the 2630–2350 cm^{-1} region. The BH_2 group has both an out-of-phase (2640–2570 cm^{-1}) and in-phase (2532–2488 cm^{-1}) BH_2 stretching bands.

Fully annotated IR and Raman spectra of selected carboxylic acids (#24–27), amides (#30–32), alcohols (#40–47), boronic acid (64), and amines (#51–53) are included in the last section for comparison.

2. ALIPHATIC GROUPS

The set of spectra in Fig. 6.2 illustrate some selected group frequencies for alkane groups.[1-4] The group frequencies are also summarized in Table 6.1. Figure 6.3 illustrates the vibrations of the methylene (CH_2)[1-4] group. For the non-cyclic $(CH_2)_n$ chain, the CH_2 out-of-phase and the in-phase stretching bands are near 2930 and 2855 cm^{-1}, respectively, and the CH_2 bend (deformation) band is near 1465 cm^{-1}. For the methylene segments $(CH_2)_n$ in the trans zig-zag form,

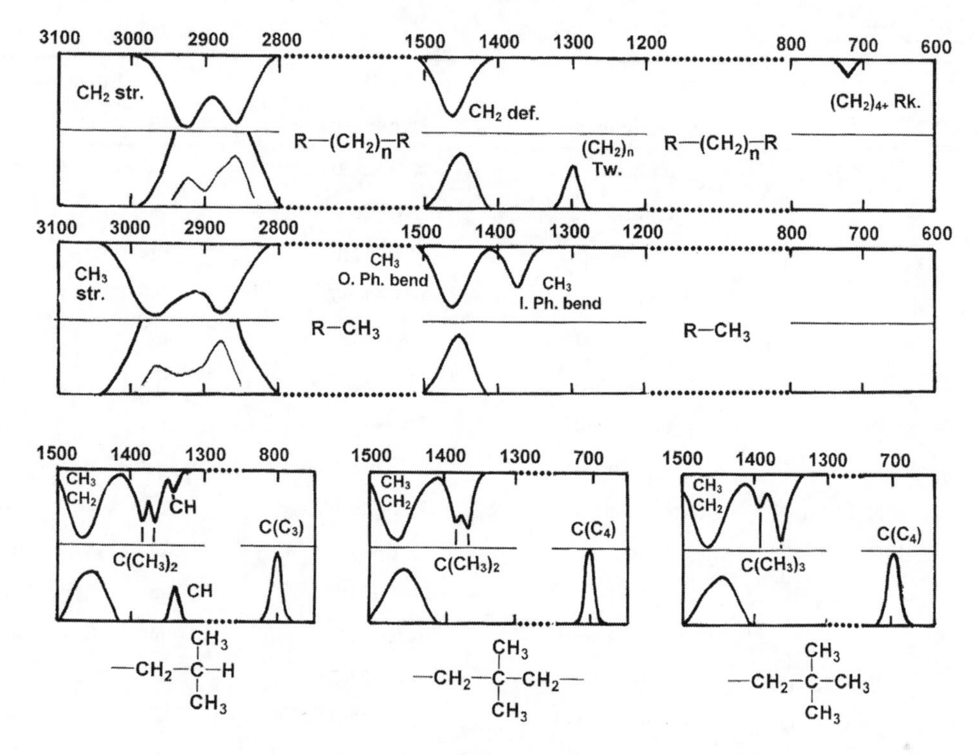

FIGURE 6.2 The generalized IR and Raman spectra of linear and branched alkane groups. The spectral components from methyl and methylene groups have been separated in the top two rows of spectra. The bottom row illustrates the sensitivity of IR and Raman spectra to branching in alkane group.

the in-phase CH_2 twist is observed near 1300 cm^{-1} in the Raman spectrum, and the in-phase $(CH_2)_n$ rock is found near 723 cm^{-1} in the IR.

Figure 6.4 illustrates the vibrations of the CH_3 group. For the CH_3 group on an alkane carbon, bands near 2960 and 2870 cm^{-1} derive from the out-of-phase and in-phase stretches and the out-of-phase bend results in a band at ca. 1465 in the IR. The in-phase bend of a single CH_3 on an alkane carbon results in a band near 1375 cm^{-1} in the IR. The presence of an adjacent group such as a N or O atom can result in a significant frequency shift in the methyl (CH_3) in-phase stretch to lower frequency between 2895 and 2815 cm^{-1}.

When there are two methyl groups on one carbon such as an iso-propyl or a gem-dimethyl group, this splits into two nearly equal intense bands near 1385 and 1365 cm^{-1} in the IR. When there are three methyl groups on one carbon, such as a tertiary-butyl group, there is further splitting into a weak band at ca. 1395 cm^{-1} and a stronger band near 1365 cm^{-1} in the IR spectrum. In addition, for these groups there are strong Raman bands at ca. 800 cm^{-1} which involves the in-phase stretch of the $C(C_3)$ group and at ca. 700 cm^{-1} for the in-phase stretch of the $C(C_4)$ group. However, although intense, this Raman band will vary considerably in frequency due to strong mechanical coupling with attached C–C, C–O, C–N or C–S groups. Table 6.2 summarizes this variation for a few selected

TABLE 6.1 Selected Group Frequencies: Aliphatics

Group		Assignment	Frequency (cm^{-1})	Intensities[1]	
				IR	R
CH_3	R–CH_3	o.ph. str.[2]	2975–2950	vs	vs
	"	i.ph. str.	2885–2860	vs	vs
	"	o.ph. bend	1470–1440	ms	ms
	"	i.ph. bend	1380–1370	m	vw
	R–$CH(CH_3)_2$	bend-bend	1385–1380	m	vw
	"	bend-open	1373–1365	m	vw
	$R(CH_3)_3$	bend-bend	1395–1385	m	vw
	"	bend-open	1373–1365	ms	vw
	aryl-CH_3	i.ph. str. +	2935–2915	ms	ms
	"	bend overtone	2875–2855	m	m
	R–O–CH_3	i.ph. str.	2850–2815	m	m
	"	i.ph. bend	1450–1430	ms	m
	R_2N–CH_3	i.ph. str	2825–2765	s	s
	O=C–CH_3	i.ph. bend	1380–1365	s	w
CH_2	R–CH_2–R	o. ph. str.[3]	2936–2915	vs	vs
	"	i. ph. str.	2865–2833	vs	vs
	"	Fermi resonance	2920–2890	w	m
	"	bend	1475–1445	ms	ms
	$(CH_2)_{>3}$	i.ph. twist	1305–1295	–	m
	$(CH_2)_{>3}$	i.ph. rock	726–720	m	–
	O=C–CH_2	bend	1445–1405	m	m
	N≡C–CH_2	bend	1445–1405	m	m
	O–CH_2	wag	1390–1340	m	m
	Cl–CH_2	wag	1300–1220	m	m
CH	R_3CH	CH bend	1360–1320	w	w
	O=C–H	CH str. +	2900–2800	m	m
	"	rk overtone	2770–2695	m	m
	"	CH rock	1410–1380	m	m

[1]*s = strong, m = medium, w = weak, v = very, br = broad, var. = variable, – = zero*
[2]*o.ph. str. = out-of-phase stretch, i.ph. str. = in-phase stretch, rk = rock*
[3]*Note: In the trans zig-zag segments of the long chain* $(CH_2)_{>6}$ *group, the Raman active* CH_2 *out-of-phase stretch band is at 2900–2880* cm^{-1} *(2883* cm^{-1} *in polyethylene). The gauche segments have this at 2936–2916* cm^{-1} *in the Raman.*

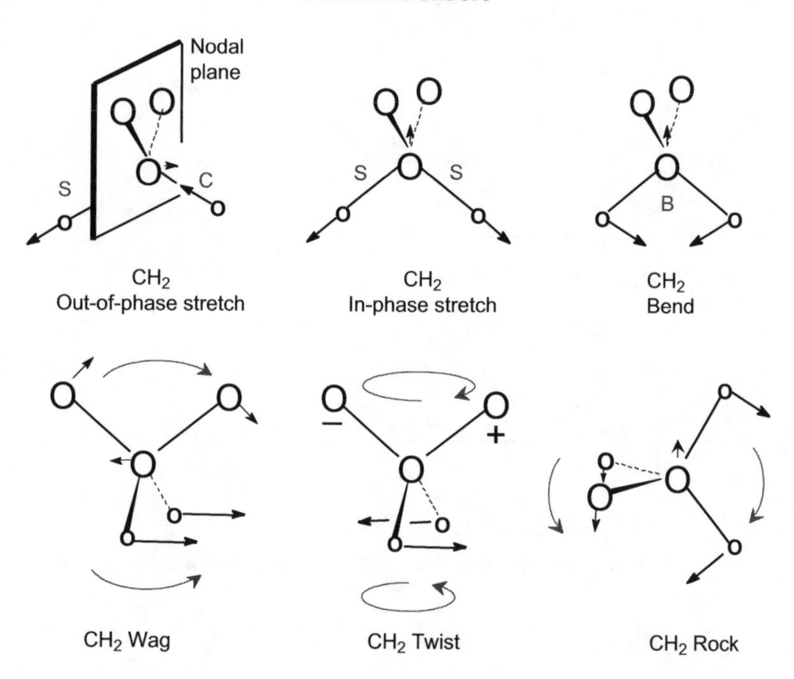

FIGURE 6.3 Vibrations of the CH_2 group. C and S stand for contract and stretch, and B and O stand for bend and open. Movement of atoms are depicted by arrows or by + and − signs if the atom movement is out of the page. Nodal planes are used to help differentiate the vibrations.

compounds. Fully annotated IR and Raman spectra of selected aliphatic compounds are included in the last section of fully interpreted spectra for comparison (#1–3).

In Fig. 6.5, the relative intensities in the IR and Raman spectra in CH stretching region are illustrated for some selected alkanes. The vibrations involving the methyl and methylene stretches (3000–2800 cm^{-1}) give rise to bands that are strong and characteristic in both the IR and Raman spectra but show greater complexity particularly in the Raman spectra due to Fermi-resonance effects.[5,6,7]

The CH stretching region of the IR and Raman spectra shows additional bands due to Fermi resonance between the methylene CH stretching vibrations and the overtone of the methylene bending (deformation) vibrations (1480–1440 cm^{-1}). Bands observed in the 2900 cm^{-1} region as a weak shoulder in the IR spectrum and as more isolated and obvious bands in the Raman spectra derive from Fermi resonance. This occurs in the IR and Raman spectra of aliphatics when:

1. There are at least two almost equivalent neighboring methylene groups.
2. The symmetry and frequency of the methylene CH_2 stretch and the CH_2 deformation overtone is suitable for Fermi resonance.

The above results in multiple Fermi-resonance doublets associated with the methylene group including the broad bands observed in the 2920–2898 cm^{-1} region and results in significantly more complicated Raman spectra of n-alkanes as observed in Fig. 6.5.

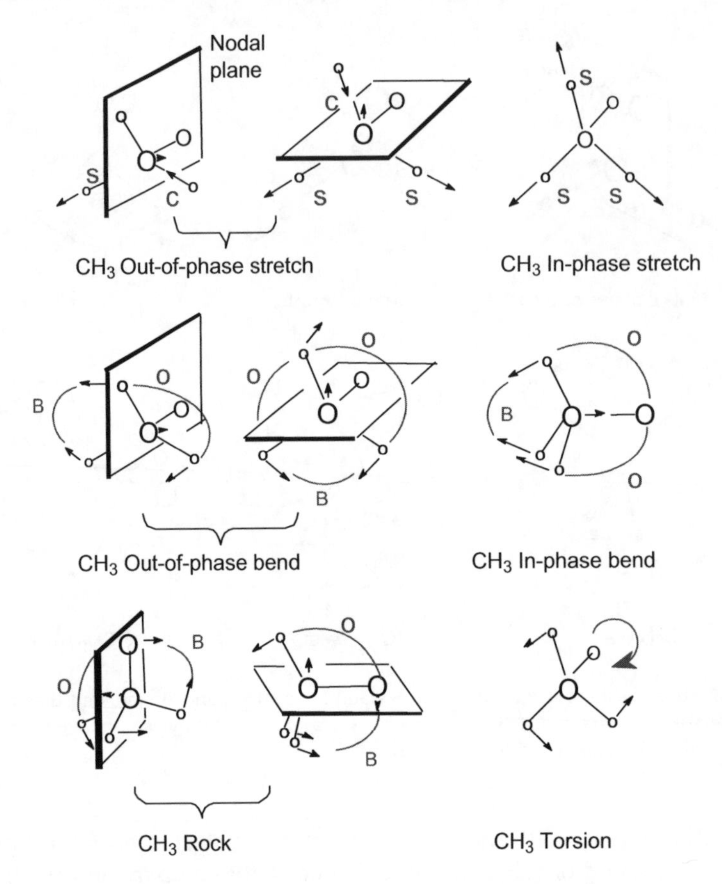

FIGURE 6.4 Vibrations of the CH_3 group. Nodal planes are used to differentiate between doubly degenerate vibrations. C and S stand for contract and stretch, and B and O stand for bend and open.

The conformational equilibrium of n-alkanes due to rotation about the C—C bond of the long chain aliphatic can also affect the IR and Raman spectra in the gas phase or in pressure induced transitions. Thus, for n-Pentane these conformations would include methylene in the all-trans (TT a planar zig-zag form), single gauche (TG), and double gauche (GG) conformers.[8] The methylene and methyl stretching bands are sensitive markers of these different conformers. The all trans conformation is typically found in the condensed state under ambient room temperature conditions such as shown in Fig. 6.5.

The relative IR intensities of the methyl and methylene bands can be used to estimate the relative concentration of the two species. In the case of the IR spectra in Fig. 6.5, the methyl (CH_3) out-of-phase stretch at ca. 2960 cm^{-1} is doubly degenerate and is therefore twice as intense at the methylene (CH_2) out-of-phase stretch at 2925 cm^{-1}. Using this relationship, the relative amount of methyl and methylene group can be rapidly estimated from the IR spectrum.

TABLE 6.2 Summary of the Raman In-Phase C–C/C–X Stretching Band for Selected Compounds, where X is Carbon, Nitrogen, Oxygen or Sulfur

X-Element	C–C–C–X (i.ph.str.)	$(C)_2C$–X (i.ph. str.)	$(C)_3C$–X (i. ph. str.)
Carbon (12)	827 cm^{-1} $CH_3CH_2CH_2CH_3$	800 cm^{-1}	733 cm^{-1}
Nitrogen (14)	868 cm^{-1}	802 cm^{-1}	748 cm^{-1}
Oxygen (16)	822 cm^{-1}	819 cm^{-1}	750 cm^{-1}
Sulfur (32)	670 cm^{-1}	631 cm^{-1}	

3. CONJUGATED ALIPHATICS AND AROMATICS

Selected IR and Raman characteristic frequencies of conjugated aliphatics and aromatics are summarized in Table 6.3.[1-4] Included are triple and cumulated double bond species, olefinics, and aromatics. Three sets of generalized IR and Raman spectra also illustrate some of the characteristic bands for these species. Below, we discuss olefinics, triple and cumulated double bonds and aromatics in more detail.

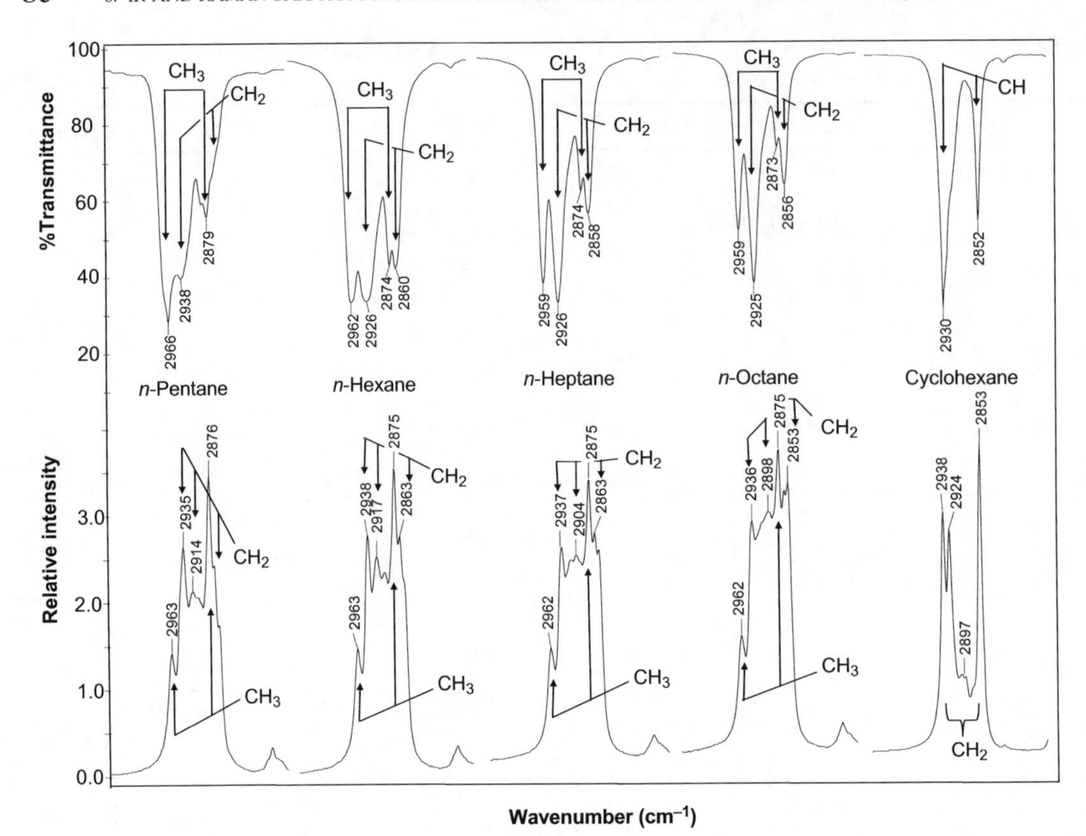

FIGURE 6.5 The infrared and Raman spectra in the CH stretching region are illustrated for *n*-pentane, *n*-hexane, *n*-heptane, *n*-octane, and cyclohexane. Additional bands due to Fermi resonance are clearly observed in the Raman spectra.

TABLE 6.3 Selected Group Frequencies: Triple Bonds, Cumulated Double Bonds, Conjugated Aliphatics and Aromatics

Group		Assignment	Frequency (cm^{-1})	Intensities[1] IR	R
Triple	C–C≡C–H	CH str.	3340–3267	s	w
Bonds	"	C≡C str.	2140–2100	w	vs
	"	CH wag	710–578	s,br	w
	C–C≡C–C	C≡C str.	2245–2100	–	s
	CH_2–C≡N	C≡N str.	2260–2240	m	vs
	"	CH_2 bend	1440–1405	m	m
	conj-C≡N	C≡N str.	2235–2185	var.	s
	S–C≡N	C≡N str.	2170–2135	ms	s

TABLE 6.3 Selected Group Frequencies: Triple Bonds, Cumulated Double Bonds, Conjugated Aliphatics and Aromatics—cont'd

Group		Assignment	Frequency (cm^{-1})	Intensities[1] IR	R
Cumulated	>C=C=CH_2	CCC o.ph. str.	2000–1900	vs	vw
Double	–N=C=O	NCO o.ph. str.	2300–2250	vs	vw
Bonds	–N=C=S	NCS o.ph. str.	2200–2000	vs	mw
C=C	C=C mono, cis, 1,1	C=C str.[2]	1660–1630	m	s
Alkane Subst.	C=C trans, tri, tetra	C=C str.[2]	1680–1665	w–o	s
	C=CH–R mono, cis, trans	CH str.	3020–2995	m	m
	C=CH_2 mono, 1,1	CH_2 o.ph. str.	3090–3075	m	m
	"	CH_2 i.ph. str.	3000–2980	m	s
	"	CH_2 bend	1420–1400	w	m
	R–CH=CH_2	trans CH i.ph. wag	995–985	s	w
	"	=CH_2 wag[3]	910–905	s	w
	R_2C=CH_2	=CH_2 wag[3]	900–885	s	w
	R–CH=CH–R trans	=CH i.ph. wag	980–965	s	–
	R–CH=CH–R cis	=CH i.ph. wag	730–650	ms	–
	R–CH=CR_2	=CH wag	840–790	m	
Aromatics	aryl CH	CH str.	3100–3000	mw	s
	ring	quadrant str.	1620–1585	var	m
	"	quadrant str.	1590–1565	var	m
	"	semicircle str.	1525–1470	var	vw
	"	semicircle str.	1465–1400	m	vw
	mono, meta, 1,3,5	2,4,6 radial i.ph str.	1010–990	vw	vs
	meta, 1,2,4 & 1,3,5	lone H wag	935–810	m	–
	para and 1,2,4	2 adj. H wag	880–795	s	–
	meta and 1,2,3	3 adj. H wag	825–750	s	–
	ortho and mono	4 & 5 adj. H wag	800–725	s	–
	mono, meta, 1,3,5	ring out-of-plane bend	710–665	s	–
	para	ring in-plane bend	650–630	–	m
	mono	ring in-plane bend	630–605	w	m

[1]*s = strong, m = medium, w = weak, v = very, br = broad, var. = variable, – = zero*
[2]*Conjugation lowers C=C frequencies 10–50 cm^{-1} (e.g., N≡C–CH=CH_2 1607 cm^{-1})*
[3]*Electron donor substituents lower the =CH_2 wag frequencies (e.g., R–O–CH=CH_2 813 cm^{-1}) and electron withdrawers raise them (e.g., N≡C–CH=CH_2 960 cm^{-1})*

3.1. Alkyl-Substituted Olefinic Groups

The set of IR and Raman spectra in Fig. 6.6 illustrate selected group frequencies for the ethylene C=C groups with alkane group substituents.[1-4] Particularly, useful Raman bands for the olefinic group include the =CH stretch, in-phase =CH_2 stretch, and the C=C stretch. In the IR spectra, the olefinic out-of-plane CH and CH_2 wag are quite important. Figure 6.7 illustrates the in-plane olefinic vibrations and Fig. 6.8 summarizes the observed and estimated C=C stretching frequencies for fifteen different compounds containing an olefinic group. The out-of-plane CH wag vibrations are shown in Fig. 6.9.

The in-plane vibrations of the olefin group are depicted in Fig. 6.7. The vinyl and 1,1-disubstituted ethylenes have a =CH_2 group with =CH_2 out-of-phase and in-phase stretching bands near 3080 and 2985 cm^{-1}, respectively. The =CH group in vinyls, cis, trans 1,2-disubstitituted ethylenes, and trisubstituted ethylenes all have a =CH stretching band at ca. 3010 cm^{-1}. The vinyl, 1,1-disubstituted, and 1,2-cis disubstituted ethylenes all have a C=C stretch with strong intensity in the Raman and medium intensity in the IR. In the case of 1,2 trans disubstituted, tri-substituted, and tetra-substituted ethylene groups, the C=C stretch at ca. 1670 cm^{-1} is quite strong in the Raman but weak or absent in the IR. The frequency of the C=C stretch will vary depending on the nature of the substituent. Conjugation effects tend to be particularly important, usually lowering the C=C stretching frequency.

The selected compounds in Fig. 6.8 vary in the cyclic olefinic structures and most importantly the C–C=C bond angle. Mechanical coupling of the stretching vibrations of bonds attached to olefinic groups is the critical variable in determining C=C stretching fequencies in cyclic olefin species.[1] The calculated frequencies are from an adapted form of the triatomic model discussed earlier and includes mechanical coupling of the C–C stretch

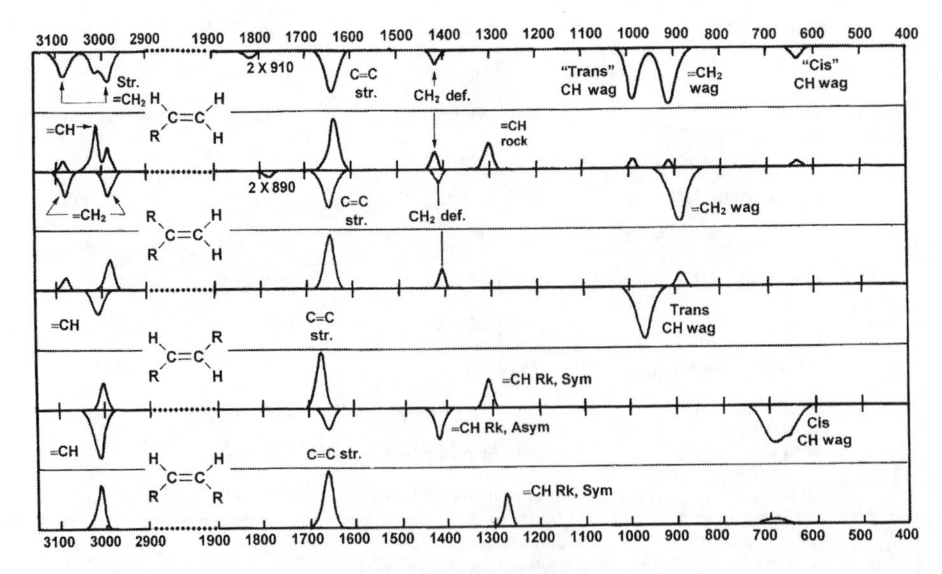

FIGURE 6.6 Generalized IR and Raman spectra of alkyl-substituted olefins.

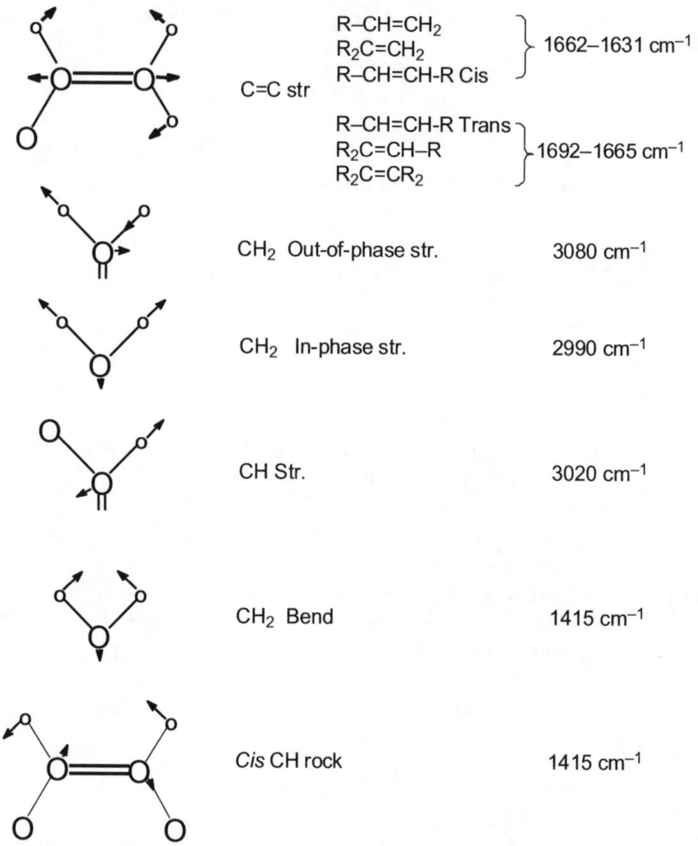

C=C str

R–CH=CH$_2$
R$_2$C=CH$_2$
R–CH=CH-R Cis
} 1662–1631 cm^{-1}

R–CH=CH-R Trans
R$_2$C=CH–R
R$_2$C=CR$_2$
} 1692–1665 cm^{-1}

CH$_2$ Out-of-phase str. 3080 cm^{-1}

CH$_2$ In-phase str. 2990 cm^{-1}

CH Str. 3020 cm^{-1}

CH$_2$ Bend 1415 cm^{-1}

Cis CH rock 1415 cm^{-1}

FIGURE 6.7 The olefin in-plane vibrations are shown above. Arrows depict the movements of atoms.

with the C=C stretch. The model uses force constants for the C=C and attached C–C of 8.9 and 4.5 mdynes/A, respectively, neglects the bending force constants, and includes the C–C=C bond angle. The approximate form of the model describing the total forces on the olefinic group for the external and internal C=C group are shown on the left side of Fig. 6.8.

Despite their simplicity, these calculations are successful in predicting trends in the frequencies in the top two rows. This data demonstrates that little or no change in the C=C force constant is required to explain the observed frequency shifts when one or both of the olefinic carbon atoms are part of a ring. No calculations are shown for the bottom row, but the experimental data indicates that for ring sizes between four and six, the effects of the indicated changes in both the external and internal C–C=C bond angles approximately cancel each other, resulting in similar C=C stretching frequencies. The critical importance that mechanical effects has in determining the C=C stretching frequencies in cyclic olefin structures has similarly been demonstrated using vibrating ball bearing and coil spring molecular models.[9]

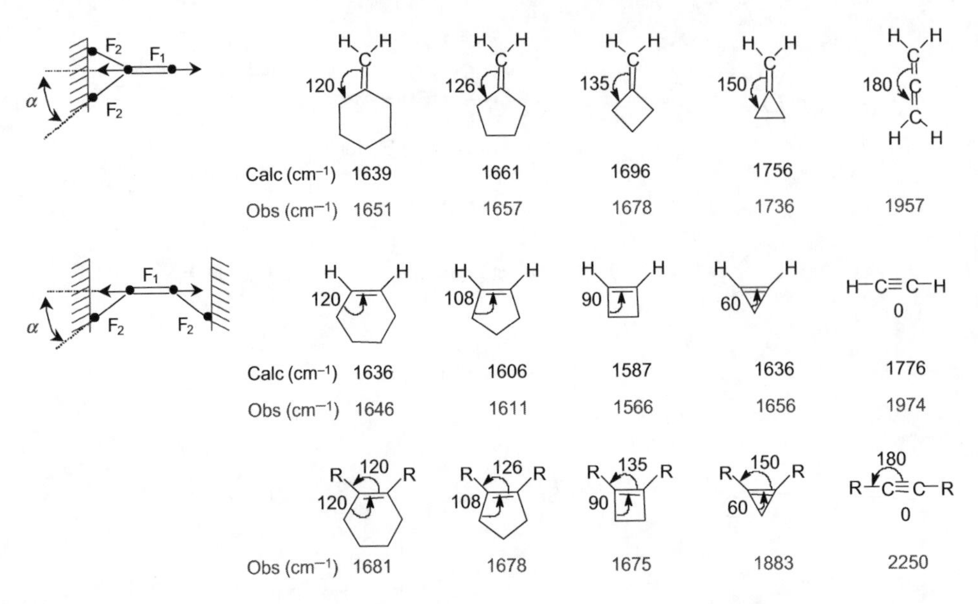

FIGURE 6.8 The cyclic C=C stretching frequencies as a function of the C−C=C bond angles. The calculated frequencies used force constants F_1 and F_2 values of 8.9 and 4.5 mdynes/A, respectively. The simple model used to calculate the frequencies is summarized on the left-hand side of the figure.

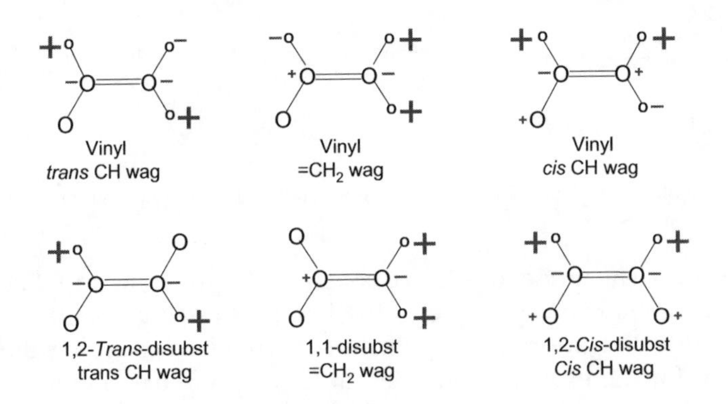

FIGURE 6.9 The out-of-plane CH wag vibrations of the olefinic group are shown above. Movement of atoms are depicted by arrows or by + and − signs.

The CH wag bands (see Fig. 6.9) are strong in the IR and highly characteristic of the alkyl substitution on the ethylene C=C group. These alkyl-substituted olefinic CH wag bands are observed near 990 and 910 cm^{-1} for vinyl groups, near 890 cm^{-1} for 1,1-disubstitution, near 970 cm^{-1} for trans 1,2-disubstitution, and roughly at 680 cm^{-1} for cis 1,2-disubstitution. The nature of the substituent can result in large but predictable frequency shifts of the olefinic CH and CH_2 wag. The $=CH_2$ wag is sensitive to electronic changes and shows a linear frequency dependence based upon experimentally determined substituent Hammet constant as well as

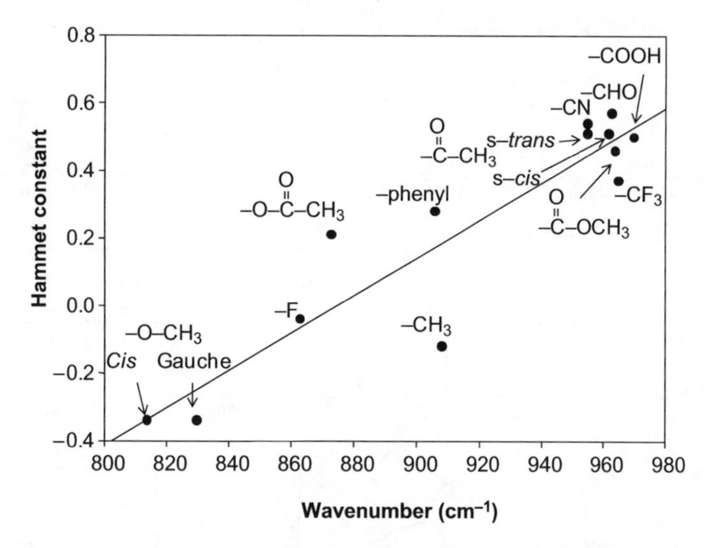

FIGURE 6.10 Correlation of the Hammet constant of substituents of mono-substituted ethylene with the CH_2 wag frequency.

the calculated substituent electronegativity and the electron density on the carbon atom.[10,11] Figure 6.10 shows the linear dependence for the mono-substituted ethylene $=CH_2$ wag as a function of the intrinsic electronic contribution to the Hammet values.[11]

These well-defined trends of the frequency dependence of the CH wag vibrations on the substituent properties are also observed in more highly substituted olefinic groups.[1] For example, substitution of an electron withdrawing ester group (–CO–O–R) in the XYC $=CH_2$ species will raise the frequency of the CH_2 wag. Similarly, substitution of an electron donating ether group (–OR) will lower the frequency. The trans CH wag shows a predictable frequency dependence upon the substituent electronegativity. An increase in the substituent electronegativity relative to an aliphatic will lower the CH wag frequency. Fully annotated IR and Raman spectra of selected olefinic compounds are included for comparison in the last section (#4–6, and 31–32).

3.2. Triple and Cumulated Double Bonds

Figure 6.11 illustrates some of the characteristic bands for C≡C, C≡N, and the X=Y=Z groups.[1-4] The C–C≡C–H group has a CH stretch band near 3300 cm^{-1} and a ≡C–H wag band near 630 cm^{-1} which are both strong in the IR and weak in the Raman. The C≡C stretch band is found at ca. 2125 cm^{-1}, which is weak in the IR but strong in the Raman. The symmetrically substituted C≡C stretching vibration is IR inactive but Raman allowed. For the dialkyl acetylenes, the Raman spectra usually has two bands near 2300 and 2230 cm^{-1} resulting from Fermi resonance splitting of the C≡C stretch. For the diaryl acetylenes, only a single C≡C stretch band is observed in the Raman near 2220 cm^{-1}. The R–C≡N group results in IR and Raman C≡N stretch bands near 2240 cm^{-1} which can be lowered by roughly 20 cm^{-1} by conjugation with a suitable substituent. The –S–C≡N

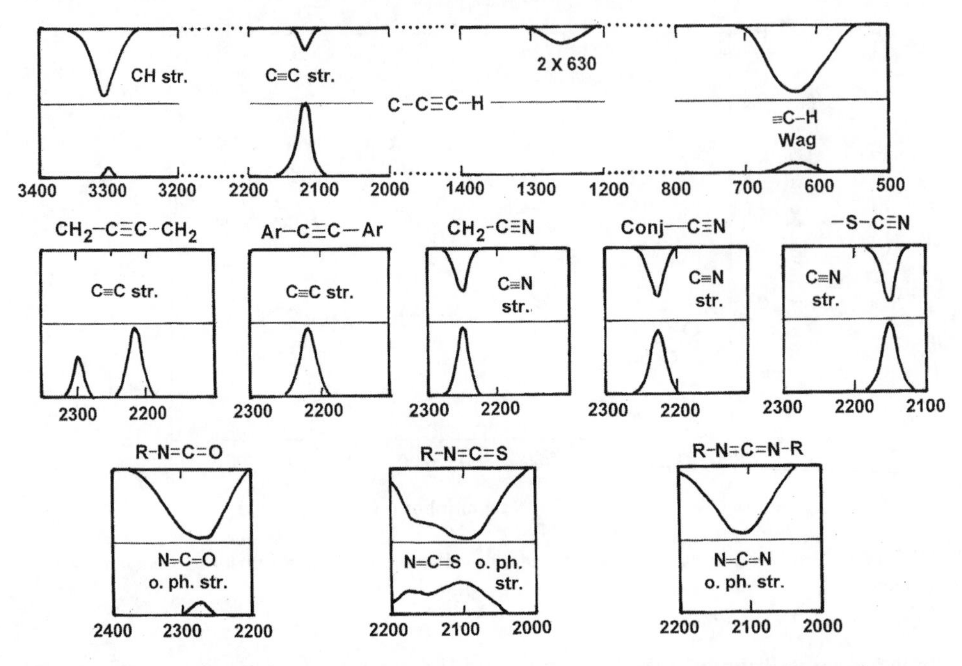

FIGURE 6.11 Generalized IR and Raman spectra of selected triple and cumulated double bond compounds.

group is characterized by IR and Raman bands near 2150 cm^{-1} which involve the C≡N stretch. The –N=C=O, –N=C =S, and –N=C=N – groups all have very strong IR bands in the 2280–2050 cm^{-1} region resulting from an out-of-phase stretch of the X=Y=Z bonds.

Fully annotated IR and Raman spectra of two selected nitrile containing compounds as well as an acetylene compound are included in the last section for comparison (#7–9).

3.3. Aromatic Benzene Rings

The twenty different normal modes of benzene are shown in Fig. 6.12[12]. A simple substitution of only one substituent results in a significant increase in the modes of vibration to thirty. Table 6.4 provides the calculated and experimental frequencies along with the unique Wilson numbers and accompanying symmetry for the modes of vibration for benzene.[13] As shown in Fig. 6.12, the benzene C–H stretching vibrations are observed in the 3100–3000 cm^{-1} region. In all other vibrations below 3000 cm^{-1} the C–H bond length remains relatively unchanged. Many of the bands observed in the 1600–1000 cm^{-1} region involve in-plane C–H bending vibrations that interact with various ring C=C vibrations. Out-of-plane C–H bending (i.e., wag) vibrations give rise to bands below ca. 1000 cm^{-1}.

In general, many of the aromatic ring vibrational modes are insensitive to the substituent groups and can be regarded as good group frequencies for the benzene ring. Some of the vibrations are substituent sensitive and result in mechanical coupling with the substituent group vibrations and can provide important structural information. However, not all of these

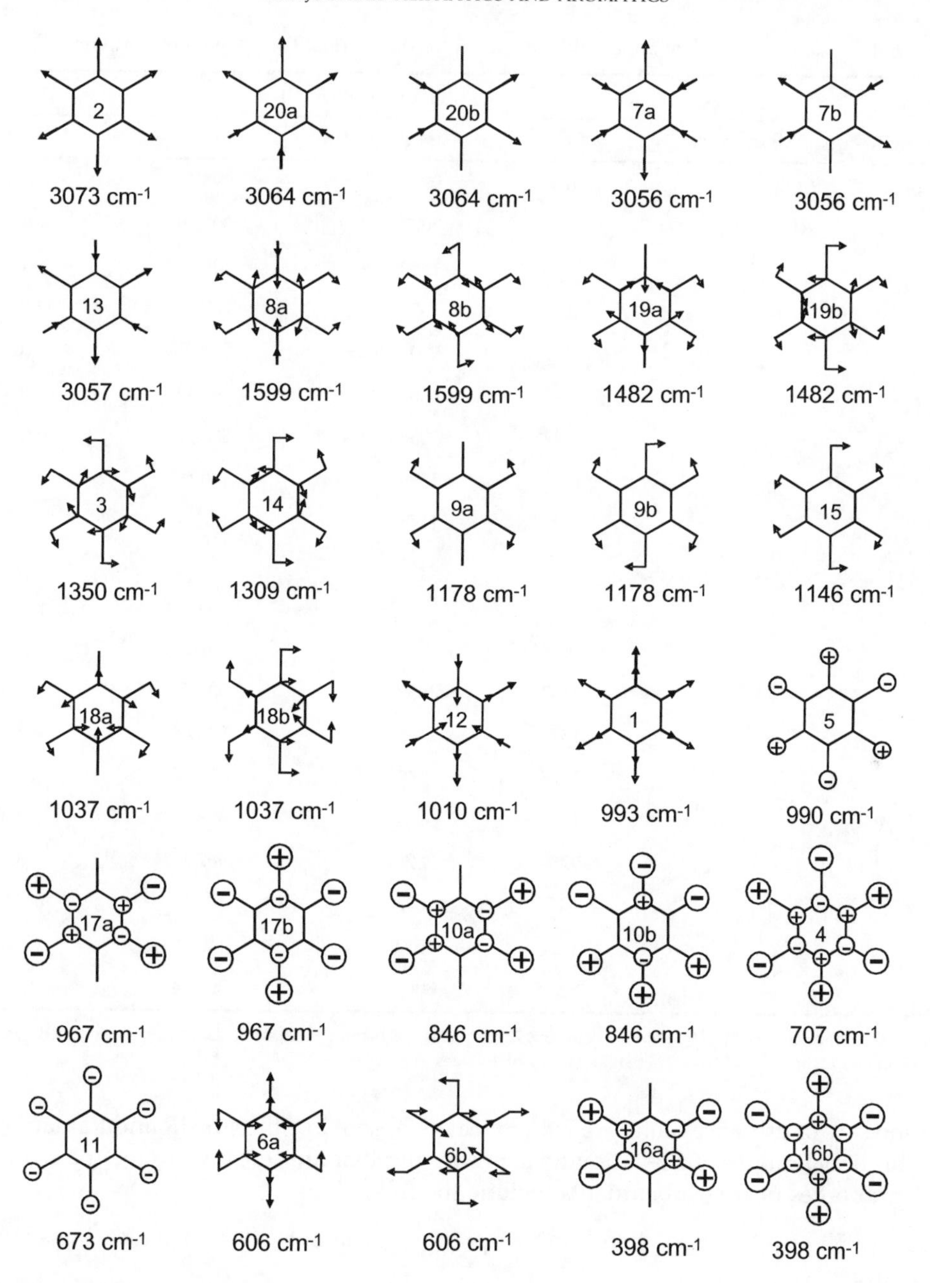

FIGURE 6.12 Approximate vibrational modes of benzene adapted from (M.A. Palafox, Int. J. Quant. Chem., 77, 661–684, 2000). The electronic structure was determined using density functional method with a 6–31G* basis set. Out-of-plane molecular vibrations are depicted using + and – to indicate out of and into the page, respectively.

TABLE 6.4 Calculated and Experimental Frequencies for the Normal Ring Modes of Benzene

		Frequency (cm^{-1})			
Wilson no.	**Symmetry**	**Calc**	**IR (Liquid)**	**IR (Gas)**	**Description**
1	a_{1g}	1026	993	993.1	δ (CCC)
2	a_{1g}	3218	3073	3073.9	ν (C–H)
3	a_{2g}	1375	1350	1350	δ (C–H)
4	b_{2g}	719	707	707	γ (CCC)
5	b_{2g}	1015	990	990	γ (C–H)
6	e_{2g}	618	606	608.1	γ (CCC)
7	e_{2g}	3192	3056	3056.7	ν (C–H)
8	e_{2g}	1664	1599	1600.9	ν (C=C)
9	e_{2g}	1203	1178	1177.8	δ (C–H)
10	e_{1g}	866	846	847	γ (C–H)
11	a_{2u}	694	673	673.9	γ (C–H)
12	b_{1u}	1013	1010	1010	δ (CCC)
13	b_{1u}	3182	3057	3057	ν (C–H)
14	b_{2u}	1378	1309	1309.4	ν (C=C)
15	b_{2u}	1178	1146	1149.7	δ (C–H)
16	e_{2u}	412	398	398	γ (CCC)
17	e_{2u}	978	967	967	γ (C–H)
18	e_{1u}	1071	1037	1038.3	δ (C–H)
19	e_{1u}	1524	1512	1483.9	ν (C=C)
20	e_{1u}	3208	3064	3064.4	ν (C–H)

The data is from (International J of Quantum Chemistry, 77, 661–684, 2000) and includes calculated frequencies used density functional method with a 6–31G* basis set and experimental values

vibrations are of use since many are either weak or absent in either the IR and Raman spectra. The vibrational modes of the aromatic benzene ring that are sensitive to the mass and electronic properties of ring substituents include the following:

1. 1300–1050 cm^{-1} region, involving in-plane movement of the ring carbons and the substituents.
2. 850–620 cm^{-1} region, involving the CH wag vibrations (intense in IR but weak in Raman) and ring out-of-plane vibrations.

Some of the characteristic group frequencies for mono- and di-substituted benzenes are illustrated in the generalized IR and Raman spectra depicted in Fig. 6.13.[1-4,14] Many different types of aromatic ring vibrations are illustrated which demonstrate how IR and

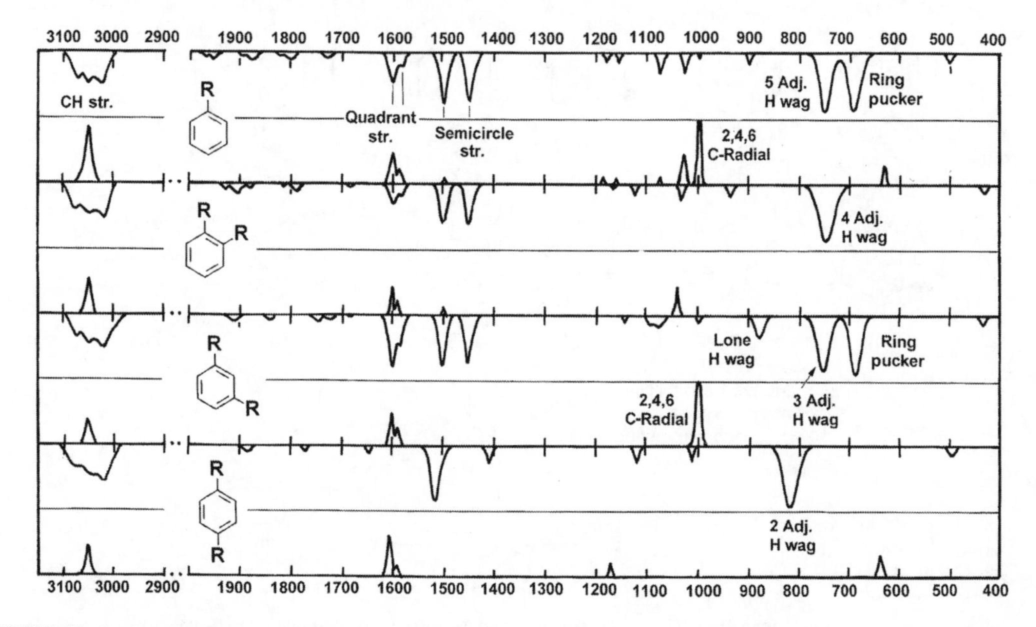

FIGURE 6.13 Generalized IR and Raman spectra of mono- and di-substituted benzenes. The complex ring vibrations can be greatly reduced using a standing wave description. This technique allows easy visualization of the form of the ring vibrations.

Raman spectra can be used to characterize aromatic substitution types. Useful group frequencies include the following:

- CH stretch bands
- Ring stretch bands
- Ring bend bands
- CH rock and CH wag bands
- Aromatic summation bands

Figure 6.14 shows the form of the 12 elementary vibrations for a six-membered ring with equal bond lengths, masses, and no substituents. The vibrational standing waves for these vibrations are also shown.[1,4] The top row illustrates the six ring stretching vibrations and the bottom row the six bending vibrations for this simple six-membered ring. The dashed lines show the nodal lines used to define the sextant , quadrant, semicircle, or whole ring vibrations. These lines separate molecular segments which vibrate out-of-phase with each other. The doubly degenerate ring modes are labeled with subscripts "a" and "b". Bands originating from ring vibrations in the IR and Raman spectra shown in Fig. 6.13 are annotated using the above description. The bottom row indicates the accompanying motion of the attached hydrogens in benzene which mechanically couple with the above ring modes.

The aryl CH stretch bands are observed in the 3100–3000 cm^{-1} region. The IR spectrum usually has several bands here while the Raman usually has only one. The ring quadrant stretch vibration has two components observed in the IR and Raman spectra near 1600

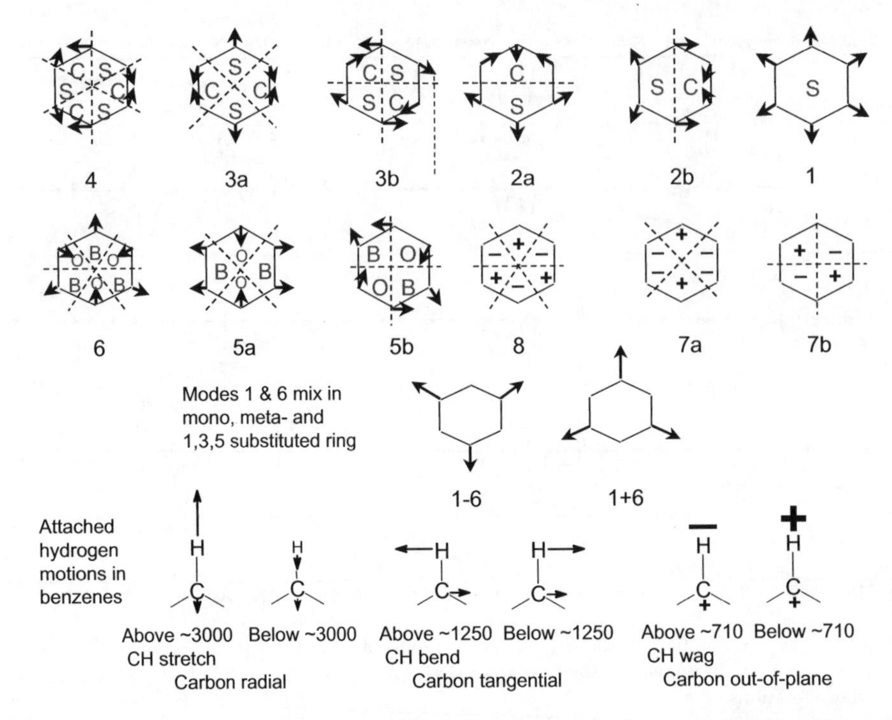

FIGURE 6.14 The twelve elementary ring vibrations for a six-membered ring with equal bond lengths, masses, and no substituents are illustrated with vibrational standing waves. The elementary stretching vibrations are shown in the top row and the in-plane and out-of-plane bending vibrations are shown in the second row. For rings of suitable symmetry, mixing of modes 1 and 6 results in vibrations shown in row three, which provide very strong Raman bands. The fourth row indicates the attached hydrogen motions in benzene which couple with the ring modes. Dotted lines are used to indicate nodal lines.

and 1580 cm^{-1}. Identical para substituents provide an exception, where no quadrant stretching bands are seen in the IR spectrum since they are forbidden by symmetry. The ring semicircle stretch has two bands in the IR near 1500 and 1450 cm^{-1} except in the para-substituted benzene where they are observed near 1515 and 1415 cm^{-1} in the IR. In mono and meta isomers, there is a very strong Raman band near 1000 cm^{-1}. This ring vibration involves movement of the 2,4,6 carbons radially in-phase and is a result of mixing of the whole ring stretch with the sextant in-plane ring bend. In the mono- and meta- substituted benzenes, there is a strong IR band near 690 cm^{-1} called the sextant out-of-plane deformation or ring pucker, where the 2,4,6 and 1,3,5 sets of carbons move oppositely out-of-plane.

3.3.1. *Aryl CH Wag*

The infrared spectrum in the region from ca. 900 to 600 cm^{-1} have intense bands that are highly characteristic of the number of hydrogen atoms present on the aromatic ring. The number of IR bands and their frequencies of vibrations involving the CH wag are significantly determined by the number of adjacent hydrogen atoms present on the benzene ring. To a lesser extent, there is some dependence of the CH wag frequency on the electron

TABLE 6.5 Selected IR Bands Useful for Substitution Determination of Aromatic Rings in the 950–680 cm^{-1} Spectral Region

	Substituent						
No. of Subst	**Pattern**	**1st CH Wag**		**2nd CH Wag**		**Ring sextant out plane def**	
1	Mono	780–720	5 Adj CH wag	NA		710–690	ring pucker
2	Ortho (1,2)	820–720	4 Adj CH wag	NA		inactive	
2	Meta (1,3)	950–880	lone H wag	870–730	2 Adj H wag	740–690	ring pucker
2	Para (1,4)	860–790	2 Adj H wag	NA		inactive	
3	1,2,3	805–760	3 adj H wag	NA		720–680	ring pucker
3	1,2,4	900–870	lone H wag	800–760	2 Adj H wag	720–680	ring pucker
3	1,3,5	865–810	lone H wag	NA		730–680	ring pucker
4	1,2,3,4	860–800	2 adj H wag	NA		inactive	
4	1,2,3,5	850–840	lone H wag	NA		720–680	ring pucker
4	1,2,4,5,	870–860	lone H wag	NA		inactive	
6	1,2,3,4,5,6	NA		NA		inactive	

donating or withdrawing characteristics of the substituent which effects the total electron density of the aromatic ring carbon atoms. Table 6.5 summarizes some of the more important and intense IR bands in this region. Complications in this spectral region can occur due to mechanical coupling of the CH wag with the bending vibrations of substituents such as $-NO_2$, carboxylic acids, and acid salts.[1-4] Some of the less intense CH wag bands are not included in table 6.5 despite the fact that they are found in a relatively small frequency range.

The mono-substituted vibrational mode shown in Fig. 6.15 is found at ca. 900 cm^{-1} and is an example of a weaker band that remains useful but is not included in Table 6.4. The range of this band is between 940 and 860 cm^{-1} and the frequency varies as a function of the substituent. Although this band can be used as a group frequency for structural determination, it is of greater use to demonstrate the linear frequency dependence of the band upon the calculated carbon electron density.[15]

The frequency of the mono-substituted benzene ring CH wag (900 cm^{-1}) shows a well-defined linear dependence on the experimentally determined Hammet values and the calculated electron density of the ring carbon atoms. The linear frequency dependence of this band upon the calculated electron density of the benzene ring para carbon atom for mono-substituted benzene ring is shown in Fig. 6.16. In many cases, the band positions for other aromatic CH wag vibrational modes listed in Table 6.5 can be estimated using an empirical expressions and a table of the mono-substituted aromatic band values (~900 cm^{-1} mode). Detailed examples of this type of analysis can be found in the literature (N.B. Colthup, Applied Spectrosc., 30, 589, 1976).[1,4,15] Fully annotated IR and Raman spectra of selected aromatic compounds are presented in the last section for comparison and include #10–15, 27, 45, 57, 64–69, 72, and 77.

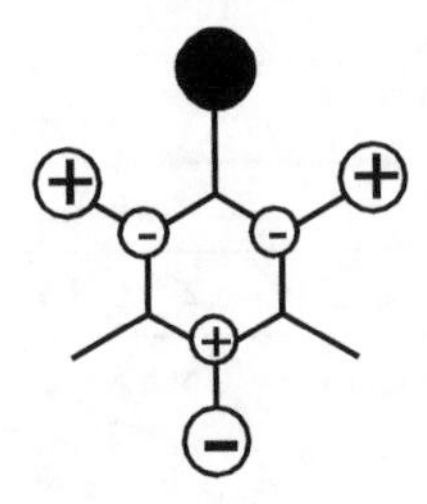

FIGURE 6.15 Approximate vibrational mode of the 900 cm^{-1} band CH wag for a mono-substituted benzene ring. The solid circle represents the substituent.

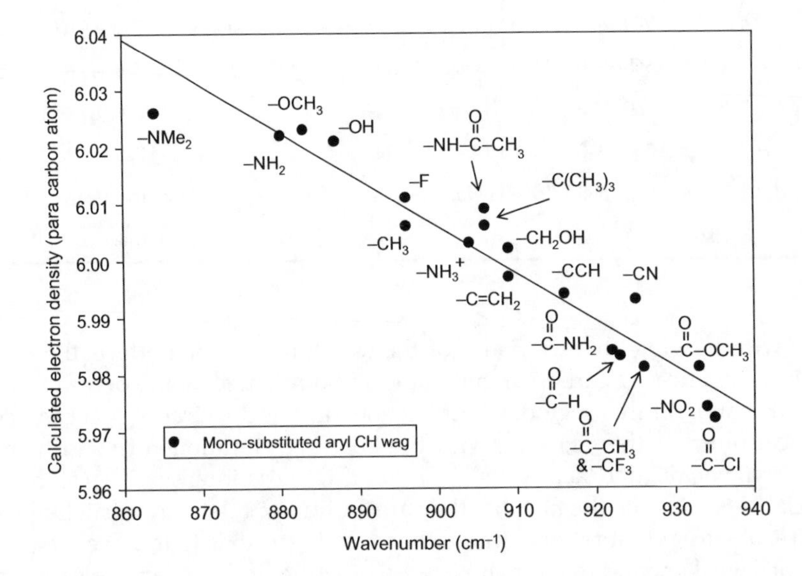

FIGURE 6.16 Correlation of the benzene ring para carbon electron density as a function of substituent with the (~900 cm^{-1} mode) CH_2 wag frequency. Data adapted from N.B. Colthup, Appl. Spectrosc., 30, 589, 1976.

3.4. Fused Ring Aromatics

Polynuclear fused ring aromatics such as naphthalenes, anthracenes, and phenanthrenes have characteristic IR and Raman bands in the same general regions as benzene derivatives.[1-4] The aryl-CH wag vibrations of these polynuclear aromatic compounds provide strong IR bands that provide important substitution information. Characteristic IR and Raman bands involving ring stretching vibrations are typically observed in the 1630–1500 cm^{-1} region.

In general, alkyl-substituted naphthalenes have a doublet near 1600 cm^{-1}, and additional bands in the 1520–1505 cm^{-1} and 1400–1390 cm^{-1}. In the Raman spectra, a very strong characteristic band is observed between 1390 and 1370 cm^{-1} and another strong band between 1030 and 1010 cm^{-1}. Some of the characteristic CH wag IR bands of naphthalenes are summarized in Table 6.6.

TABLE 6.6 Selected IR bands Useful for Substitution Determination of Naphthalene Rings in the 900–700 cm^{-1} Spectral Region

	Substituent						
No. of Subst	**Pattern**	**1st (cm^{-1})**	**CH Wag**	**2nd (cm^{-1})**	**CH Wag**	**3rd (cm^{-1})**	**CH Wag**
1	1 (mono)	810–775	3 Adj CH wag	780–760	4 adj H wag	NA	
1	2 (mono)	875–823	lone H wag	825–800	2 adj H wag	760–735	4 adj H wag
2	1,2 (di)	835–800	2 adj H wag	800–726	4 adj H wag	NA	

3.5. Heterocyclic Aromatic Six-Membered Ring Compounds

Six-membered heterocyclic aromatic compounds are benzene-like rings where one or more of the ring carbons are replaced with nitrogen atoms. Important examples of these compounds include pyridine which involves replacement with one nitrogen atom and triazine which involves substitution with three nitrogen atoms. Because, the six-membered heterocyclic rings typically have an aromatic double bonds in the ring analogous to benzenes, the resultant IR and Raman spectra have many of the same type of group frequencies. However, substitution of an OH or SH group in the ortho or para position relative to the ring nitrogen atom typically results in tautomerization and the formation of a C=O or C=S bond, respectively.[1,3,4] Such tautomerization also results in characteristic changes in the bands associated with the ring vibrations. Structural examples of these type of tautomerization is shown in Fig. 6.17 for 2-hydroxypyridine and cyanuric acid.

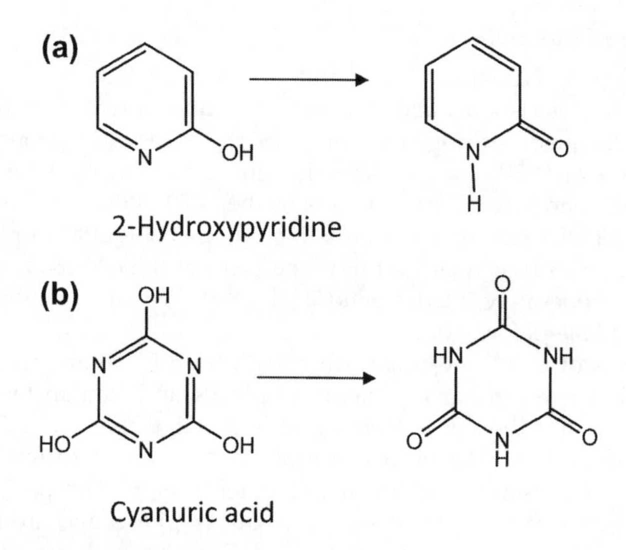

FIGURE 6.17 The resultant keto form resulting from tautomerization of 2-hydroxypyridine and Cyanuric Acid. The keto tautomer of 2-hydroxypyridine results in a carbonyl band between ca. 1650 and 1680 cm^{-1}. The keto tautomer of cyanuric acid results in a carbonyl band cluster between 1700 and 1780 cm^{-1}.

3.5.1. *Pyridines*

The IR and Raman spectra of pyridine compounds have characteristic bands due to the CH stretch, ring stretching and deformation, and substituent informative CH wag vibrations. A fully annotated IR and Raman spectrum of pyridine (#15) is included in the last section for comparison. Analogous to benzene, the CH stretching bands of pyridine compounds is found in the 3100–3000 cm^{-1} region. The quadrant ring stretch vibrations typically give rise to moderate to strong IR and Raman bands in the 1620–1555 cm^{-1} region. The ring semi-circle stretch also results in characteristic IR and Raman bands in the 1500–1410 cm^{-1} region. The Raman spectra of pyridine compounds is typically dominated by a very strong band in the 1050–980 cm^{-1} region that involves the 2,4,6 carbon radial stretch. The quadrant ring in-plane bend also results in a moderate Raman band in the 850–750 cm^{-1} region which is correlated with the substituent position. Similarly, the strong adjacent CH wag observed in the IR in the 850–740 cm^{-1} region is correlated with the pyridine substitution. Some of the characteristic bands for pyridine compounds are listed in Tables 6.7 and 6.8.[1-4]

Protonation of the pyridine nitrogen to form the pyridinium salt results in multiple bands from the N^+–H stretch between 3300 and 1900 cm^{-1} that have moderate IR intensity but are weak in the Raman spectrum. The multiple band structure derives from Fermi resonance. Potentially useful Raman bands with variable intensities include the NH in-plane bend at ca. 1250–1240 cm^{-1} and the NH wag at 940–880 cm^{-1}.

The N-oxides of pyridines have a strong IR band in the 1300–1200 cm^{-1} region involving the N–O stretch. This band is moderate to weak in the Raman spectra. IR and Raman bands involving the N–O stretch and the ring in-plane bend occurs between 880 and 830 cm^{-1}.

3.5.2. *Triazines and Melamines*

The triazine ring in s-triazine, alkyl- and aryl-substituted triazines, and melamine compounds have characteristic IR and Raman bands due to ring stretch and deformation vibrations.[1-4,16,17] The quadrant ring stretching vibrations result in strong IR and moderate Raman bands in the 1600–1500 cm^{-1} region. The semi-circle ring stretching vibrations result in multiple strong IR and weak Raman bands in the 1450–1350 cm^{-1} region. Particularly, intense Raman bands of triazine compounds include the N-Radial in-phase stretch in the 1000–980 cm^{-1} region and the quadrant in-plane bend in the 690–660 cm^{-1} region. Lastly, in the IR there is a moderately intense band in the 860–775 cm^{-1} region that involves the ring sextant out-of-plane deformation.

Melamine consists of a triazine ring with substitution of an amino group on all three ring carbon atoms. The amino group has multiple IR and Raman bands between 3500 and 3100 cm^{-1} involving the NH_2 stretch and a doublet between 1650 and 1620 cm^{-1} from the NH_2 deformation. The added complexity is due to differences in hydrogen bonding in the solid crystalline state. Measurements of melamine in DMSO solution results in the expected doublet from the NH_2 out-of-phase and in-phase stretch. The triazine ring itself exhibits the expected IR and Raman bands at 1550 cm^{-1} deriving from the quadrant ring stretch and 1470–1430 cm^{-1} from the semi-circle ring stretch. The unique characteristic of Raman bands include the strong band at 984 cm^{-1} from

TABLE 6.7 Selected In-plane Stretching and Bending Vibrations for Pyridine and Mono-Subsituted Pyridine

Pyridine	Assignment	Region (cm^{-1})	Intensities[1]	
			IR	R
Pyridine	Aryl CH str.	3100–3000	m	m
	Quadrant str.	1615–1570	s	m
	Semi-circle str	1490–1440	s	mw
	2,4,6 carbon radial str.	1035–1025	m	vs
	Ring breath/str.	995–985	m	s
	Quadrant in-plane bend	660–600	–	m
2-Mono- substituted	Quadrant str.	1620–1570	s	m
	Quadrant str.	1580–1560	s	m
	Semi-circle str.	1480–1450	s	m
	Semi-circle str.	1440–1415	s	w
	CH in-plane rk	1050–1040	m	s
	2,4,6 carbon radial str.	1000–985	m	vs
	Quadrant in-plane bend	850–800	w	ms
3-Mono- substituted	Quadrant str.	1595–1570	m	ms
	Quadrant str.	1585–1560	s	m
	Semi-circle str.	1480–1465	s	m
	Semi-circle str.	1430–1410	s	w
	2,4,6 carbon radial str.	1030–1010	m	vs
	Quadrant in-plane bend	805–750	w	m
4-Mono- substituted	Quadrant str.	1605–1565	m	ms
	Quadrant str.	1570–1555	m	m
	Semi-circle str.	1500–1480	m	m
	Semi-circle str.	1420–1410	s	w
	2,4,6 carbon radial str.	1000–985	m	vs
	Quadrant in-plane bend	805–785	w	ms

[1] *s = strong, m = medium, w = weak, v = very, var. = variable, – = zero*

the N-Radial in-phase stretch and the quadrant in-plane ring bend at 675 cm^{-1}. A sharp, moderate IR band is observed between 825 and 800 cm^{-1} which involves the traizine ring sextant out-of-plane deformation. This vibration involves an out-of-plane movement of all three carbon atoms and the three nitrogen atoms in opposite directions.

TABLE 6.8 Selected Out-of-plane CH wag and Out-of-plane Sextant Deformation Vibrations for Pyridine and Mono-Subsituted Pyridine

			Intensities[1]	
Pyridine	**Assignment**	**Region (cm^{-1})**	**IR**	**R**
Pyridine	5 adjacent H wag	760–74	s	–
	out-of-plane sextant def.	710–700	s	–
2-Mono- substituted	4 adjacent H wag	780–740	s	–
3-Mono- substituted	3 adjacent H wag	820–770	s	–
	out-of-plane sextant def	730–690	m-s	w
4-Mono- substituted	2 adjacent H wag	850–790	s	–

[1]*s = strong, m = medium, w = weak, v = very, var. = variable, – = zero*

Tautomerizing the triazine ring by making one of the double bonds external to the ring results in the iso form in which there are fewer than three double bonds in the ring. As discussed above, substitution of the carbon atom with an OH or SH results in formation of the iso form of the triazine ring. In addition to the formation of a C=O or C=S bond, the IR active band that derives from the sextant out-of-plane bend will be found in the 795–750 cm^{-1} region. Thus, the sextant out-of-plane bend vibration found near 800 cm^{-1} can be used to discriminate between normal triazine rings and the iso form of the triazine ring. A fully annotated IR and Raman spectrum of ammeline (#34) and trimethyl isocyanurate (#35) are included in the last section for comparison.

3.5.3. *Five–membered Ring Heterocyclic Compounds*

Pyrroles, furans, and thiophenes are five-membered ring aromatics containing two conjugated double bonds and a single nitrogen, oxygen or sulfur atom in the ring respectively. Additional compounds exist in which an additional carbon atom (position 2 or 3) are replaced with nitrogen or oxygen atoms. The double bonds are somewhat delocalized in these five-membered rings and the ring vibrations can be described using a standing wave description. Figure 6.18 depicts selected approximate ring vibrations for a five-membered ring containing one heteroatom.[1] However, the ring carbon atoms are not identical and changing the substituent can significantly affect the IR and Raman spectra.

Table 6.9 lists some of the characteristic ring stretching vibrations for pyrroles, furans, and thiophene compounds.[1-4,18] The quadrant and semi-circle ring stretching vibrations typically mix with the CH rock vibrations. The quadrant stretching vibration consists primarily of the C=C stretch while the first semi-circle stretch consists mainly of the C–X–C in-phase stretch with some C–C contraction. The second semi-circle stretch consists mainly of the C–X–C out-of-phase stretch.

Important five-membered ring heteroaromatic vibrations not depicted in Fig. 6.18 include the ring CH stretch, the –CH=CH– ring double bond cis CH wag and NH stretch in pyrrole type compounds. The NH stretch of pyrroles are typically observed between 3450 and

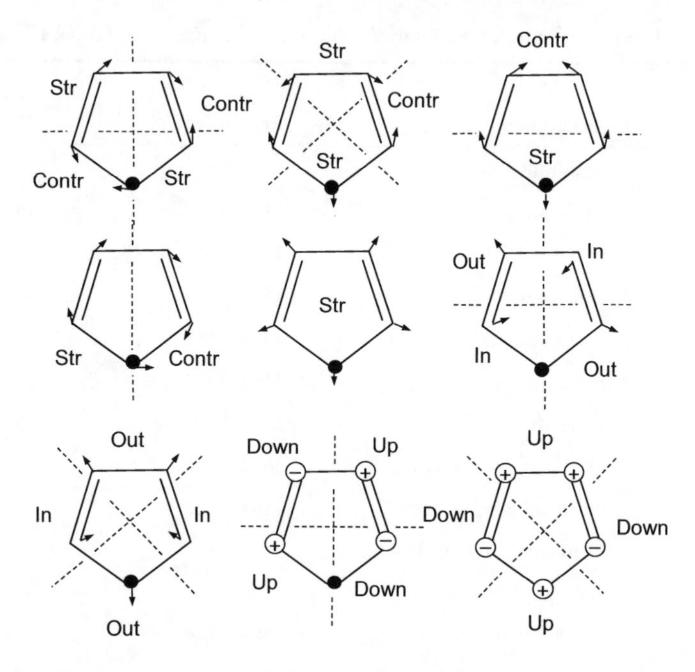

FIGURE 6.18 Approximate ring vibrations for a five-membered ring containing one heteroatom (solid circle) and two double bonds. The quadrant stretch predominantly involves the double bond stretching vibration.

TABLE 6.9 Selected In-plane Stretching Vibrations for Selected Pyrroles, Furans and Thiophenes

Pyridine	Assignment	Region (cm^{-1})	Intensities[1]	
			IR	R
Pyrrole	NH str.	3500–3000	s	m–w
	=CH str.	3135, 3103	m	s
	Quadrant str. + CH rk	1530	s	–
	Quadrant str. + CH rk	1468	m	s
	Semi-circle str + CH rk	1418	m	–
	Semi-circle str. + CH rk	1380	s	m
	Ring in-phase str.	1143	m	vs
1-subst pyrrole	Quadrant str.	1560–1540	w–m	m
	Quadrant str.	1510–1490	m	vs
	Semi-circle str.	1390–1380	s–m	s

(Continued)

TABLE 6.9 Selected In-plane Stretching Vibrations for Selected Pyrroles, Furans and Thiophenes—cont'd

			Intensities[1]	
Pyridine	**Assignment**	**Region (cm^{-1})**	**IR**	**R**
2-subst pyrrole	Quadrant str.	1570–1545	w–m	m
	Quadrant str.	1475–1460	m	vs
	Semi-circle str.	1420–1400	s–m	s
3-subst pyrrole	Quadrant str.	1570–1560	w–m	m
	Quadrant str.	1490–1480	m	vs
	Semi-circle str.	1430–1420	s–m	s
Furan	=CH str.	3156, 3121, 3092	m	s–m
	Quadrant str. + CH rk	1590	s	w
	Quadrant str. + CH rk	1483	s	vs
	Semi-circle str + CH rk	1378	s	s
	Ring in-phase str.	1138	–	vs
2-subst furan	Quadrant str.	1560–1540	w–m	m
	Quadrant str.	1510–1490	m	vs
	Semi-circle str.	1390–1380	s–m	s
Thiophene	=CH str.	3110, 2998	m	s–m
	Quadrant str. + CH rk	1588	s	–
	Quadrant str. + CH rk	1408	s	vs
	Semi-circle str + CH rk	1357	s	s
	Ring in-phase str.	1031	–	vs
	Quad bend	832	s	vs
2-subst thiophene	Quadrant str.	1540–1514	w–m	m
	Quadrant str.	1455–1430	m	vs
	Semi-circle str.	1361–1345	s–m	s

3200 cm^{-1}. However, for five-membered rings with more than one nitrogen atom in the ring such as imidazoles, a strongly hydrogen-bonded NH–N = complex occurs resulting in a broad complex set of bands in the 3000–2600 cm^{-1} observed in both the IR and Raman spectra. Pyrroles and furans which have nitrogen or oxygen atom in the ring, respectively, have moderate to strong IR and Raman bands involving the =CH stretch between 3180 and 3090 cm^{-1}. In the case of thiophenes which has a sulfur atom in the ring, the =CH stretching vibration is somewhat lower in frequency at 3120–3060 cm^{-1}. In general, the =CH stretching vibrations of five-membered ring heterocycles are slightly higher than observed for their six-membered ring analogues. Five-membered ring heterocyclic compounds with a unsubstituted –CH =CH– group are characterized by a strong IR band in the 800–700 cm^{-1} region which derives from the in-phase cis CH wag.

4. CARBONYL GROUPS

The generalized IR and Raman spectra of the carbonyl region are shown in Fig. 6.19 and the carbonyl group frequencies are summarized in Table 6.10.[1-4] The C=O stretching vibration responsible for these bands typically give rise to strong IR bands and weak Raman bands. In addition to the carbonyl band itself, much useful information can be obtained from the bands of the attached groups. The column of spectra on the left side shows examples of unconjugated carbonyl species including ketones (near 1715 cm^{-1}), aldehydes (near 1730 cm^{-1}), esters (near 1740 cm^{-1}), and acid chlorides (near 1800 cm^{-1}). The central column of spectra shows examples of carboxylic acid, carboxylic acid salt, anhydride, and cyclic anhydride while the third column of spectra shows examples of various amides.

4.1. Review of Selected Carbonyl Species

Carboxylic acid dimers are characterized by an IR band near 1700 cm^{-1}, and a Raman band near 1660 cm^{-1} which derive from the out-of-phase and the in-phase C=O stretch in the dimer, respectively. Weak, characteristic bands from the C–OH bend are also observed

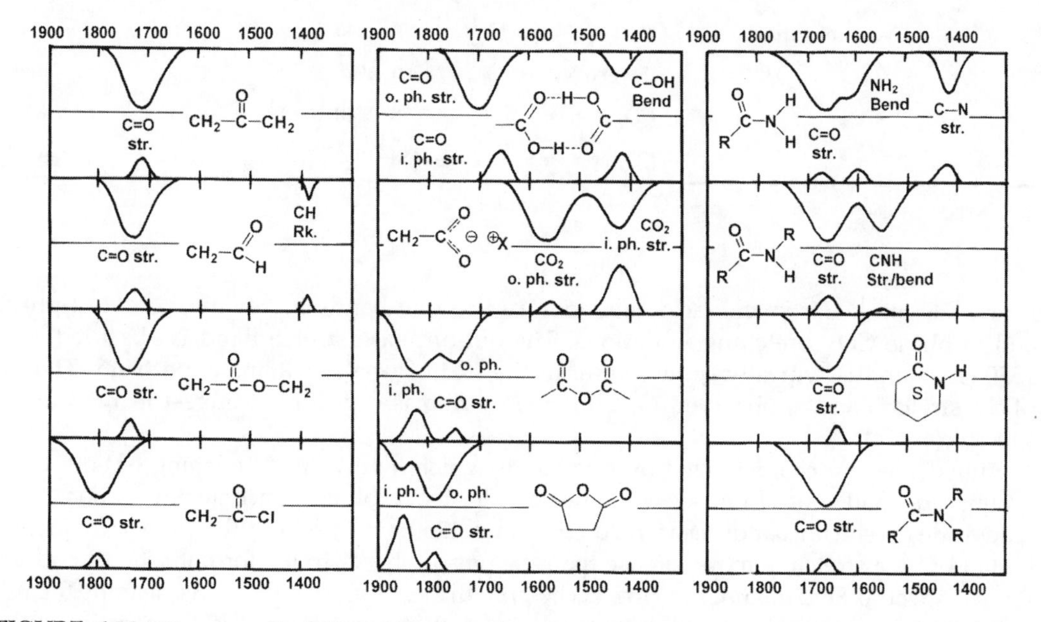

FIGURE 6.19 The generalized IR and Raman spectra of selected carbonyl compounds between 1900 and 1400 cm^{-1}.

TABLE 6.10 Selected Group Frequencies: Carbonyls

Group		Assignment	Frequency (cm^{-1})	Intensities[1]	
				IR	R
C=O	R–CO–H	C =O str.	1740–1720	s	m
	conj-CO–H	C =O str.	1710–1685	s	w
	R–CO–R	C =O str.	1725–1705	s	m
	conj-CO–R	C =O str.	1700–1670	s	m
	conj-CO-conj	C =O str.	1680–1640	s	m
	H–CO–O–R	C =O str.	1725–1720	s	m
	R–CO–O–R	C =O str.	1750–1735	s	m
	Conj-CO–O–R	C =O str.	1735–1715	s	m
	γ lactones	C =O str.	1795–1760	s	m
	R–CO–OH dimer	C =O o.ph. str.	1720–1680	s	–
	"	C =O i.ph. str.	1670–1630	–	m
	R–CO–N	C =O str.	1695–1630	s	m
	R–CO–Cl	C =O str.	1810–1775	s	m
	R–CO–O–CO–R	C =O i.ph. str.	1825–1815	s	m
	"	C =O o.ph. str.	1755–1745	ms	m
	Cyclic anhydride	C =O i.ph. str.	1870–1845	m	s
	" "	C =O o.ph. str.	1800–1775	s	mw
	R–CO_2^-	CO_2 o.ph. str.	1650–1540	s	w
	"	CO_2 i.ph. str.	1450–1360	ms	s

[1]*s = strong, m = medium, w = weak, v = very, br = broad, var. = variable, – = zero*

near 1420 cm^{-1}. Carboxylic acid salts give rise to two bands involving the out-of-phase and in-phase CO_2 stretching vibrations. The out-of-phase stretch band is observed near 1570 cm^{-1} and is typically quite strong in the IR but weak in Raman spectrum. The in-phase stretch band is observed near 1415 cm^{-1} and is typically strongest in the Raman spectrum.

Anhydrides are characterized by two bands which involve the stretching of both C=O groups in- and out-of-phase with respect to each other. Unconjugated, non-cyclic anhydrides result in bands near 1820 cm^{-1} and 1750 cm^{-1}, where the higher frequency, C=O in-phase stretch is more intense than the out-of-phase stretch for both the IR and the Raman spectra. Five-membered ring cyclic anhydrides give rise to bands near 1850 cm^{-1} and 1780 cm^{-1}, where the lower frequency out-of-phase stretch band is strongest in the IR spectrum, while the higher frequency in-phase band is strongest in the Raman spectrum.

This significant difference between the IR intensity of the in-phase and out-of-phase C=O stretch for the non-cyclic anhydrides and the cyclic anhydrides is directly related to the relative orientation of the two carbonyl groups. An integrated IR absorbance ratio of the two carbonyl bands can be used to predict the dihedral angle of the C–O bonds.[1]

Amides are characterized by strong IR bands from the C=O stretch near 1670 cm^{-1}. The Raman bands for the amide carbonyl stretch are typically quite weak. The primary amide group (O=C–NH_2) also has bands near 1610 cm^{-1} involving the NH_2 bend as well as near 1410 cm^{-1} involving the C–N stretch. The non-cyclic secondary amide group (O=C–NH–C) not only has a C=O stretch at 1670 cm^{-1}, but another strong band in the IR spectrum near 1550 cm^{-1} involving the CNH bend and the C–N stretch. In contrast, the cyclic secondary amide (O=C–NH–C) group does not have a 1550 cm^{-1} band. Amides with no NH have only the C=O stretch in this region. Fully annotated IR and Raman spectra of selected carbonyl compounds are included in Chapter 8 for comparison (see #16–29 for ketones, esters and anhydrides, 30–39 for amides, ureas, and related compounds).

4.2. Factors that Effect Carbonyl Frequencies

The carbonyl stretching frequencies are dependent upon mass, mechanical, and force constant effects. The mass effect is typically the least important factor in understanding carbonyl frequency dependence upon structure. This arises because the carbonyl stretching vibration involves mostly a change in the C=O bond length with only a slight movement of the attached atoms (bending). This is an excellent example of a high frequency C=O oscillator attached to a low frequency C–X oscillator (see unequal oscillators discussed earlier). The effect of changing mass on the carbonyl frequency when replacing a carbon atom with a heavier atom is quite slight due to the mass effect alone. The mass effect is more important when decreasing the substituent mass from carbon (12) to hydrogen (1) and should decrease the C=O frequency about 17 cm^{-1} when comparing a ketone to an aldehyde. However, even in this simple example, the higher C=O frequency of the aldehyde is counter to that expected by the mass effect, and demonstrates the importance of changes in the force constants.

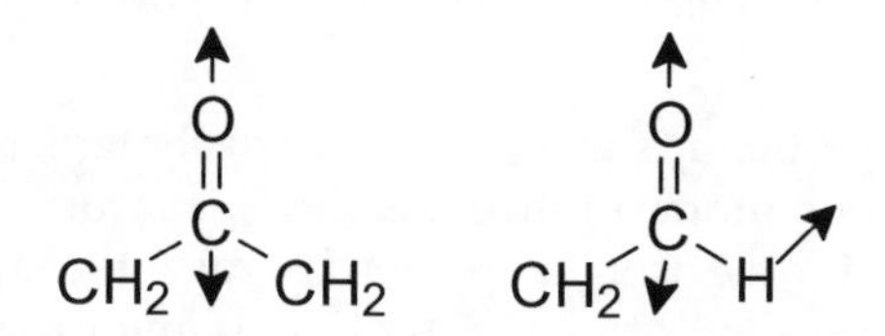

The above shows the approximate carbonyl stretching vibration for a ketone and an aldehyde. The very slight movements of the carbon atoms which results in changes in the carbonyl carbon bond angle are not shown. The aldehyde CH bend and C=O stretch are similar in energy and therefore couple more strongly than either the C–C stretch or the C–C bend.

The characteristic frequencies found in strained ring carbonyls is an excellent example of the importance of mechanical effects in carbonyl compounds and is analogous to that observed in cyclic olefin structures. Once again, the bond angle is the critical variable in determining the carbonyl stretching frequency. Shown below are the structures, important bond angles, as well as the observed and calculated frequencies based upon a simple mechanical model for the carbonyl stretching frequencies of four cyclic carbonyl compounds.

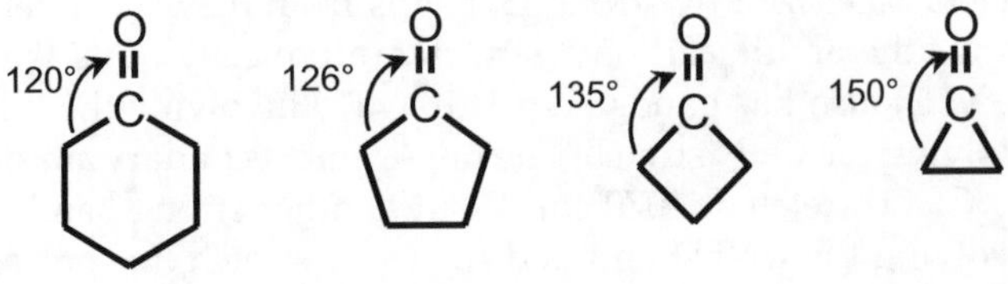

	120°	126°	135°	150°
Calc (cm^{-1})	1714	1737	1774	1836
Obs (cm^{-1})	1715	1740	1782	1822

The cyclic C=O stretching frequencies as a function of the C–C=O bond angles is shown above.

The same mechanical triatomic model is used for the above calculations as was used for the cyclic olefin structures earlier, where F_{C-C} and $F_{C=O}$ are 4.5 and 11.1 mdynes/A, respectively. The good agreement between the calculated and observed frequencies indicates that most of the shifts derive from mechanical effects and not from changes in force constants.

Force constant effects on carbonyl stretching frequencies include mesomeric, inductive, and field effects.[1-4] Carbonyl inductive effects derive from the polar nature of the carbonyl C=O bond. As shown below, the more the oxygen atom can attract electrons, the weaker the carbonyl C=O force constant becomes.

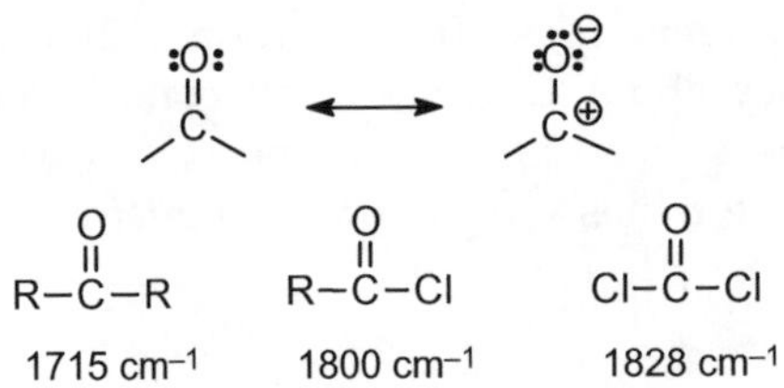

R–C(=O)–R	R–C(=O)–Cl	Cl–C(=O)–Cl
1715 cm^{-1}	1800 cm^{-1}	1828 cm^{-1}

The polarity of the carbonyl bond is a consequence of the greater electronegativity of the oxygen relative to the carbon atom. Electronegative substituents on the carbon atom will compete with the oxygen for electrons and thereby raise the carbonyl frequency. Substitution with strongly electron-withdrawing chlorine atoms provides a classic example of this. Field effects are another example of inductive effects in the carbonyl group. In this case, the presence of an electron rich chlorine spatially near to the oxygen atom favors the non-polar C=O group and raises the frequency. Thus, IR can discriminate between the cis and gauche rotational isomers in α-chlorocarbonyl compounds ($Cl{-}CH_2{-}C({=}O){-}X$).

Mesomeric and conjugation effects can be understood by examining resonance structures. Mesomeric effects derive from substituting the carbonyl group with lone pair electron containing heteroatoms such as chlorine, oxygen, or nitrogen.

$$-\overset{\overset{:O:}{\|}}{C}-\ddot{\underset{..}{Cl}}: \longleftrightarrow -\overset{\overset{:\ddot{O}:^{\ominus}}{|}}{C}=\ddot{\underset{..}{Cl}}^{\oplus} \qquad 1800\ cm^{-1}$$

$$-\overset{\overset{:O:}{\|}}{C}-\ddot{\underset{..}{O}}-CH_2- \longleftrightarrow -\overset{\overset{:\ddot{O}:^{\ominus}}{|}}{C}=\overset{\oplus}{\underset{..}{O}}-CH_2- \qquad 1740\ cm^{-1}$$

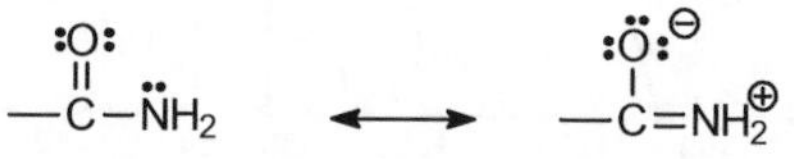

$$-\overset{\overset{:O:}{\|}}{C}-\ddot{N}H_2 \longleftrightarrow -\overset{\overset{:\ddot{O}:^{\ominus}}{|}}{C}=NH_2^{\oplus} \qquad 1660\ cm^{-1}$$

As shown above, the non-bonding electrons rearrange, resulting in donation of electrons to the oxygen atom and thereby weakening the carbonyl C=O bond. The mesomeric effect is not important for acid chlorides since the resonance structure placing a positive charge on the chlorine atom is not favored. Conversely, the mesomeric effect is the dominant determinant of the carbonyl frequency in amides.

Conjugation of a carbonyl with either a vinyl or a phenyl group typically lowers the carbonyl frequency by 20–30 cm^{-1}. This is also a result of electron rearrangement, analogous to the mesomeric effect discussed above and can easily be understood by use of resonance structures. Shown below are the approximate frequencies for an unconjugated ketone, single, and double conjugated ketone.

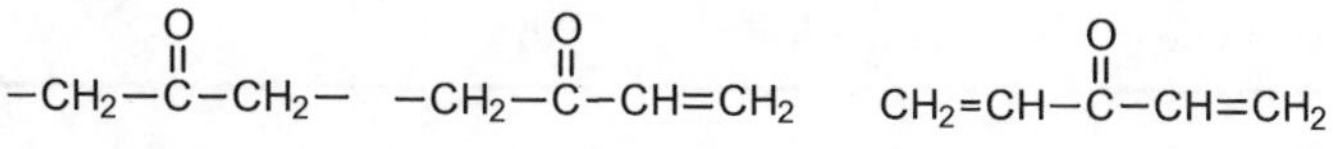

$$-CH_2-\overset{\overset{O}{\|}}{C}-CH_2- \qquad -CH_2-\overset{\overset{O}{\|}}{C}-CH=CH_2 \qquad CH_2=CH-\overset{\overset{O}{\|}}{C}-CH=CH_2$$

$$1715\ cm^{-1} \qquad 1685\ cm^{-1} \qquad 1655\ cm^{-1}$$

Lastly, interaction effects are key in understanding the source of two bands in anhydrides and carboxylic acid salts; these have their origin in both mechanical interactions and electron redistribution. Anhydrides have two carbonyl groups that vibrate in- and out-of-phase to each other resulting in two bands in the IR and Raman spectra. Comparison of the generalized IR spectra of cyclic and non-cyclic anhydrides shows that the relative intensity easily enables rapid identification. Similarly, carboxylic acid salt have two bond-and-a-half C–O bonds which vibrate in- and out-of-phase to give two bands (see Fig. 6.19). The different relative intensities for these two vibrations derive from the total dipole moment change. The electron redistribution that occurs during the in- and out-of-phase vibrations for these compounds is termed as an interaction force constant and also contributes to the observed frequencies.

5. C–O AND C–N STRETCHES

The IR and Raman spectra in Fig. 6.20 illustrate some group frequencies for vibrations involving C–O or C–N stretches interacting with connected C–C vibrations.[1-4] Table 6.11

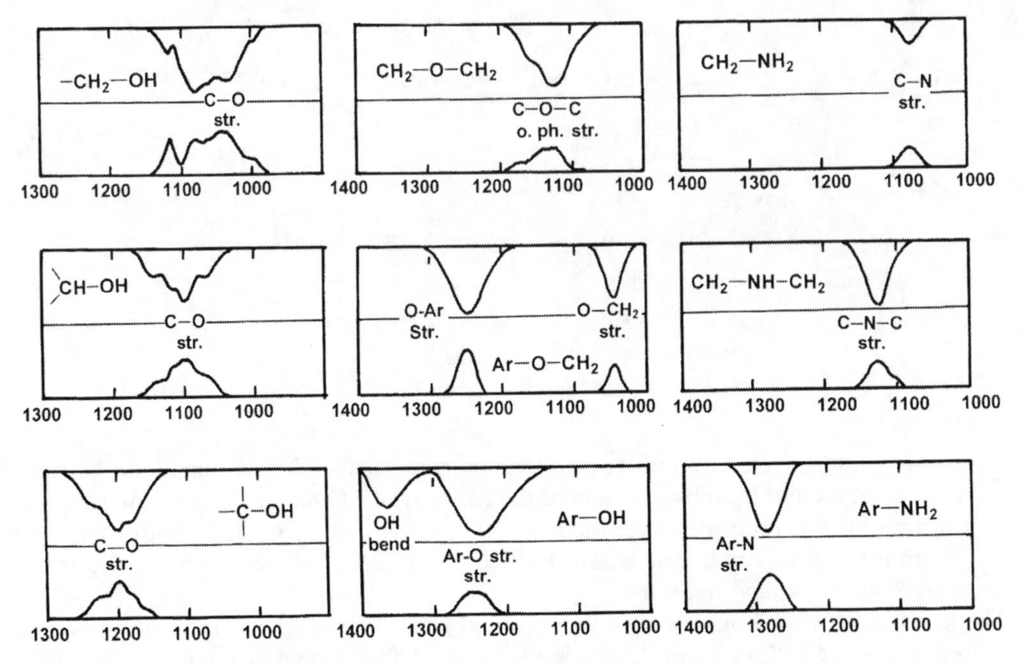

FIGURE 6.20 The generalized IR and Raman spectra of selected C–O and C–N stretching bands.

TABLE 6.11 Selected Group Frequencies: Alcohols and Ethers

Group		Assignment	Frequency (cm^{-1})	Intensities[1] IR	R
OH	R–OH······O	OH str.	3400–3200	sbr	vw
	O=C–OH dimer	OH str.	3200–2600	sbr	–
	"	OH wag	960–875	m	–
	P–OH······O	OH str.	2800–2100	s br	vw
	S–OH······O	OH str.	3100–2200	sbr	vw
C–O	CH_2–O–CH_2	COC o.ph. str	1270–1060	s	w
	"	COC i.ph. str.	1140–800	m	s
	C=C–O–CH_2	COC o.ph. str.	1225–1200	s	w
	Ar–O–CH_2	Ar–O str.	1310–1210	s	m
	"	O–CH_2 str.	1050–1010	m	m
	CH_2–OH	C–O str.	1090–1000	s	mw
	"	C–O str.	900–800	mw	s

TABLE 6.11 Selected Group Frequencies: Alcohols and Ethers—cont'd

Group	Assignment	Frequency (cm^{-1})	Intensities[1]	
			IR	R
R_2CH–OH	C–O str.	1150–1075	m	mw
"	C–O str.	900–800	mw	s
R_3C–OH	C–O str.	1210–1100	s	mw
"	C–O str.	800–750	mw	s
Ar–OH	C–O str.	1260–1180	s	w
O=C–O–C	C–O str.	1300–1140	s	w
O=C–OH	C–O str.	1300–1200	s	w
Epoxy	ring i.ph. str.	1270–1245	m	s
"	ring o.ph. str.	935–880	s	m
"	ring o.ph. str.	880–830	s	m

[1]*s = strong, m= medium, w = weak, v = very, br = broad, var. = variable, – = zero*

summarizes some of the important group frequencies for alcohols and ethers. Typically, vibrations involving these groups tend to give band clusters rather than single bands. The bands involving the C–O and C–N stretches of alcohols and amines show a systematic frequency dependence on their substitution (i.e., primary, secondary...).

Primary alcohols and amines have bands at approximately 1050 cm^{-1} and near 1075 cm^{-1} respectively. Secondary alcohols have bands at roughly 1100 cm^{-1}, while aliphatic secondary amines have IR bands near 1135 cm^{-1}. Tertiary alcohols usually have IR bands near 1200 cm^{-1}.

Aliphatic ethers have characteristic IR bands near 1120 cm^{-1} while alkyl–aryl ethers have two sets of bands near 1250 cm^{-1} involving aryl-O stretch and 1040 cm^{-1} involving the O–CH_2 stretch. Phenols have IR bands near 1240 cm^{-1} and aromatic-NH_2 aniline species have IR bands roughly at 1280 cm^{-1} involving aryl-O and aryl-N stretches, respectively.

Fully annotated IR and Raman spectra of selected alcohol and ethers are included in Chapter 8 (alcohols #40–47 and ethers 48–50) for comparison.

6. N=O AND OTHER NITROGEN CONTAINING COMPOUNDS

The generalized IR and Raman spectra illustrating group frequencies for some N=O type compounds are shown in Fig. 6.21.[1-4] Table 6.12 summarizes the selected group frequencies for selected nitrogen containing compounds. As shown in Fig. 6.21 and in Table 6.12, NO_x containing compounds provide strong IR and Raman strong bands that have excellent group frequencies.

The unconjugated nitro group (C–NO_2) are illustrated in the top spectra in Fig. 6.21 and have NO_2 stretching bands near 1560 and 1375 cm^{-1}. The out-of-phase stretch at ~1560 cm^{-1}

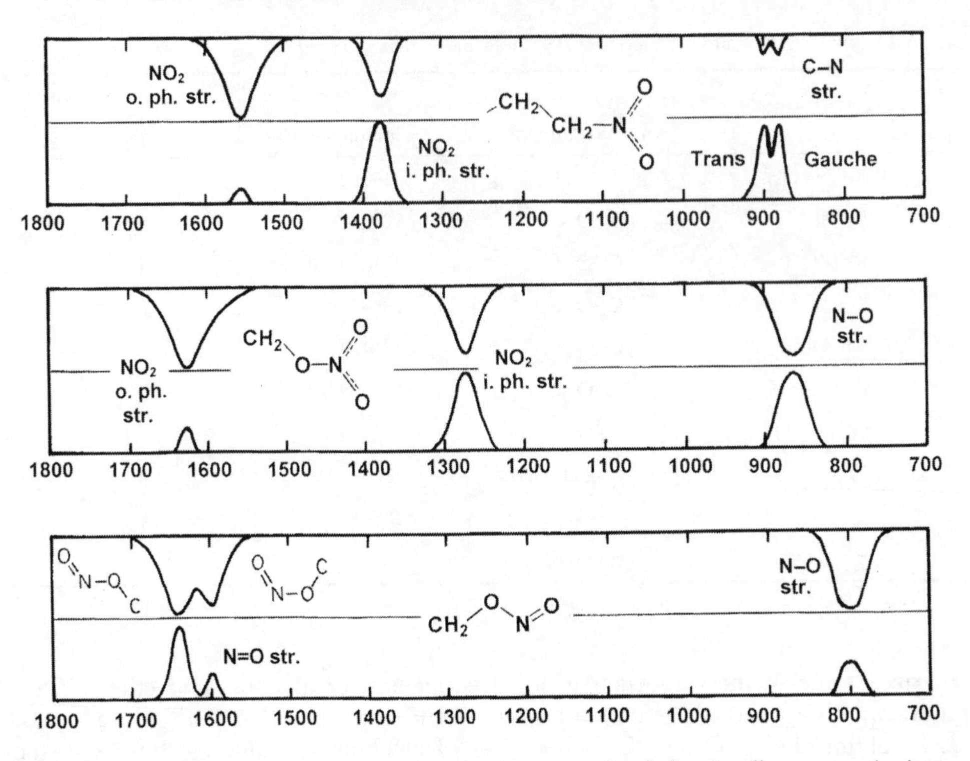

FIGURE 6.21 The generalized IR and Raman spectra of important bands for nitroalkanes, organic nitrates, and organic nitrites.

TABLE 6.12 Selected Group Frequencies: Nitrogen Compounds

Group	Assignment	Frequency (cm^{-1})	Intensities[1] IR	Intensities[1] R
CH_2–NH_2	NH_2 o.ph. str.	3500–3300	m	vw
"	NH_2 i.ph. str.	3400–3200	m	m
"	NH_2 bend	1630–1590	m	vw
"	NH_2 wag	900–600	sbr	w
CH_2–NH–CH_2	NH str.	3450–3250	vw	w
"	C–N–C o.ph. str.	1150–1125	m	mw
Ar–NH_2	C–N str.	1380–1260	sbr	m
C–NH_3^+ ······X^-	NH_3 str.	3200–2700	s	vw
"	NH_3 o.ph. bend	1625–1560	mw	vw
"	NH_3 i.ph. bend	1550–1505	w	vw

TABLE 6.12 Selected Group Frequencies: Nitrogen Compounds—cont'd

Group	Assignment	Frequency (cm^{-1})	Intensities[1]	
			IR	R
$C_2NH_2^+ \cdots\cdots X^-$	NH_2 str.	3000–2700	sbr	w
"	NH_2 bend	1620–1560	mw	w
$C_3NH^+ \cdots\cdots X^-$	NH str.	2700–2300	s	w
O =C–NH_2	NH_2 o.ph. str.	3475–3350	s	w
"	NH_2 i.ph. str.	3385–3180	s	w
"	NH_2 bend	1650–1620	ms	w
O =C–NH–C	NH str.	3320–3270	ms	w
"	CNH str. bend	1570–1515	ms	w
O =C–NH–C cyclic	NH str.	3300–3100	ms	w
>C =N	C =N str	1690–1630	mw	ms
>C =N–OH	OH str.	3300–3150	s	w
"	N–O str.	1000–900	s	s
CH_2–NO_2	NO_2 o.ph. str.	1600–1530	s	mw
"	NO_2 i.ph. str.	1380–1310	s	vs
Ar–NO_2	NO_2 o.ph. str.	1555–1485	s	–
"	NO_2 i.ph. str.	1357–1318	s	vs
C–O–NO_2	NO_2 o.ph. str.	1640–1620	s	mw
"	NO_2 i. ph. str.	1285–1220	s	s
"	N–O str.	870–840	s	s
C–O–N =O	*s*-trans N =O str.	1681–1640	s	s
"	*s*-cis N =O str.	1625–1600	s	s

[1]*s = strong, m = medium, w = weak, v = very, br = broad, var. = variable, – = zero*

is more intense in the IR spectrum while the in-phase stretch at ~1375 cm^{-1} is more intense in the Raman spectrum. In nitroalkanes, the C–N stretch gives rise to bands in the 915–865 cm^{-1} region that are intense in the Raman and medium–weak in the IR spectra. These bands are sensitive to the rotational isomer present in the CH_2–CH_2–NO_2 group. When the C–C group is trans to the CNO_2 species the C–N band is at ~900 cm^{-1} and when the gauche orientation is present, it is at ~ 880 cm^{-1}.

The covalent nitrate group (R–O–NO_2^-) shown in the middle spectra in Fig. 6.21 has bands near 1620 and 1275 cm^{-1} involving the NO_2 stretch. The out-of-phase stretch at ~1620 cm^{-1} is more intense in the IR spectrum while the in-phase stretch at ~1275 cm^{-1} is more intense in the Raman spectrum. The N–O stretch also gives rise to a prominent band near 870 cm^{-1}.

The covalent nitrite group (R—O—N=O) is characterized by N=O stretching bands between 1600 and 1640 cm^{-1} and a N—O stretch near 800 cm^{-1}. The N=O bands are sensitive to the rotation isomer present in the CH_2—O—NO group. The N=O stretch bands are typically observed near 1640 and 1600 cm^{-1} for the *s*-trans and *s*-cis isomers, respectively. Here *s*-cis indicates the single (O—C) bond is cis to the N=O bond and *s*-trans indicates the single (O—C) bond is trans to the N=O bond. The fully annotated IR and Raman spectra of selected nitrogen containing compounds including amines (#51—56), C=N type groups (#57—59), N=O type groups (60—62), and an azo group (63) are found in Chapter 8 for comparison.

7. C-HALOGEN AND C—S CONTAINING COMPOUNDS

The generalized IR and Raman spectra illustrating group frequencies for the stretching vibrations of C—Cl, C—Br, C—S—C, and C—S—S—C groups is shown in Fig. 6.22.[1-4] Table 6.13 shows some selected group frequencies of halogenated compounds. All of these have rotational isomers where the atom that is trans to the halogen or sulfur is either

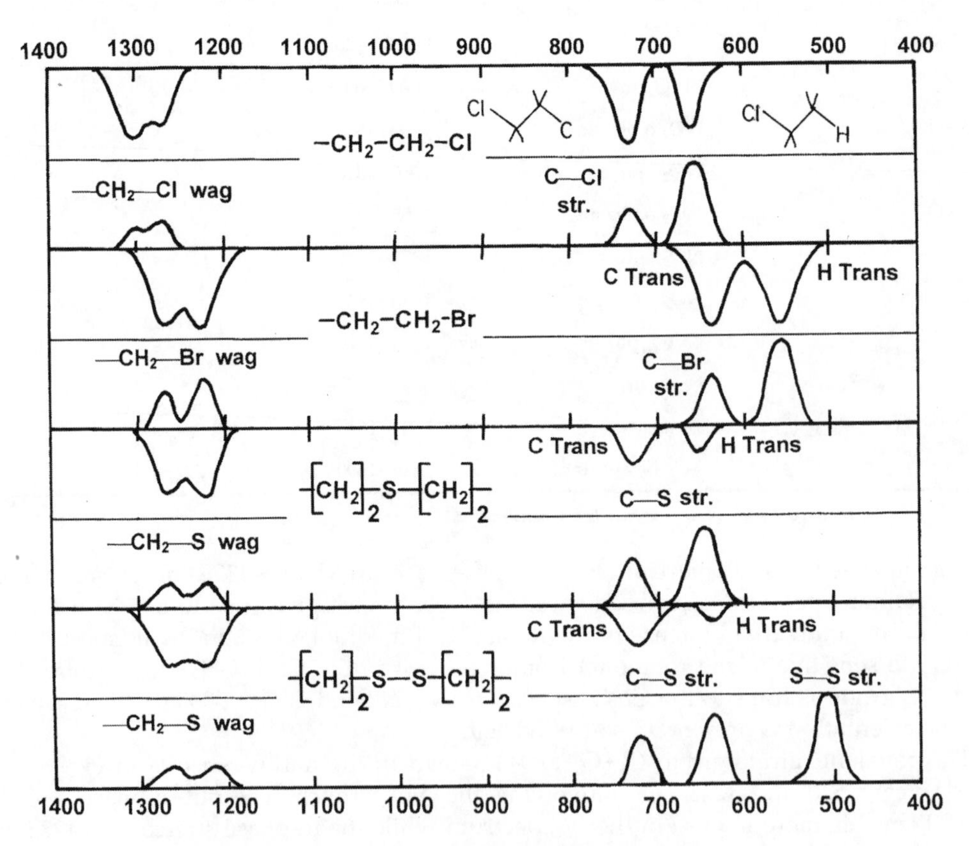

FIGURE 6.22 The generalized IR and Raman spectra for selected vibrations involving the C—Cl, C—Br, C—S—C, and C—S—S—C groups.

TABLE 6.13 Selected Group Frequencies: Halogen Compounds

Group	Assignment	Frequency (cm^{-1})	Intensities[1]	
			IR	R
CF_2 and CF_3	CF str.	1350–1120	vs	mw
Ar–F	CF str.	1270–1100	s	m
C–Cl	C–Cl str.	830–560	s	s
C–CH_2–CH_2–Cl	trans C–Cl str.	730–720	s	s
"	gauche C–Cl str.	660–650	s	s
CH_2–Cl	CH_2 wag	1300–1240	s	s
C–Br	C–Br str.	700–515	s	vs
C–CH_2–CH_2–Br	trans C–Br str.	650–640	s	vs
"	gauche C–Br str.	565–560	s	vs
CH_2–Br	CH_2 wag	1250–1220	s	m

[1]*s = strong, m = medium, w = weak, v = very, br = broad, var. = variable, – = zero.*

a carbon or a hydrogen atom. Both the IR and Raman spectra have isolated bands C–S and C-Halogen stretching bands which are due to the rotational isomers. For chlorinated aliphatics such as C–CH_2–CH_2–Cl, the bands involving the C–Cl stretch appear near 730 and 660 cm^{-1}. The aliphatic sulfide group (–CH_2–CH_2–S–CH_2–CH_2) is quite similar with C–S stretching bands near 720 and 640 cm^{-1}. For brominated aliphatics such as C–CH_2–CH_2–Br, the bands involving the C–Br stretch appear at much lower frequency near 635 and 560 cm^{-1}.

The C–F stretch of organic fluorine compounds results in strong IR bands but weak to moderately intense Raman bands. Monofluorinated compounds have a strong IR and a weak to moderate Raman band between 1100 and 1000 cm^{-1}. Difluorinated compounds (–CF–$_2$) have two very strong IR bands between 1250 and 1050 cm^{-1} involving the CF_2 out-of-phase stretch. These bands are weak to moderate in the Raman spectrum. Trifluorinated compounds (–CF_3) have multiple strong bands between 1350 and 1050 cm^{-1} involving the CF_3 out-of-phase stretch.

Both the IR and Raman have some characteristic bands such as the CH_2 wags and the disulfide S–S stretch, which are not sensitive to rotational isomers. The disulfide group (R–S–S–R) is characterized by a very strong Raman band near 500 cm^{-1} that involves the S–S stretch. In addition, the CH_2–Cl group has a CH_2 wag band cluster near 1285 cm^{-1}, whereas the CH_2–Br and CH_2–S groups have these bands near 1240 cm^{-1}.

In aryl-halides, the C–X stretching vibration couples with the aryl ring vibrations resulting in strong IR and moderate Raman bands. Fluorine-substituted aromatics (Ar—F) have a moderately strong IR band between 1270 and 1100 cm^{-1} involving the aryl ring C–F stretch. This same band is found between 1100 and 1030 cm^{-1} for the aryl-Cl species and between 1075 and 1025 cm^{-1} for aryl bromo species. These band regions can be further defined from high to lower frequency for para, meta, and ortho substitution of the Cl and

Br aryl halides. Iodobenzenes (para) have a characteristic aryl-I band between 1061 and 1057 cm^{-1}.

The fully annotated IR and Raman spectra of selected chlorine, bromine, and fluorine containing compounds are included in Chapter 8 for comparison (#62–69, 99).

8. S=O, P=O, B–O/B–N AND Si–O COMPOUNDS

The generalized IR and Raman spectra illustrating group frequencies for some S=O, SO_2, P=O, and Si–O type compounds are shown in Fig. 6.23. Table 6.14 summarizes the selected group frequencies for the chosen S=O, P=O, B–O, and Si–O containing compounds.[1-4]

Compounds containing S=O generally have strong characteristic IR and Raman bands. The characteristic IR and Raman bands of a sulfoxide (R_2SO), sulfone (R_2SO_2), and a dialkyl sulfate (RO–SO_2–OR) are shown in the first column of spectra in Fig. 6.23. The frequency of these bands can be correlated with the substituent electronegativity. In general, electronegative substituents tend to raise the S=O frequency since they favor the S=O resonance structure. As illustrated in the top left spectra, the intensity of the S=O stretch near 1050 cm^{-1} is more intense in the IR than the Raman spectrum.

For both sulfones and dialkyl sulfates, the SO_2 group results in two bands due to the in- and out-of-phase stretches. The higher frequency out-of-phase SO_2 stretch is stronger in the IR and the in-phase SO_2 stretch is more intense in the Raman spectra. Table 6.14 also summarizes

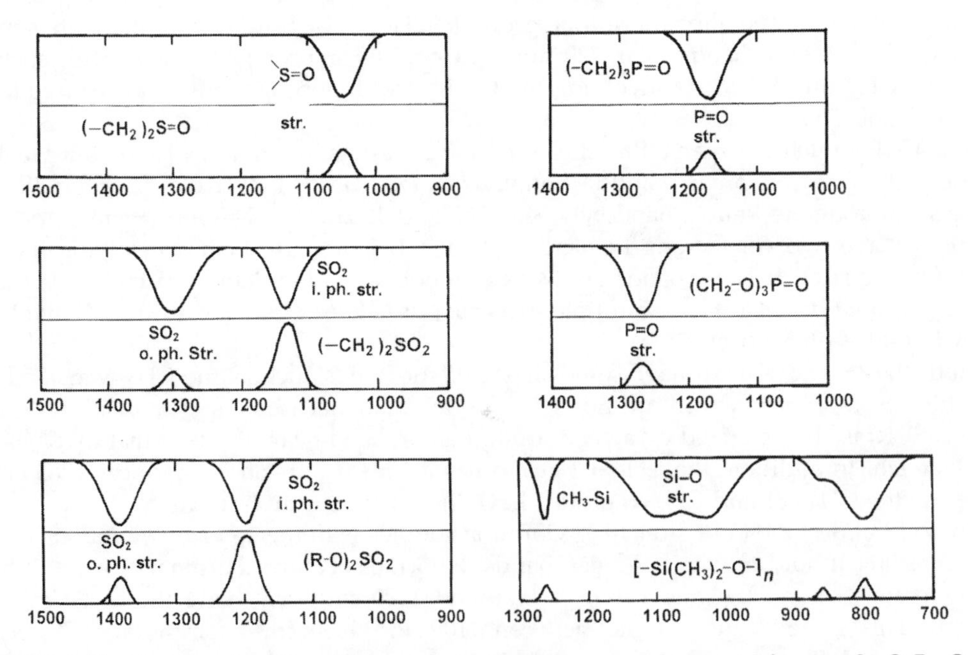

FIGURE 6.23 The generalized IR and Raman spectra illustrating group frequencies for some S=O, P=O, and Si–O type compounds.

important bands for sulfonamides (R–SO_2–N), sulfonates (R–SO_2–OR), sulfonyl chlorides, sulfonic acids (anhydrous: R–SO_2–OH), and sulfonic acid salts (R–SO_3^-).

The characteristic IR and Raman bands of a phosphine oxide (R_3P =O) and phosphorus ester ((R–O)$_3$P =O) are shown in the second column of spectra in Fig. 6.23. The P=O stretching frequency depicted have been shown to correlate with the sum of the substituent

TABLE 6.14 Selected Group Frequencies: Sulfur, Phosphorus, Silicon, and Boron Compounds

Group		Assignment	Frequency (cm^{-1})	Intensities[1]	
				IR	R
Sulfur Cpds	SH	SH str.	2590–2540	w	s
	C–SO_2–C	SO_2 o.ph. str.	1340–1290	s	w
	"	SO_2 i.ph. str.	1165–1120	s	s
	C–SO_2–N	SO_2 o.ph. str.	1380–1310	s	m
	"	SO_2 i.ph. str.	1180–1140	s	s
	C–SO_2–O–C	SO_2 o.ph. str.	1375–1335	s	ms
	"	SO_2 i.ph. str.	1195–1165	s	s
	C–SO_2–Cl	SO_2 o.ph. str.	1390–1361	s	w
	"	SO_2 i.ph. str.	1181–1168	s	s
	C–SO_2–OH anhydrous	OH str.	3100–2200	sbr	w
	" "	SO_2 o.ph. str.	1352–1342	s	w
	" "	SO_2 i.ph. str.	1165–1150	s	s
	C–SO_3^- H_3O^+ (hydrate)	OH str.	2800–2100	sbr	–
	" "	SO_3 o.ph. str.	1230–1120	s	–
	" "	SO_3 i.ph. str.	1085–1025	m	–
	C_2S =O	S =O str.	1065–1030	s	w
Phosphorus Cpds	PH	PH str.	2450–2270	ms	mw
	P =O	P =O str.	1320–1140[2]	s	m
	P–OH	OH str.	2800–2100	sbr.	vw
	P–O–CH_2	P–O–C str.	1050–970	s	mw
	P–O–Ar	O–Ar str.	1240–1160	s	mw
	"	P–O str.	995–855	s	w
Silicon Cpds	SiH	Si–H str.	2250–2100	vs	m
	Si–O	Si–O str.	1100–1000	vs	w
	Si–CH_3	CH_3 i.ph. bend	1270–1250	vs	w

(Continued)

TABLE 6.14 Selected Group Frequencies: Sulfur, Phosphorus, Silicon, and Boron Compounds—cont'd

				Intensities[1]	
Group		**Assignment**	**Frequency (cm^{-1})**	**IR**	**R**
Boron Cpds	BH	B–H str	2640–2350	s	mw
	BH (complete octet)	B–H str	2400–2200	s	mw
	B–H–B (Bridge)	B–H–B str.	2220–1540	s	mw
	B–OH	OH str	3300–3200	s	w
	B–O	B–O str	1380–1310	s	var
	B–N	B–N str.	1465–1330	s	var
	B-Phenyl	C–B–C/ring str.	1440–1430	s	

[1]*s = strong, m = medium, w = weak, v = very, br = broad, var. = variable, – = zero*
[2]*Electronegative substituents raise the P =O frequency (example: R_3P =O ~1160 cm^{-1}, $(R–O)_3P$ =O ~ 1270 cm^{-1})*

electronegativities. In general, the IR spectrum provides a better selection of strong characteristic bands. Table 6.14 summarizes selected important group frequencies for phosphorus compounds including the PH stretch, the OH stretch of the P–OH group, and the P–O–C stretch of the P–O–CH_2 group. The P–O-phenyl group gives rise to two bands involving the phenoxy C–O stretch and the P–O stretch.

The characteristic IR and Raman bands of a siloxane polymer is shown in the bottom right hands corner in Fig. 6.21. In general, IR spectroscopy provides a much better selection of strong, characteristic group frequency bands of silicon compounds. For silicones, the Si-CH_3 group has a good IR in-phase CH_3 bending band near 1265 cm^{-1} and the Si–O stretch has a strong, broad IR band in the 1100–1000 cm^{-1} region.

Table 6.14 also summarizes selected important group frequencies for boron containing compounds. The B–O stretch results in a strong characteristic IR band between 1380 and 1310 for both organic and inorganic compounds containing the B–O species. Compounds containing B–N species also result in strong characteristic IR bands between 1465 and 1330 cm^{-1}. Little information on the characteristic Raman bands B–O and B–N containing species are available in the literature.

The interpreted IR and Raman spectra for sulfur containing compounds (70–76, 83, 84), phosphorus containing compounds (77–80, 88, 89), silicon containing compounds (81, 93, 108), and boron containing compounds (64, 90) are included in Chapter 8 for comparison.

9. INORGANICS

Both IR and Raman spectroscopy provides important structural information on a multitude of ionic inorganic compounds.[19] Both techniques are sensitive to the crystalline form of the inorganic species. Table 6.15 summarizes selected IR and Raman bands for a few common inorganic compounds. The interpreted IR and Raman spectra of several selected inorganic compounds (#82–94) are included in Chapter 8 for comparison.

TABLE 6.15 Selected Group Frequencies: Common Inorganic Compounds

Inorganics	IR bands (cm^{-1})	Raman bands (cm^{-1})
NH_4^+	3100 s, 1410 s	3100 w, 1410 w
NCO^-	2170 vs, 1300 m, 1210 m	2170 m, 1300 s, 1260 s
NCS^-	2060 vs	2060 s
CN^-	2100 m	2080 s
CO_3^-	1450 vs, 880 m, 710 w	1065 s
HCO_3^-	1650 m, 1320 vs	1270 m, 1030 s
NO_3^-	1390 vs, 830 m, 720 w	1040 s
NO_2^-	1270 vs, 820 w	1320 s
SO_4^{-2}	1130 vs, 620 m	980 s
HSO_4^-	1240 vs, 1040 m, 870 m	1040 s, 870 m
SO_3^-	980 vs	980 s
PO_4^{-3}	1030 vs, 570 m	940 s
$(-SiO_2^-)_x$	1100 vs, 470 m	–
TiO_2	660 vs, 540 vs	–

s = strong, m = medium, w = weak, v = very, – = zero.

Chapter 6: Problems

1. Explain the frequency dependence observed for CH_3 in-phase bending band in the following series: B—CH_3 = 1310 cm^{-1}, C—CH_3 = 1380 cm^{-1}, N—CH_3 = 1410 cm^{-1}, O—CH_3 = 1445 cm^{-1}, and F—CH_3 = 1475 cm^{-1}.
2. Explain the observed splitting in the CH_3 in-phase bend IR band in isopropyl or gem-dimethyl groups and *t*-butyl groups.
3. Explain the observed frequency dependence of the C=CH_2 wag vibration: (N≡C)$_2$—C=CH_2 = 985 cm^{-1}, N≡C—CH =CH_2 = 960 cm^{-1}, R—C=CH_2 = 910 cm^{-1}, R—O—C=CH_2 = 813 cm^{-1}. Hint draw possible resonance structures and consider mesomeric electron donation to the =CH_2 carbon. How would you expect the IR intensity to vary?
4. Trans-substituted olefins (R—C=C—R) show the following systematic increase in the observed CH wag. An isolated trans alkene (T) has a CH wag at 965 cm^{-1}, a conjugated trans—trans diene (TT) at 986 cm^{-1}, the trans—trans—trans alkene (TTT) at 994 cm^{-1}, and the trans—trans—trans—trans alkene (TTTT) at 997 cm^{-1}. This is a result of vibrational interaction with conjugated diene groups. Using diagrams of the CH wag explain this progression. Hint: mechanical interaction (see N.B. Colthup, Appl Spectrosc., 25, 368, 1971).
5. The P=O, S=O, and SO_2 stretching frequencies are relatively unaffected by conjugation or strained ring. How does the geometry relate to this? The inductive effect has a significant impact on the P=O, S=O, and SO_2 stretching frequencies. Where the inductive effect is: $X_3P{=}O \leftrightarrow X_3P^{+}{-}O^{-}$ How could changing the substituents change the frequency (i.e., what simple parameter would explain the frequency shifts)?
6. Draw the five expected ring stretching modes for a five-membered ring. Use standing wave to help. Use your vibrations to assign IR bands observed for the following ring stretch vibrations:

Rings	Region 1 (IR:var, R:var)	Region 2 (IR:m, R:vs)	Region 3 (IR:ms, R:s)
1-Substituted pyrrole	1560—1540	1510—1490	1390—1380
2-Substituted pyrrole	1570—1545	1475—1460	1420—1400
2-Substituted Furans	1605—1560	1515—1460	1400—1370
2-Substituted Thiophenes	1535—1514	1454—1430	1361—1347

7. The aromatic sextant stretch typically is IR and Raman forbidden and is therefore not observed. This band has been assigned for certain Cl-substituted aromatics to occur between 1330 and 1305 cm^{-1} which is below that observed for either the quadrant or semi-circle ring stretches. Draw the two different extremes of the sextant ring stretch. The lower frequency is due to an interaction force

constant. Explain this concept as it relates to the observation for the aromatic sextant stretch.

8. How does the interaction force constant explain the observed differences between 2,4-pentanedione ($CH_3-CO-CH_2-CO-CH_3$ at 1725 and 1707 cm^{-1}) and an anhydride ($CH_3-CO-O-CO-CH_3$ at 1833 and 1764 cm^{-1}).

References

1. Colthup, N. B.; Daly, L. H.; Wiberley, S. E. *Introduction to Infrared and Raman Spectroscopy*, 3rd ed.; Academic: New York, NY, 1990.
2. Bellamy, L. J.; *The Infrared Spectra of Complex Molecules*, 3rd ed.; Chapman and Hall: New York, NY, 1975; vol. 1.
3. Lin-Vien, D.; Colthup, N. B.; Fateley, W. G.; Grasselli, J. G. *Infrared and Raman Characteristic Frequencies of Organic Molecules*; Academic: San Diego, 1991.
4. Socrates, G. *Infrared Characteristic Group Frequencies*, 2nd ed.; John Wiley: New York, NY, 1994.
5. Jordanov, B.; Tsankov, D.; Korte, E. H. *J. Mol. Struct.* **2003**, 651–653, 101–107.
6. Atamas, N. A.; Yaremko, A. M.; Seeger, T.; Leipertz, A.; Bienko, A.; Latajka, Z.; Ratajczak, H.; Barnes, A. J. *J. Mol. Struct.* **2004**, *708*, 189–195.
7. Amorim da Costa, A. M.; Marques, M. P. M.; Batista de Carvalho, LA. E. *Vib. Spectrsoc.* **2002**, *29*, 61–67.
8. Balabin, R. M. *J. Phys. Chem. A* **2009**, *113*, 1012–1019.
9. Colthup, N. B. *J. Chem. Educ.* **1961**, *38* (8), 394–396.
10. Colthup, N. B.; Orloff, M. K. *Spectrochim. Acta, Part A* **1971**, *27A*, 1299.
11. Determined by the author using Hammet values from Domingo, L. R.; Perez, P.; Contreras, R. *J. Org. Chem.* **2003**, *68*, 6060–6062.
12. Tasumi, M.; Urano, T.; Nakata, M. *J. Mol. Struct.* **1986**, *146*, 383–396.
13. Palafox, M. A. *Int. J. Quantum Chem.* **2000**, *77*, 661–684.
14. Versanyi, G. *Vibrational Spectra of Benzene Derivatives*; Academic: NY, 1969.
15. Colthup, N. B. *Appl. Spectrosc.* **1976**, *30*, 589.
16. Larkin, P. J.; Makowski, M. P.; Colthup, N. B. *Spectrochim. Acta A* **1999**, *55*, 1011–1020.
17. Larkin, P. J.; Makowski, M. P.; Colthup, N. B.; Flood, L. *Vibr. Spectrosc.* **1998**, *17*, 53–72.
18. El-Azhary, A. A.; Hilal, R. H. *Spectrochim. Acta, Part A* **1997**, *53*, 1365–1373.
19. Nakamoto, K. *Infrared and Raman Spectra of Inorganic and Coordination Compounds, Part A: Theory and Applications in Inorganic Chemistry, Part B: Applications in Coordination, Organometallic, and Bioinorganic Chemistry*, 5th ed.; John Wiley: New York, NY, 1997.

CHAPTER

7

General Outline and Strategies for IR and Raman Spectral Interpretation

Spectroscopists must develop an analytical toolbox suitable for their needs. This includes the ability to measure a suitable spectrum, understand the basic chemistry of the compound, and systematically interpret the spectrum. In this chapter, the IR and Raman spectrum will be discussed briefly in terms of spectral regions rather than functional groups.

Both Raman and FT-IR spectroscopies provide important molecular structural information. However, the successful application of both techniques relies heavily on the skill of the spectroscopist. There are common skills used for solving structural problems using vibrational spectroscopy. In general, tools and skills that can be called upon include the following components:

- Interpersonal and communication skills—Define the problem and get background information on the sample.
- Instrumentation—Select most appropriate technique and sample preparation.
- Data analysis techniques—Experience is important in selecting the most appropriate tool for your application.
- Spectral interpretation skills.
- Spectral database tools (spectral libraries, interpretation software).
- *Ab initio*-based normal mode analysis (ultimate structural verification).

1. TOOLS OF THE TRADE

IR and Raman spectroscopies are used to study a very wide range of sample types and may be examined either in bulk or in microscopic amounts over a wide range of temperatures and physical states (e.g., gases, liquids, latexes, powders, films, fibers, or as a surface or embedded layer). Because IR and Raman spectroscopies have a very broad range of applications, it is important to be knowledgeable about the advantages and limitations of various available techniques in order to select the most appropriate technique for the problem facing the spectroscopist. Often a broad-based knowledge of spectroscopy is required including an understanding of the instrument, the data analysis as well as spectral interpretation skills.

Basic instrumental techniques for bulk analysis include mid-IR (most commonly FT-IR) as well as Raman spectroscopy. Raman spectroscopy has various laser excitation sources and instrument designs (such as FT-Raman or grating-based instruments). When examining smaller samples micro-analysis techniques must be employed such as microscopy (IR and Raman, imaging, and confocal), and micro-ATR for IR.

Once a suitable sampling technique has been selected to provide the spectrum (or spectra) needed, a data analysis technique to provide information needed to solve the problem is required. For identification and structural verification purposes, this can include spectral databases and interpretation. Chemometric data analysis may be employed for developing quantitative models or analysis of large spectral data sets such as those encountered in imaging techniques.

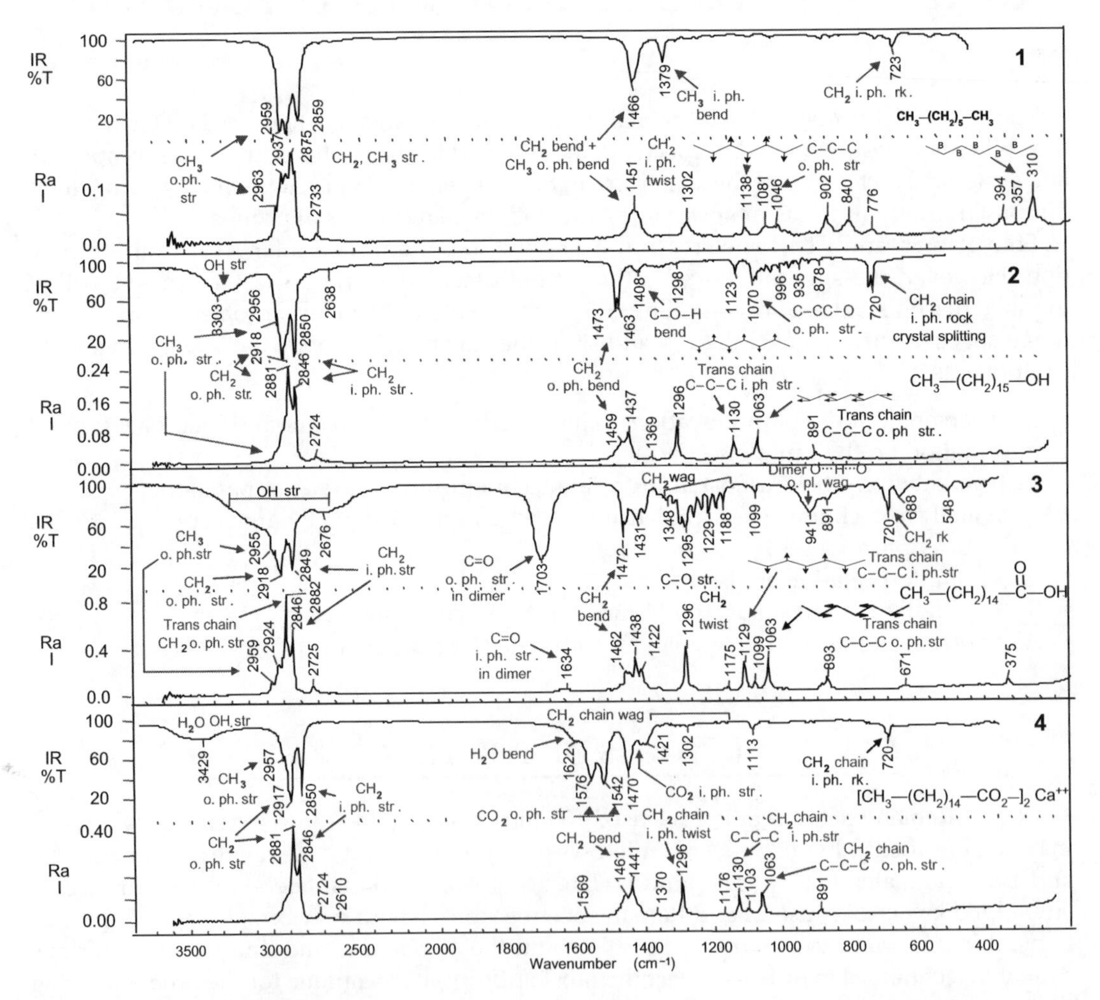

FIGURE 7.1 The FT-IR and FT-Raman spectra of *n*-heptane, 1-hexadecanol, palmitic acid, and calcium palmitic acid salt.

Success will often depend on your selection of the most suitable technique for the problem at hand. As a rule, mid-IR provides a better general workhorse technique than Raman spectroscopy because it contains a greater selection of bands characteristic of functional groups. This is illustrated in the Fig. 7.1 of the IR and Raman spectra for several organic species. The Raman spectra have characteristic bands for the aliphatic backbone, but do not have strong bands associated with important functional groups of the alcohol, carboxylic acid, and carboxylic acid salt. IR has strong characteristic bands for the aliphatic methyl and methylene groups as well as the alcohol, carboxylic acid, and carboxylic acid salt groups. The Raman spectrum of each of the four species is very similar with bands characteristic of the methyl, methylene, and skeletal carbon backbone.

In the four compounds selected (*n*-heptane, 1-hexadecanol, palmitic acid, and calcium palmitic acid salt), the Raman spectrum does not provide strong bands for the hydroxyl, carbonyl, or carboxylic acid. Moderate-to-strong Raman bands can occur for carbonyl groups that are more polarizable (e.g., amide carbonyls or the very strong carbonyl observed in isocyanurate).

There are functional groups for which Raman is the preferred method while some functional groups provide useful bands in both types of spectra. In general, IR provides a broader representation of characteristic bands for a greater variety of functional groups. The Raman spectra have characteristic bands for the aliphatic backbone, but do not have strong bands associated with important functional groups of the alcohol, carboxylic acid, and carboxylic acid salt.

2. IR SAMPLE PREPARATION ISSUES

The quality of the information that can be derived from a FT-IR spectrum is directly related to the quality of the spectrum itself. Therefore selection of the appropriate sample technique and the resultant spectrum quality is critical. Operators must develop requisite sample-preparation skills necessary to measure photometrically accurate spectra. In the case of neat liquids, the liquid film should be uniformly thick (ca. 5–8 μm) without holes or voids to insure a good, quantitative IR spectrum. Similarly, films (cast or pressed) should also be uniformly thick without holes or voids. Lastly, solid-powdered samples can be prepared as a uniform dispersion in either a Nujol or a KBr matrix. In both cases, the sample powder must be sufficiently grounded and mixed with the support matrix to insure a good, artifact-free IR spectrum.

A wide variety of IR sample preparation techniques exist because there is not a universal sampling technique that will provide high-quality IR spectra for all samples. The nature of the particular sample will often dictate the selection of the technique. As an example, the IR spectra of potato starch measured using a Nujol mull, KBr disc, water cast film, and ATR sampling are shown in Fig. 7.2.

Starch provides an excellent example of the importance of selecting an appropriate IR sample-preparation technique suitable for the sample of interest. Both the Nujol mull and KBr disc sample preparations are inappropriate for starch and result in poor-quality IR spectra. The Nujol mull provides an extremely false spectrum and results in very limited spectral information that could be used for identification and interpretation. The KBr disc

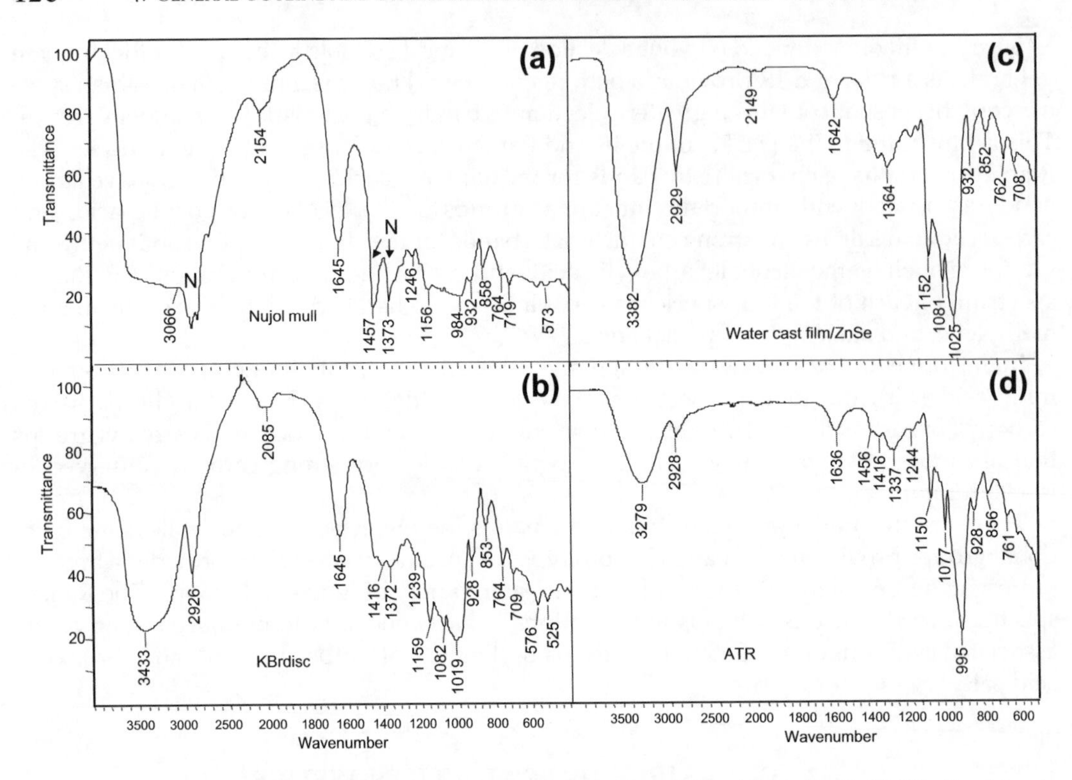

FIGURE 7.2 IR Spectra of potato starch: Comparison of (a) Nujol mull (N = Nujol signal), (b) KBr disc, (c) water cast film on ZnSe sample preparation, and (d) ATR sampling.

spectrum, while false, is a significant improvement over the Nujol mull preparation and could be used for identification. Dissolving the starch in water and preparing a cast film on ZnSe results in a photometrically accurate IR spectrum. The resultant IR spectrum is of an amorphous film. Lastly, using ATR, an excellent IR spectrum is obtained of the starch powder. Note that the ATR technique results in changes in peak position, shape, and also results, as is typical of ATR, in a wavelength dependence of the band intensities.

Five different general criteria should be kept in mind for suitable IR sample preparation for both samples and library reference spectrum.

1. *Suitable band intensities in the spectrum*. The strongest band in the spectrum should have intensity in the range of 5–15% transmittance. The resulting preparation must be uniformly thick and homogenously mixed without holes or voids to insure a good quality spectrum.
2. *Baseline*. The baseline should relatively be flat. The highest point in the spectrum should lie between 95 and 100% transmittance.
3. *Atmospheric compensation*. Water vapor and carbon dioxide bands should be minimized. Since the sample scan is ratioed against a previously measured background, spectral subtraction of water vapor/carbon dioxide, reference spectrum should be used.

4. *Spurious bands.* Any band that does not belong to the sample is spurious and must be accounted for or better yet eliminated. Use of an IR window contaminated with a previous sample is a very common source of spurious bands.
5. *X-axis scale.* Plots should employ a 2*X*-axis expansion and not a simple linear *X*-axis. This provides an expansion of the important fingerprint region. Some spectral databases such as the Aldrich collection utilize a 4*X*-axis expansion.

2.1. Select Suitable Sampling and Know the Limitations

The sample and problem at hand will determine if a bulk or microscopy measurement is used. The sampling by Raman spectroscopy is straightforward with fluorescence interference representing the main limitation. In the case of IR spectroscopy, sample preparation choices must be made to produce the most photometrically accurate spectrum. Sample preparation for bulk analysis includes KBr disc, Nujol mull, ATR, DRIFTS, and cast films. For a KBr disc preparation, the entire spectrum is available, but KBr is hydroscopic. Water will be introduced in the KBr disc preparation and this will affect the ability to interpret the OH/NH stretching region. For a Nujol mull sample preparation, there will be interferences from the mulling oil aliphatic groups. ATR can be considered a more universal sampling choice but the IR band intensities exhibit a wavelength dependence. Some band shape distortions will occur. Less commonly used techniques include DRIFTS where the analyst must be concerned about the photometric accuracy of the spectrum. Lastly, a solvent cast films can be used. Issues with this technique include preparing a uniform film, eliminating much or all of the solvent, and the resultant crystalline state of the material.

3. OVERVIEW OF SPECTRAL INTERPRETATION

Spectral interpretation requires a systematic, knowledge-based approach for optimal success. Both the sample IR and Raman spectrum and any reference spectrum must be of good quality. Before beginning to interpret the IR and Raman spectrum, the analyst should get a suitable amount of background information. Spectral interpretation is a critical skill that is learned with practice and experience. Resources to use include electronic and paper libraries, empirical characteristic frequencies for chemical functional groups as well as outside vendor and interior libraries. The quality of the available reference library is critical. The IR and Raman libraries ideally should capture knowledge of the spectroscopy of the band assignments of molecules included in the database in order to facilitate interpretation of other molecules.

3.1. Hierarchy of Quality for Spectral Reference Libraries

Spectral reference libraries have a hierarchy of quality. The usefulness of a spectral library is defined in part by how accurate it is and how much known information is captured in it. For example, the resolution of both the sample and reference libraries should match. In the case of IR spectrum, the sample preparation should always be indicated. In the case of Raman spectral reference libraries, a non-resonant excitation frequency should be employed and the final processed spectrum should include a background intensity correction.

The following is a general hierarchy of spectral library quality:

- Spectrum incorrectly identified with wrong structure. (This is the worst case scenario.) An example of this can include an IR spectrum dominated by water vapor and carbon dioxide.
- A false IR spectrum with trade name only (useless).
- A photometrically accurate IR or Raman spectrum with trade name only (almost useless).
- A photometrically accurate IR or Raman spectrum with structure (Best case scenario of commercial libraries).
- A fully interpreted photometrically accurate IR and Raman spectrum with structure (Best case scenario of in-house libraries).

3.2. Computer-Based Libraries and Software Tools

Computer-based systems that provide guidance in spectral identification or group frequency analyses are a useful tool, but have limitations that must be understood. The usefulness of a spectral library is defined in part by how accurate it is and how much known information is captured in it. The spectral library must contain structures similar to the unknown in order to be useful. The critical decision to be made is whether the measured spectrum can be considered consistent with the reference spectrum.

A common first step for structural identification is to employ a spectral library search. A fundamental limitation is that libraries contain no knowledge of the chemistry of the system and are limited by the size and scope of spectral libraries. Libraries are most useful if the compound of interest, with suitable sample preparation/presentation, exists in the database. A reference spectrum and sample of the same species, but different crystallinity, in the library will negatively affect the accuracy of the identification.

In general, a number of automatic search methodologies exist that can be used to compare the spectrum of an unknown sample against many different spectra in the library. These include a simple peak matching to discrimination analysis using Euclidean distance, wavelength correlation (or other algorithms). A search simply reports the best-known match from the spectral library. The unknown may be from an entirely different class of materials.

Another approach is to use an empirically based, computer-assisted structural interpretation to aid in a group frequency analysis of the IR and Raman spectrum. The analyst should also use their background knowledge of the chemistry of the sample to help screen such results. One approach involves software that incorporates numerous correlation tables organized by functional group fragments. The analysis of spectrum results in suggested fragments with highlights of important bands for that fragment. These fragments can be combined to provide a hypothetical structure for confirmation.

Some software allows the addition of functional group band assignments by the analyst. This can enable the software to focus on the type of molecules commonly encountered in the particular lab. Approximate intensities are often included and include the frequency range. This is a more flexible analysis tool than the "brute force" spectral library approach.

The most rigorous approach to structural verification involves using *ab initio* determined structure with normal mode analysis. The calculations are based on an isolated molecule and result in the theoretical IR and Raman band frequencies, relative intensities, and corresponding vibrations. Interactions encountered in the condensed phase such as hydrogen bonding and various environments from different crystalline structures are not taken into account. Therefore, the computational results are most appropriate for gas-phase spectra. The size of the molecule limits this approach because of the computationally intensive analysis employed. Analysis of the Cartesian displacement vectors describing the vibrations from the normal mode analysis can provide important insight into group frequencies and mechanical interactions of the vibrations in the molecule. This approach provides the most rigorous structural verification approach.

3.3. Define the Problem that Needs to be Solved

The first step involves defining the problem. Ideally, this should be done before the spectrum of a sample is measured. A complete unknown is rare. Get all the background chemistry information available so that you have a good sample history.

Obviously, the analyst should be clear if the sample is research or product related. If this is a research sample, obtain the reaction scheme. Ask the submitter what are the expected products and impurities and if this is an intermediate or final product. If the sample is product related then the analyst should determine the product area/general area of chemistry. Examples can include paper industry related, polymer additives, polymers, and pharmaceutical API's. IR and Raman spectroscopy are commonly employed for contaminant analysis. In this case, find out what the source matrix is and if this is believed to be a contaminant. In some instances, an identification of extracted materials is needed. In this case, determine the matrix from which the material was extracted. What solvent was used? What filters were used? The analyst must know what spectral signals might be expected from these possible contaminants.

Questions to be asked to the submitter can include the following: Where did the sample come from exactly? Was there any sample manipulation prior to analysis? Has there been similar work done in the past? What other techniques are being used in the analysis?

3.4. Examine the IR and Raman Spectrum

The first step of interpreting the IR and Raman spectrum is to identify any common impurities that could be present. Some common impurities include silicone oil, water, plasticizers such as phthalates, flow aids such as Mg stearate and commonly encountered polymers (e.g., PE, Nylon, Teflon), and long-chain aliphatics.

Look for possible solvents that may have been introduced by any extraction procedures and not fully removed from an isolated reaction product. The analyst should also look for inorganic sulfates, sulfites, or nitrates that may be present as a result of treatment of a reaction mixture with sulfuric acid or nitric acid.

Next, determine if the spectrum represents single component or multiple components. Almost all organic and many inorganic species have IR and Raman bands. A complex spectrum can represent either a complex molecule with a wide range of functional groups or multiple components. Multiple component identification is challenging and is a skill that improves with practice. If you have assigned multiple species, review if that is consistent with the chemical background. If possible estimate relative amounts of components present.

Be aware of orientation/crystallinity effects since vibrational spectroscopy has excellent environmental sensitivity. This can result in very different IR and Raman spectra for amorphous versus crystalline states. Most of the observed spectral differences are a result of changes in hydrogen-bonding environments. Because of this the analyst should try to ensure that the reference spectra followed the same sample-preparation method and has the same crystalline state as the unknown. Some sample preparations (e.g., cast films) can result in a different crystalline state from the sample of interest.

The last step for structural identification and verification is to assign all main peaks in the IR and Raman spectrum. A standard systematic way to begin analyzing the IR/Raman spectrum is to begin at the high wavenumber end of the spectrum (4000 cm^{-1}) and look for the presence and absence of characteristic absorptions as you move to lower wavenumbers. The most common and distinctive bands can then be organized into several spectral regions. Whenever possible, assignments should utilize confirming bands. If possible, measuring a full spectral region (4000–400 cm^{-1}) is important because, for example, the confirming bands for many inorganic species occurs below 650 cm^{-1} (*vide infra*). When possible use the basic knowledge about the sample and the chemistry associated with it. Check for group frequencies associated with suspected structural components. Computer-based spectral libraries and structural interpretation software may be valuable tools. Remember to keep in mind the limitations and spectral contributions from any sample-preparation techniques used. Try to assign the most intense bands first. In most cases, you will not be able to fully and accurately assign group frequencies to all of the bands in the spectrum.

4. INTERPRETATION GUIDELINES AND MAJOR SPECTRA–STRUCTURE CORRELATIONS

This section is to be used as a general guide to interpret IR and Raman spectra systematically. As a first check, examine the overall spectral appearance. If the spectrum is very simple with only a few vibrational bands then focus on a low molecular weight organic or inorganic species. Look at inorganic salts such as sulfate, carbonate, or nitrate or simple common organic solvents and water. For organics and hydrocarbons look for bands in the IR and Raman spectra in the region between 3200 and 2700 cm^{-1}, because these are excellent group frequencies associated with C–H-type stretches.

After a quick evaluation of the spectral appearance and determining, if hydrocarbons are present, begin to interpret the vibrational spectrum. Start from 4000 cm^{-1} and work systematically to lower wavenumbers. The presence and absence of functional groups needs to be noted.

4.1. Hydroxy (OH), Amino (NH), and Acetylene (≡CH) Groups: 3700–3100 cm^{-1}

Bands in this region are typically due to various OH and NH stretching vibrations. Hydrogen bonding will have a large effect on the amino and hydroxyl stretching vibrations. The OH stretch is medium to strong in intensity in the IR spectrum but is generally weak in the Raman spectrum. IR spectroscopy will be excellent for NH stretching vibrations. Raman spectroscopy will be poor for OH and acetylene CH stretching vibrations but moderate for NH stretching vibrations.

In the condensed phase, the hydroxyl OH stretching band is generally intermolecularly hydrogen bonded with a broad band appearing between 3550 and 3230 cm^{-1}. In some cases due to crystalline structure or steric hindrance, the hydroxyl group is not hydrogen bonded and resultant sharp and moderately strong free OH stretching bands are observed between 3670 and 3580 cm^{-1}. The OH stretches for water alcohols and phenols are generally observed at 3400 cm^{-1} but can be found between 3700 and 3300 cm^{-1}. The observed OH stretch depends upon the strength of the hydrogen bond. Acidic protons for the hydroxyl species of carboxylic acid dimers and phosphoric acids result in very broad bands centered at ca. 3000 cm^{-1}.

IR bands for hydrogen-bonded amino groups are generally not as broad as those observed for hydrogen-bonded hydroxyl groups. Primary amine (NH) groups result in a single weak band between 3500 and 3300 cm^{-1} while a secondary amine (NH_2) group results in a weak doublet between 3550 and 3250 cm^{-1} with a spacing of about 70 cm^{-1}. A secondary amide NH results in a stronger band between 3470 and 3200 cm^{-1} while a primary amide (NH_2) group results in a strong doublet between 3550 and 3150 cm^{-1} with a spacing of about 100–200 cm^{-1}.

The acetylene ≡CH group has a moderate-to-strong sharp IR band between 3340 and 3280 cm^{-1}. The band is quite weak in the Raman spectrum.

Lastly, weak overtones from carbonyl bands can occur between 3500 and 3400 cm^{-1} region. Typically, the overtone band frequencies will be ca. 20 cm^{-1} less than twice the carbonyl stretching band wavenumber and the intensity will be significantly less intense than the carbonyl stretching band.

4.2. Unsaturated Aryl and Olefinic Groups: 3200–2980 cm^{-1}

Both IR and Raman spectroscopy provide moderately strong bands for these groups. If sharp bands are observed between 3000 and 3200 cm^{-1} then it is likely an unsaturated group is present. Aromatics have multiple weak bands from CH stretch vibrations between 3100 and 3000 cm^{-1}. Most olefins have characteristic CH stretching bands in the 3100–2980 cm^{-1} region. Observation of isolated bands at 3010 cm^{-1} and 3040 cm^{-1} most likely indicates a simple olefinic group. Pyroles and furans will typically have bands in the 3180 and 3090 cm^{-1} region. Three-membered rings such as cyclopropyl and epoxy rings will also have bands in 3100–2990 cm^{-1} and 3060–2990 cm^{-1} regions, respectively.

4.3. Aliphatic Groups: 3000–2700 cm^{-1}

Well-defined, strong characteristic IR and Raman bands for C–H stretching vibrations characteristic of aliphatics are typically observed. Both methyl and methylene groups (CH_3, CH_2) result in doublets at slightly different frequencies. Methyl CH stretching bands typically have bands between 2975– 2950 cm^{-1} and 2885–2865 cm^{-1}. Methylene CH stretching bands typically have bands between 2940–2915 cm^{-1} and 2870–2840 cm^{-1}. The relative intensities of the bands can be used to estimate the relative amount of methyl to methylene groups. Heteroatoms adjacent to these groups result in characteristic group frequencies:

The methoxy group has one band at ca. 2835 cm^{-1}, methyl and methylene groups next to the nitrogen atom in tertiary amines result in bands at ca. 2800 cm^{-1}, and aldehydes result in two peaks near 2730 cm^{-1} (due to Fermi resonance).

4.4. Acidic Protons: 3100–2400 cm^{-1}

Broad bands in this region are typically due to acidic and strongly hydrogen-bonded hydrogen atoms. In general, IR is a much better technique than Raman spectroscopy for these groups. Hydrogen bonding is an important determinant in the frequencies for these groups. The main OH stretching band for carboxylic acid dimers is found at ca. 3000 cm^{-1} with additional overtone/combination bands at ca. 2650 cm^{-1} and 2550 cm^{-1}. Primary, secondary, and tertiary amine salts absorb in this region and have distinctive band structure. The $N{-}H^+$ stretch for these groups are characterized by moderate-to-strong IR and Raman bands. Primary amine salts for the $-NH_3^+$ have an absorption band(s) between 3350 and 3100 cm^{-1}. Amines with $> NH_2^+$, $>NH^+-$, and $C{=}NH^+-$ have bands between 2700 and 2250 cm^{-1}. Phosphoric acids have broad absorption bands in this region.

4.5. SH, BH, PH, and SiH: 2600–2100 cm^{-1}

Both IR and Raman can be very useful for these species. The SH stretch for mercaptans and thiophenols occurs in the region 2590–2540 cm^{-1} and is generally quite strong in the Raman spectrum, but weak in the IR. BH stretches occur in the region 2630–2350 cm^{-1} and are moderate to strong in the IR spectrum, but are generally weak in the Raman spectrum. PH stretches occur in the region 2440–2275 cm^{-1} and are moderately intense in the IR spectrum and moderate to weak in the Raman spectrum. SiH stretches occur in the region 2250–2100 cm^{-1} and are strong in both the IR and Raman spectrum.

4.6. Triple Bonds and Cumulated Double Bonds: 2300–1900 cm^{-1}

Both IR and Raman can be very useful for these species. The IR and Raman intensities for the $X{\equiv}Y$ and $X{=}Y{=}Z$ stretching vibrations are variable. Refer to chapter 6 (section 3.2) to use the table and idealized spectra (Figure 6.11 and table 6.3). Contributions from atmospheric carbon dioxide generate a false signal in this region of the IR spectrum.

4.7. Carbonyl-Containing Species: 1900–1550 cm^{-1}

IR is excellent for carbonyl species while Raman is quite variable. The carbonyl C=O stretching vibration results in strong characteristic IR bands. Raman bands for this vibration are typically moderate to weak with some structures resulting in a strong C=O stretch. This band is easily identified in the IR spectrum because of its intensity and its lack of interference from most other group frequencies. Carbonyl groups are present in many different compounds such as ketones, carboxylic acids, aldehydes esters, amides, acid anhydrides, acid halides, lactams, lactones, urethanes, and carbamates. Bands above 1775 cm^{-1} are often due to carbonyl compounds such as anhydrides (doublet), acid halide, or strained ring carbonyl (such as lactones and carbonates in five-membered rings). Bands between

1750 and 1700 cm^{-1} fall in the middle of the carbonyl range. Carbonyl compounds in this region can include aldehydes, esters, carbamates, ketones, and carboxylic acids. Bands below 1700 cm^{-1} fall in the low end of the carbonyl range. Such carbonyl compounds can include amides, ureas, and carboxylic acid salts. Since conjugation can lower carbonyl frequencies, this range will also include conjugated aldehydes, ketones, esters, and carboxylic acids. In general, bands below 1600 cm^{-1} are due to the out-of-phase stretch of carboxylic acid salts. The remaining carbonyl signals are due to C=O stretches and occur above 1600 cm^{-1}. Fermi resonance sometimes gives rise to two bands in the C=O stretching region even though only one C=O bond is present. See Chapter 2 for an explanation of this effect.

4.8. Olefinic (C=C), Imino (C=N), and Azo (N=N) Compounds: 1690–1400 cm^{-1}

The stretching vibrations of these groups are characterized by strong Raman band which makes Raman an excellent technique for these types of compounds. The intensities of the IR absorptions are variable. The C=C stretching vibrations result in variable IR intensities. Most olefin C=C stretching bands occur between 1680 and 1600 cm^{-1}. In IR this results in a narrow, weak absorption band. Conjugation with another double bond lowers the frequency and often increases the IR band intensity.

Imino (C=N) stretching vibrations result in strong Raman and moderately strong IR bands. Most C=N stretching vibrations occur between 1690 and 1630 cm^{-1}. Azo (N=N) stretching vibrations give rise to strong, well-defined Raman bands, but a very poor quality IR bands. Most Azo N=N stretching bands occur between 1580 and 1400 cm^{-1}. The nature of the compound is very important in analyzing spectra of azo species. Symmetrically substituted *trans* azo compounds give rise to strong Raman bands. Although the signal is forbidden in the IR, non-symmetrically substituted *trans* azo compounds can give rise to strong IR signals.

4.9. Organic Nitrates (N=O): 1660–1450 cm^{-1}

These species include nitro (R–NO_2), organic nitrates (R–NO_2), organic nitrites (R–O–N=O), and nitroso groups. The N=O stretching vibrations result in strong IR bands and generally moderate Raman bands. Aliphatic and aromatic nitro compounds have two IR strong bands between 1580–1475 cm^{-1} and 1390–1320 cm^{-1}. Organic nitrites have two strong IR bands between 1660–1615 cm^{-1} and 1300–1270 cm^{-1}. Only the lower band is observed by Raman spectroscopy. Inorganic nitrate salts (NO_3^-) provide a broad band associated with the out-of-phase NO_3^- between 1410 and 1350 cm^{-1} (IR strong, Raman moderate).

Organic nitrites provide strong IR and Raman bands with generally two bands resulting because of rotational isomers. The *cis* form exhibits a band in the region 1625–1610 cm^{-1} while the *trans* form exhibits a band in the region 1680–1650 cm^{-1}. Nitroso groups are characterized by strong IR and Raman bands. In the solid and liquid state, nitroso groups exist as dimers in either the *cis* or *trans* state. The aliphatic *cis* form results in two bands: 1425–1330 cm^{-1} and 1345–1320 cm^{-1}. The *trans* form results in a single band in the region 1290–1175 cm^{-1}. The aromatic *cis* form results in two bands: 1400–1390 cm^{-1} and 1410 cm^{-1}. The *trans* form results in a single band in the region 1300–1250 cm^{-1}.

4.10. Amine NH Deformation Vibrations for Amines, Amine Salts, and Amide Compounds: 1660–1500 cm^{-1}

NH deformation bands result in moderate IR bands and weak Raman bands. Primary amine NH_2 in-phase deformation occurs in the region 1660–1590 cm^{-1}. Amine salts result in a NH_2^+ in-phase deformation band found between 1620 and 1560 cm^{-1} while a NH_3^+ out-of-phase deformation results in two sets of bands between 1635–1585 cm^{-1} and 1585–1560 cm^{-1}. In secondary carbamates and mono-substituted amides, the CNH group results in a strong IR band between 1570 and 1510 cm^{-1}. Note that water also has a band near 1640 cm^{-1} from the OH deformation.

4.11. Aromatic and Hetero-aromatic Rings: 1620–1420 cm^{-1}

Sharp bands in this region involve the aromatic ring quadrant and semi-circle stretching vibrations. The IR intensities for these bands are variable in intensity. The Raman intensities are moderate for the quadrant stretch and quite weak for the semi-circle stretching bands. Typically two sets of bands are observed in this region due to the quadrant and semi-circle ring stretch vibrations. These may occur as single bands or as a multiple component envelop of bands.

In general, the quadrant stretch will result in bands near 1600 and 1580 cm^{-1} while the semi-circle stretch will result in bands near 1500 and 1460 cm^{-1} The relative intensity of the IR bands will vary with the aromatic substituent pattern and type of substituent. Six-membered ring heteroaromatics resulting from substituting an aromatic carbon atom with a nitrogen atom result in similar characteristic ring-stretching bands between 1600 and 1500 cm^{-1}. These include single nitrogen-substituted aromatic molecules such as pyridines, quinolines, and isoquinolines, two nitrogen-substituted aromatic molecules such as pyrimidines and quinazolines, and three nitrogen-substituted molecules such as triazine species. Simple aromatic compounds will also generate a series of weak IR absorption bands between 2000 and 1700 cm^{-1} which derive from multiple combination bands.

4.12. Methyl and Methylene Deformation Vibrations: 1500–1250 cm^{-1}

These alkane deformation vibrations results in moderate IR bands and weak to moderate Raman bands. When these groups have identical substituents then the CH_2 "scissors" deformation and CH_3 out-of-phase deformation are found in the same frequency region, ca. 1480–1430 cm^{-1}. When the CH_2 and CH_3 are on hydrocarbons, this band is near 1460 cm^{-1}. When the CH_2 (methylene) is near an unsaturated group, the deformation band is found near 1440 cm^{-1}. If the methylene is adjacent to chlorine, bromine, iodine, sulfur, phosphorus, nitrile, nitro, or carbonyl, this band occurs between 1450 and 1405 cm^{-1}. The CH_3 in-phase deformation is a useful moderately intense IR band, but provides only a weak Raman band. The CH_3 in-phase ("umbrella") deformation is strongly dependent upon the electronegativity of the adjacent atom. The more electronegative the atom, the higher the frequency with a range of 1470–1250 cm^{-1}. Examples include O–CH_3 at ca. 1450 cm^{-1}, C–CH_3 at ca. 1450 cm^{-1}, and Si–CH_3 at ca. 1265 cm^{-1}. In aliphatic groups, steric hindrance to the vibrations due to the presence of two or three adjacent methyl groups can result in highly characteristic bands. Here $C{-}CH_3$ results in a single band at 1378 cm^{-1}, while the isopropyl $C(CH_3)_2$ has an equal intensity doublet at 1385 and 1368 cm^{-1}, and the t-butyl $C(CH_3)_3$ has a non-equally intense doublet at 1395 and 1368 cm^{-1}.

4.13. Carbonate, Nitrate, Ammonium, and B–O Type Compounds: 1480–1310 cm^{-1}

Inorganic carbonate, nitrate, and ammonium ions provide a broad, intense IR band near 1400 cm^{-1} but only a weak Raman band. IR bands result from the out-of-phase XY_3 stretch for carbonate- and nitrate-containing compounds. Strong sharp Raman bands in the 1040–1100 cm^{-1} region result from the in-phase XY_3 stretch. The ammonium ion deformation band is due to a NH_4^+ bend and is found in the 1480–1390 cm^{-1} region.

The B–O stretch provides a strong characteristic IR band. Inorganic boric acid has a strong broad band near 1410 cm^{-1} due to the BO_3 out-of-phase stretch. Organic borates [$B(RO)_3$], boronates [$PhB(RO)_2$], metaborates (cyclic six-membered B–O ring), boronic acid esters, and anhydrides are all characterized by strong IR bands in the region 1390–1310 cm^{-1} involving the B–O stretch.

4.14. Organic and Inorganic SO_x Type and Thiocarbonyl (C=S) Compounds: 1400–900 cm^{-1}

In general, both IR and Raman provide strong characteristic bands associated with SO_x and C=S stretching vibrations. Inorganic SO_x type compounds include sulfites (SO_3^{2-}), sulfates (SO_4^{2-}) as well as more complex S_XO_Y type compounds (thiosulfate, pyrosulfite, dithionate). Sulfites have a strong broad IR band and a strong doublet Raman band between 1010 and 900 cm^{-1} involving the SO_3 out-of-phase and in-phase stretching vibrations. Sulfates are characterized by a very strong IR band between 1130 and1080 cm^{-1} involving the SO_4^{-2} out-of-phase stretch and a strong Raman band between 1065 and 955 cm^{-1} involving the SO_4^{-2} in-phase stretch. S_XO_Y type compounds are characterized by strong IR and Raman bands between 1250 and 1000 cm^{-1} involving complex S=O stretching vibrations.

Organic SO_X type compounds include >S=O, >SO_2 as well as sulfonic acids (–SO_3H), and sulfonic acid salts (–SO_3^- M^+) (See chapter 6, section 8). In general, the stretching frequencies involving the S=O group of these types of compound show a dependence on the electronegativity of the substituents. Sulfoxides (R_2>S=O) are characterized by a strong IR band and a moderate-to-weak Raman band between 1070 and 1030 cm^{-1} from the S=O stretch. Similarly, sulfinic acids (–S(=O)–OH) are characterized by a very strong IR and moderate-to-strong Raman band in the region 1090–990 cm^{-1} due to the S=O stretch. The S=O stretch is shifted to higher frequencies for sulfinic acid esters (R–S(=O) –O–R) at 1140–1125 cm^{-1} (strong IR; moderate intensity Raman) and dialkyl sulfites ($(RO)_2S{=}O$) in the region 1220–1170 cm^{-1} (strong IR; strong, polarized Raman). Sulfones (R– SO_2–R) are characterized by two vibrations, an out-of-phase and in-phase stretch of the SO_2 group. In most cases, strong IR bands result for both in-phase and out-of-phase vibrations while a strong Raman band only occurs for the in-phase SO_2 stretch. Sulfones (>SO_2) have strong characteristic bands at 1360–1290 cm^{-1} and 1200–1120 cm^{-1}. Sulfanamides (R–SO_2–N⟨)

have strong bands at 1360–1315 cm^{-1} and 1180–1140 cm^{-1}. The higher frequencies are due to the influence of the more electronegative nitrogen atom.

Sulfonyl halides (—SO_2—X) have strong bands at 1385–1360 cm^{-1} and 1190–1160 cm^{-1}. The out-of-phase SO_2 stretch results in a moderate-to-strong Raman band. Covalent sulfonates (R—SO_2—OR) and organic sulfates (—O—SO_2—O—) have bands at 1420–1330 cm^{-1} and 1200–1145 cm^{-1}. Organic sulfate salts (RO—SO_2—O^- M^+) are characterized by a strong IR band in the region 1315–1210 cm^{-1} and a strong Raman band in the region 1140–1050 cm^{-1}.

Organic sulfonic acids (—SO_3H) and salts (—SO_3^- M^+) are characterized by strong IR and Raman bands involving the SO_2 and SO_3 stretching vibrations. Sulfonic acids (—SO_3H), if not dried, are easily ionizable with small amounts of water resulting in a hydrated form (—SO_3^- H_3O^+). Anhydrous sulfonic acids (R—SO_2—OH) have strong IR and Raman bands at 1355–1340 cm^{-1} and 1200–1100 cm^{-1} from the SO_2 out-of-phase and in-phase stretches, respectively. Hydrated sulfonic acids (RSO_3^- H_3O^+) have bands at 1230–1120 cm^{-1} and 1120–1025 cm^{-1} from the SO_3 out-of-phase and in-phase stretches, respectively. Both SO_3 vibrations result in strong IR bands. However, a moderate-to-strong Raman band is observed only for the SO_3 in-phase stretch. Sulfonic acid salts (—SO_3^- M^+) have bands at 1250–1140 cm^{-1} and 1070–1030 cm^{-1}. The out-of-phase SO_3 stretch results in very strong IR bands but weak Raman bands. However, the in-phase SO_3 stretch results in strong IR and Raman bands.

Thiocarbonyl compounds (>C=S) are generally characterized by a strong IR and Raman band between 1250 and 1200 cm^{-1} that involve the C=S stretch.

4.15. P=O-Containing Compounds: 1350–1080 cm^{-1}

The P=O stretching vibration for organic phosphorus compounds results in a strong IR band between 1350 and 1140 cm^{-1} while the Raman band intensity is moderate to weak. The frequency of the P=O stretch shows a strong dependence on the substituent electronegativity and hydrogen bonding effects. Inorganic PO_x type compounds include dibasic phosphate (HPO_4^{2-}), monobasic phosphate ($H_2PO_4^-$), phosphate (PO_4^{3-}). The phosphate out-of-phase stretch has strong IR and Raman bands between 1180 and 1000 cm^{-1}. The dibasic phosphate (HO—PO_3^{2-}) has strong IR and Raman bands between 1150 and 1000 cm^{-1} from the PO_3 out-of-phase stretch. The monobasic phosphate ($(HO)_2$ PO_2^-) has strong IR and Raman bands between 1200 and 950 cm^{-1} from the PO_2 out-of-phase stretch. The P=O band can appear as a doublet, possibly due to Fermi resonance or, in some cases, rotational isomerism.

4.16. Fluorinated Alkane Groups: 1350–1000 cm^{-1}

The C—F stretch of organic fluorine compounds results in strong IR bands but weak to moderately intense Raman bands. Monofluorinated compounds have a strong IR and a weak to moderate Raman band between 1100 and 1000 cm^{-1}. Difluorinated compounds (—CF_2—) have two very strong IR bands between 1250 and 1050 cm^{-1} involving the CF_2 out-of-phase stretch. These bands are weak to moderate in the Raman spectrum. Trifluorinated compounds (—CF_3) have multiple strong bands between 1350 and 1050 cm^{-1} involving the CF_3 out-of-phase stretch. Fluorine substituted aromatics (Ar—F) have a moderately strong IR band between 1270 and 1100 cm^{-1} involving the aryl ring C—F stretch.

4.17. C–O Stretching Vibrations: 1300–750 cm^{-1}

Bands in the region 1300–750 cm^{-1} are associated with the C–O stretch of species such as alcohols, ethers, esters, carboxylic acids, and anhydrides. Ethers are characterized by C–O–C out-of-phase and in-phase stretches. In general, the out-of-phase C–O–C stretch results in strong IR bands in the range 1270–1060 cm^{-1} but weak Raman bands. The in-phase C–O–C stretch results in strong Raman bands and moderate IR bands at 1140–800 cm^{-1}. Alcohols are characterized by C–O type stretch vibrations in the region 1200–750 cm^{-1}. Primary alcohols are characterized by a strong IR band between 1090 and 1000 cm^{-1} and a strong Raman band between 900 and 800 cm^{-1}. Secondary alcohols are characterized by multiple strong IR bands between 1150 and 1075 cm^{-1} and a strong Raman band between 900 and 800 cm^{-1}. Tertiary alcohols are characterized by multiple strong IR bands between 1210 and 1100 cm^{-1} and a strong Raman band between 800 and 750 cm^{-1}.

Carbonyl compounds such as esters, carboxylic acids, and anhydrides include a C–O group that is often used to confirm the functional group. The C–O stretch of carboxylic acids is characterized by moderately intense IR bands in the range 1380–1210 cm^{-1} with the corresponding Raman bands having highly variable intensity. Similarly, esters result in strong IR bands between 1300 and 1100 cm^{-1} and corresponding Raman bands of variable intensity. Anhydrides are characterized by strong IR and Raman bands between 1310 and 980 cm^{-1} involving the C–O–C stretching vibrations.

4.18. Si–O and P–O Containing Compounds: 1280–830 cm^{-1}

The Si–O group is characterized by strong IR bands and variable Raman bands. Compounds containing P–O–C group are characterized by bands in the range 1090–920 cm^{-1} which involve the out-of-phase stretching vibration. These characteristic bands are typically strong in the IR and moderate to weak in the Raman spectrum. The P–O–P and P–OH groups are characterized by strong IR bands but only moderate-to-weak Raman bands.

The P–O–P group is characterized by a strong IR band between 1025 and 870 cm^{-1} involving the P–O–P out-of-phase stretch. The P–OH group is characterized by bands involving the P–O stretch between 1000 and 900 cm^{-1} with strong IR intensity, but moderate-to-weak Raman intensity.

The siloxane Si–O–Si and Si–OH groups are characterized by strong IR bands at 1090–1010 cm^{-1} and 955–830 cm^{-1}, respectively, involving the Si–O stretch. Raman spectra of siloxanes do not provide useful bands in this region. Alkoxysilanes, Si–O–R, are characterized by an out-of-phase stretch (1110–1000 cm^{-1}) which results in a strong IR band and an in-phase stretch (1070–990 cm^{-1}) which results in moderate-to-strong intensity IR and Raman bands.

4.19. CH Wag of Olefinic and Acetylenic Compounds: 1000–600 cm^{-1}

The CH wag vibration provides important medium-to-strong IR bands for characterizing alkenes and acetylenes. These bands are typically weak in the Raman spectrum. The position of these bands is dependent on the properties (electron withdrawing) of the substituents.

Vinyl groups, $-CH=CH_2$, have strong IR bands between 1000–940 cm^{-1} and 960–810 cm^{-1} (995–980 cm^{-1} and 915–905 cm^{-1}, respectively, for hydrocarbons). Vinylidene groups, $>C=CH_2$, exhibit strong IR bands between 985 and 700 cm^{-1} (895–885 cm^{-1} for hydrocarbons).

Trans-di-substituted ethylenes, $-CH=CH-$, have strong IR bands between 980 and 890 cm^{-1} (1000–955 cm^{-1} for hydrocarbons). Vibrational interaction with conjugated diene groups results in a systematic increase in the observed CH wag. An isolated trans alkene (T) has a CH wag at 965 cm^{-1}, a conjugated trans–trans diene (TT) at 986 cm^{-1}, the trans–trans–trans alkene (TTT) at 994 cm^{-1}, and the trans–trans–trans–trans alkene (TTTT) at 997 cm^{-1}.

Cis-di-substituted ethylenes have strong IR bands between 800 and 600 cm^{-1} (730–650 cm^{-1} for hydrocarbons). Tri-substituted alkenes, $>CH=CH-$, absorb between 850 and 790 cm^{-1}. The CH wag of acetylenic, $C\equiv CH$, have moderate-to-strong IR bands between 700 and 578 cm^{-1}.

4.20. Aromatic In-plane 2,4,6 Radial Carbon In-phase Stretch: 1290–990 cm^{-1}

Very strong Raman bands are observed from aromatic ring vibrations involving the 2,4,6 radial carbon in-phase stretch. A very strong Raman band between 1010 and 990 cm^{-1} is observed for mono-, 1,3 di-, and 1,3,5 tri-substituted benzenes. Only a weak to moderately intense sharp IR band is observed.

4.21. Aromatic CH Wag: 900–700 cm^{-1}

The aromatic ring CH wag vibrations results in strong IR absorption bands but only weak Raman bands. The frequencies of the aromatic CH wag are largely dependent upon the number of adjacent hydrogen atoms since the adjacent CH wag vibrations are mechanically coupled to one another. For this reason, the IR bands from the in-phase CH wag vibrations are quite useful in distinguishing aromatic ring substitutions.

Mono-substituted benzenes have a band at ca. 750 cm^{-1} involving the five adjacent CH wag (overall region between 820 and 728 cm^{-1}). Ortho di-substituted benzenes also have a band at ca. 750 cm^{-1} involving the four adjacent CH wag (overall region between 790 and 728 cm^{-1}). The three adjacent CH wag vibrations of meta- or 1,2,3 tri-substituted benzenes have a band at ca. 782 cm^{-1} (overall region between 825 and 750 cm^{-1}). The two adjacent CH wag vibrations of para, 1,2,4 tri-substituted, or 1,2,3,4 substituted benzenes have a band at ca. 817 cm^{-1} (overall region between 880 and 795 cm^{-1}). The lone CH wag vibration (meta, or 1,2,4,… or 1,3,5,… or 1,2,3,5,… or 1,2,4,5,… or penta) substituted benzenes have a band at ca. 860 cm^{-1} (overall region between 935 and 810 cm^{-1}). The aryl CH adjacent classification can also be extended to pyridines and napthalenes.

4.22. Halogen-Carbon Stretch: 850–480 cm^{-1}

Strong IR and Raman bands result for the halo-alkanes involving the C–X stretching vibrations (X=F, Cl, Br, I). Fluoroalkanes are summarized above. The C–Cl stretching vibrations occur between 860 and 505 cm^{-1}. Compounds with more than one chlorine atom will have multiple bands deriving from the in-phase and out-of-phase $C{-}Cl_x$ ($X = 2,3,4$) which may also couple with other groups. The C–Br stretching vibrations occur between 680 and 485 cm^{-1}. Compounds with more than one bromine atom will have multiple bands deriving from the in-phase and out-of-phase $C{-}Br_2$ which may also couple with other groups. The C–I stretching vibrations occur between 610 and 485 cm^{-1}.

4.23. OH, NH, NH_2 Wag: 900–500 cm^{-1}

In the condensed state, moderately intense (but broad) IR bands are found between 900 and 500 cm^{-1} due to the X–H wag of hydrogen-bonded water, amines, amides, and alcohols. Hydrogen bonding significantly influences the frequency for these characteristic IR bands. No significant Raman bands from these vibrations are observed. Alcohols exhibit bands that are generally centered around 650 cm^{-1}. Water absorbs below 800 cm^{-1}.

Primary amines have bands between 900 and 770 cm^{-1} and secondary amines between 750 and 680 cm^{-1}. Amides have bands between 750 and 550 cm^{-1}.

4.24. Metal Oxides: 800–200 cm^{-1}

Metal oxides of various types have strong IR and Raman bands from the metal-O-metal group.

CHAPTER

8

Illustrated IR and Raman Spectra Demonstrating Important Functional Groups

The IR and Raman spectra of each compound are presented in a stacked format with the spectrum number given in the upper right-hand corner of each IR spectrum. The sample preparation and type of crystal used for the IR measurement is listed in parenthesis after the compound name in the spectral index below. The IR measurements were made using a Bio-Rad FTS-575C spectrometer. All Raman measurements were made using 180° backscattering on a BioRad FT-Raman spectrometer (575C with Raman III accessory) and background corrected using a KBr white light reflectance measurement. A melting point or NMR tube was used for the Raman measurements. A single bounce diamond ATR accessory was used for the IR measurements of the polymers. The assignment shown on all of the spectra are consistent with that discussed in the text. Page numbers of spectra are shown on the right hand side.

1. ALIPHATIC

2. C=C DOUBLE BONDS

3. TRIPLE BONDS

4. AROMATIC RINGS

5. KETONES, ESTERS, AND ANHYDRIDES

6. AMIDES, UREAS, AND RELATED COMPOUNDS

7. ALCOHOLS

8. ETHERS

9. AMINES AND AMINE SALTS

10. C=N COMPOUNDS

11. N=O COMPOUNDS

12. AZO COMPOUND

13. BORONIC ACID COMPOUND

14. CHLORINE, BROMINE, AND FLUORINE COMPOUNDS

15. SULFUR COMPOUNDS

16. PHOSPHORUS COMPOUNDS

17. SILOXANE COMPOUNDS

18. INORGANIC COMPOUNDS

19. POLYMERS AND BIOPOLYMERS

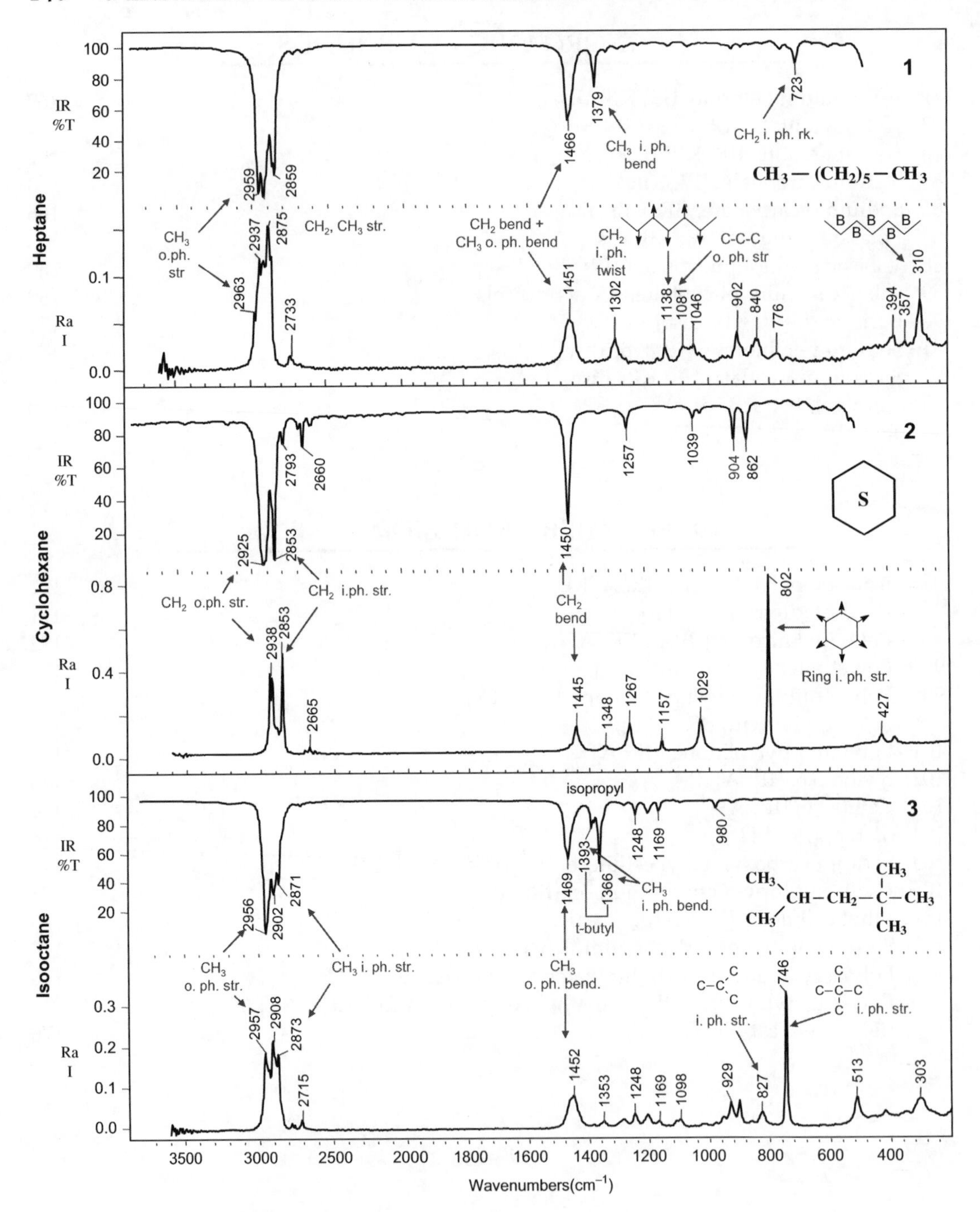
Heptane
IR %T
Ra I
1
2959
2859
1466
1379
723
CH3 i. ph. bend
CH2 i. ph. rk.
CH3—(CH2)5—CH3
CH3 o.ph. str
2937
2875
CH2, CH3 str.
CH2 bend + CH3 o. ph. bend
CH2 i. ph. twist
C-C-C o. ph. str
2963
2733
1451
1302
1138
1081
1046
902
840
776
394
357
310
Cyclohexane
2
2793
2660
1257
1039
904
862
S
2925
2853
1450
CH2 o.ph. str.
CH2 i.ph. str.
CH2 bend
802
2938
2853
2665
1445
1348
1267
1157
1029
Ring i. ph. str.
427
Isooctane
3
isopropyl
1393
1366
1248
1169
980
CH3 i. ph. bend.
t-butyl
1469
2956
2902
2871
CH3 o. ph. str.
CH3 i. ph. str.
CH3 o. ph. bend.
i. ph. str.
746
2957
2908
2873
2715
1452
1353
1248
1169
1098
929
827
513
303
Wavenumbers(cm−1)

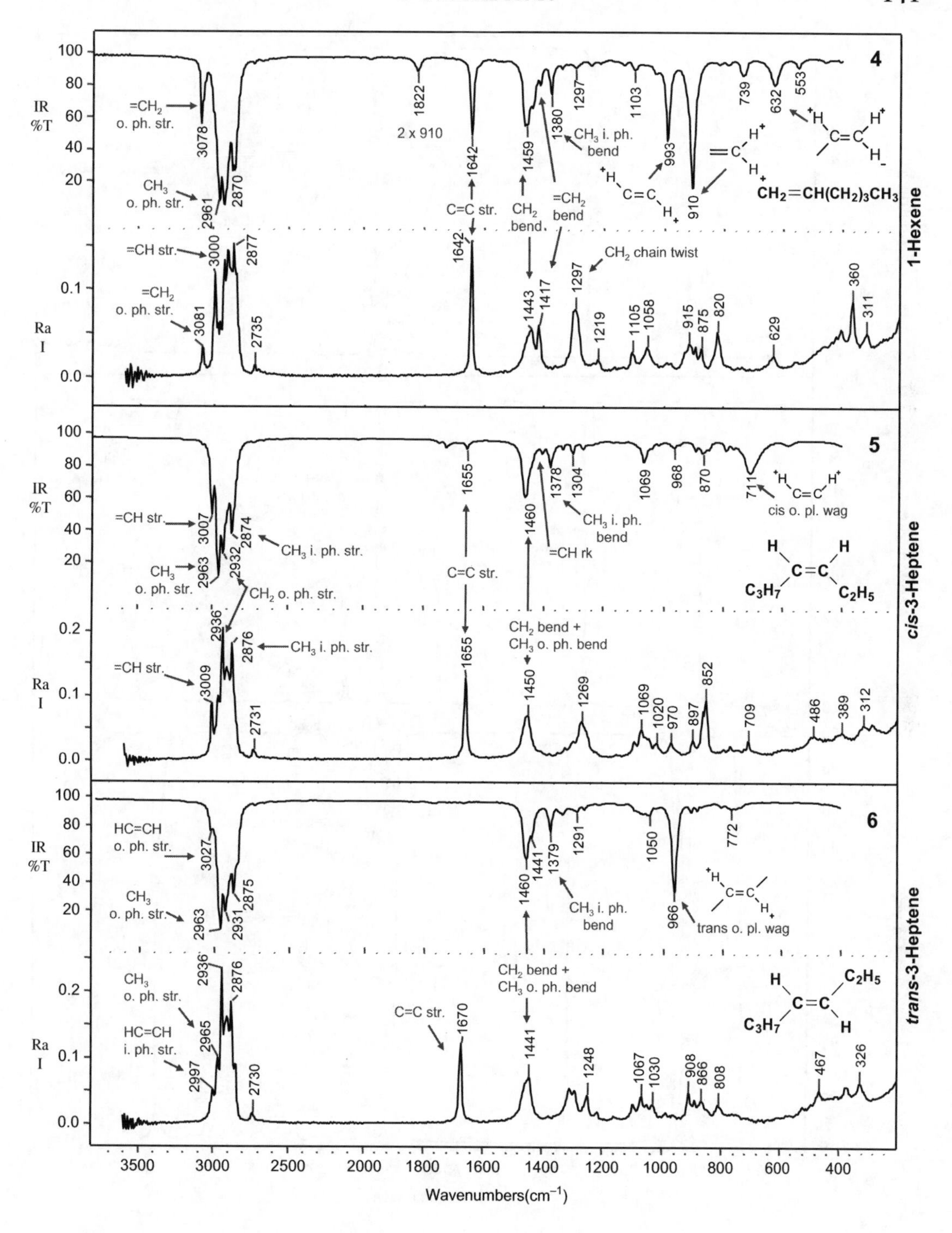
4
1-Hexene
$CH_2=CH(CH_2)_3CH_3$
=CH₂ o. ph. str.
CH₃ o. ph. str.
2 x 910
C=C str.
CH₂ bend.
=CH₂ bend
CH₃ i. ph. bend
=CH str.
CH₂ chain twist
5
cis-3-Heptene
=CH str.
CH₃ o. ph. str.
CH₃ i. ph. str.
CH₂ o. ph. str.
CH₂ bend + CH₃ o. ph. bend
=CH rk
cis o. pl. wag
6
trans-3-Heptene
HC=CH o. ph. str.
HC=CH i. ph. str.
trans o. pl. wag
Wavenumbers(cm⁻¹)
IR %T
Ra I

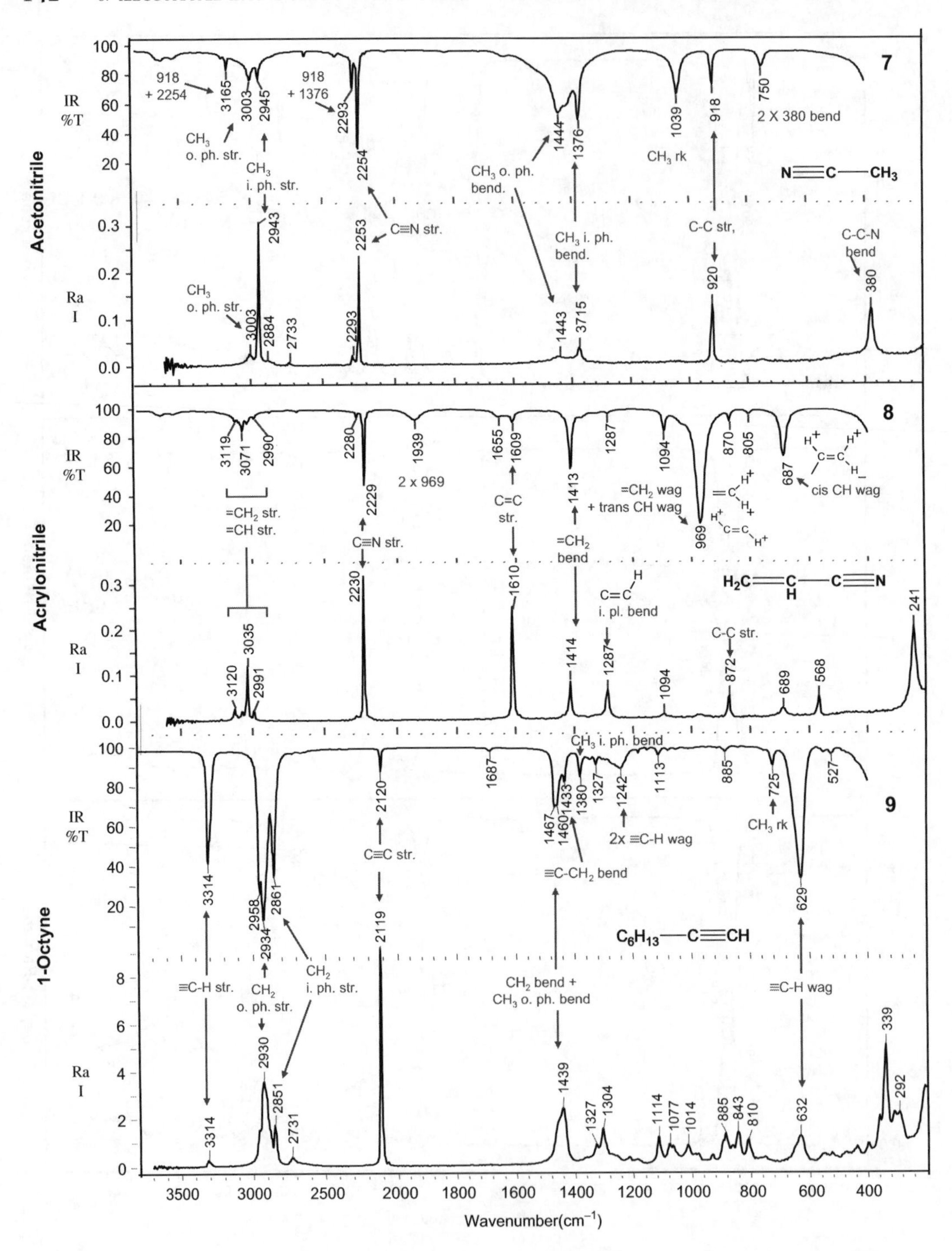
Acetonitrile
7
IR
%T
Ra
I
918
+ 2254
3165
3003
2945
918
+ 1376
2293
2254
CH3
o. ph. str.
CH3
i. ph. str.
C≡N str.
CH3 o. ph.
bend.
1444
1376
1039
918
750
2 X 380 bend
CH3 rk
N≡C—CH3
2943
2253
CH3 i. ph.
bend.
C-C str,
C-C-N
bend
920
380
CH3
o. ph. str.
3003
2884
2733
2293
1443
3715
Acrylonitrile
8
3119
3071
2990
2280
1939
1655
1609
1287
1094
870
805
2229
2 x 969
1413
=CH2 str.
=CH str.
C=C
str.
=CH2 wag
+ trans CH wag
969
687
cis CH wag
C≡N str.
=CH2
bend
H2C=CH—C≡N
2230
1610
C=C
i. pl. bend
241
3035
3120
2991
1414
1287
C-C str.
1094
872
689
568
1-Octyne
9
CH3 i. ph. bend
1687
2120
1467
1460
1433
1380
1327
1242
1113
885
725
527
CH3 rk
2x ≡C-H wag
C≡C str.
≡C-CH2 bend
3314
2958
2934
2861
2119
629
C6H13—C≡CH
≡C-H str.
CH2
o. ph. str.
CH2
i. ph. str.
CH2 bend +
CH3 o. ph. bend
≡C-H wag
339
2930
2851
2731
3314
1439
1327
1304
1114
1077
1014
885
843
810
632
292
Wavenumber(cm−1)

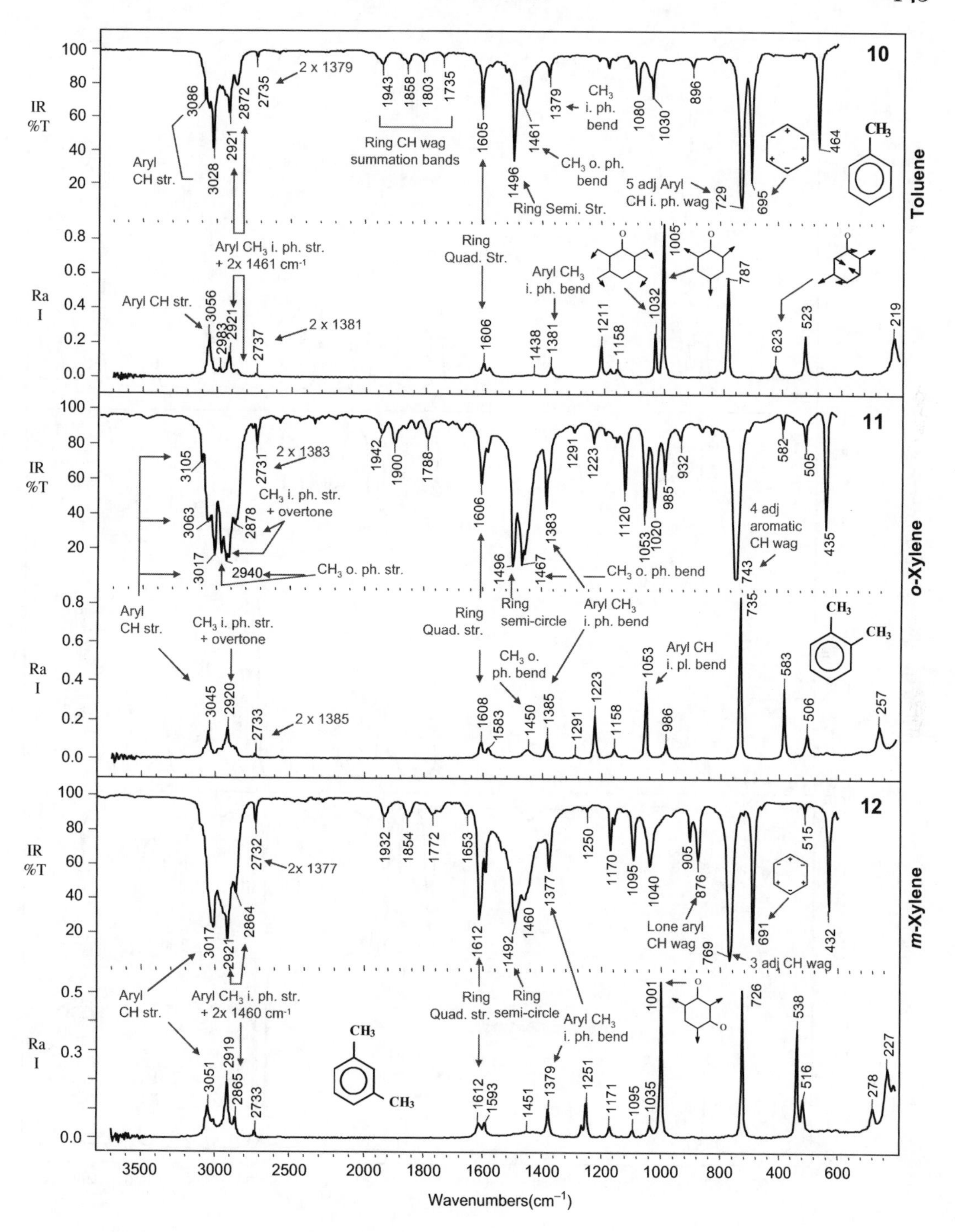
10
Toluene
Aryl CH str.
Ring CH wag summation bands
2 x 1379
CH3 i. ph. bend
CH3 o. ph. bend
Ring Semi. Str.
5 adj Aryl CH i. ph. wag
Aryl CH3 i. ph. str. + 2x 1461 cm-1
Ring Quad. Str.
Aryl CH3 i. ph. bend
2 x 1381
11
o-Xylene
2 x 1383
CH3 i. ph. str. + overtone
CH3 o. ph. str.
CH3 o. ph. bend
4 adj aromatic CH wag
Aryl CH str.
Ring Quad. str.
Ring semi-circle
Aryl CH3 i. ph. bend
CH3 o. ph. bend
Aryl CH i. pl. bend
2 x 1385
12
m-Xylene
2x 1377
Lone aryl CH wag
3 adj CH wag
Aryl CH str.
Aryl CH3 i. ph. str. + 2x 1460 cm-1
Ring Quad. str.
Ring semi-circle
Aryl CH3 i. ph. bend
IR %T
Ra I
Wavenumbers(cm−1)

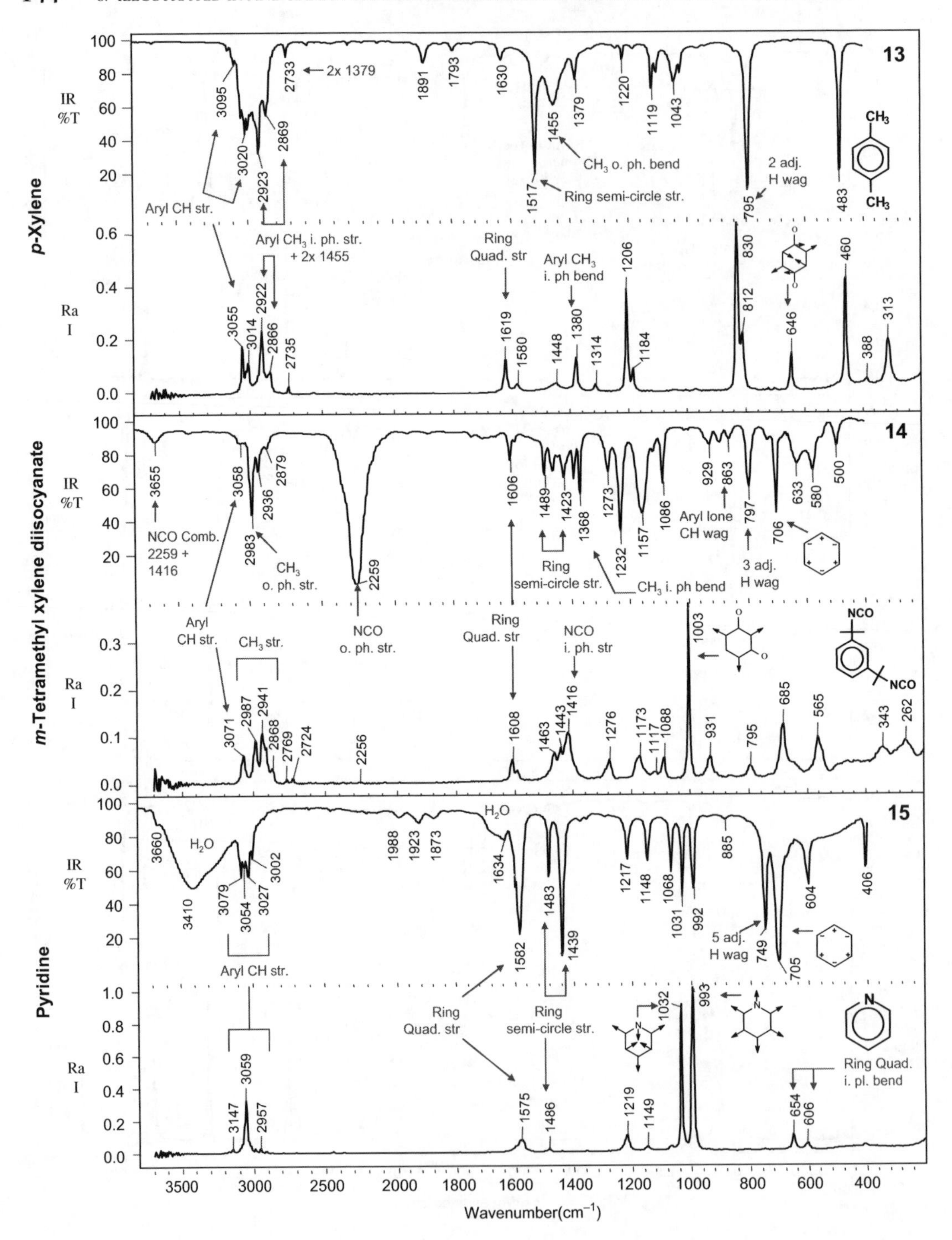

p-Xylene
m-Tetramethyl xylene diisocyanate
Pyridine
IR %T
Ra I
13
14
15
Aryl CH str.
CH3 o. ph. bend
Ring semi-circle str.
2 adj. H wag
Aryl CH3 i. ph. str. + 2x 1455
Ring Quad. str
Aryl CH3 i. ph bend
NCO Comb. 2259 + 1416
CH3 o. ph. str.
Ring semi-circle str.
CH3 i. ph bend
Aryl lone CH wag
3 adj. H wag
Aryl CH str.
CH3 str.
NCO o. ph. str.
NCO i. ph. str
NCO
H2O
Aryl CH str.
5 adj. H wag
Ring Quad. str
Ring semi-circle str.
Ring Quad. i. pl. bend
Wavenumber(cm−1)

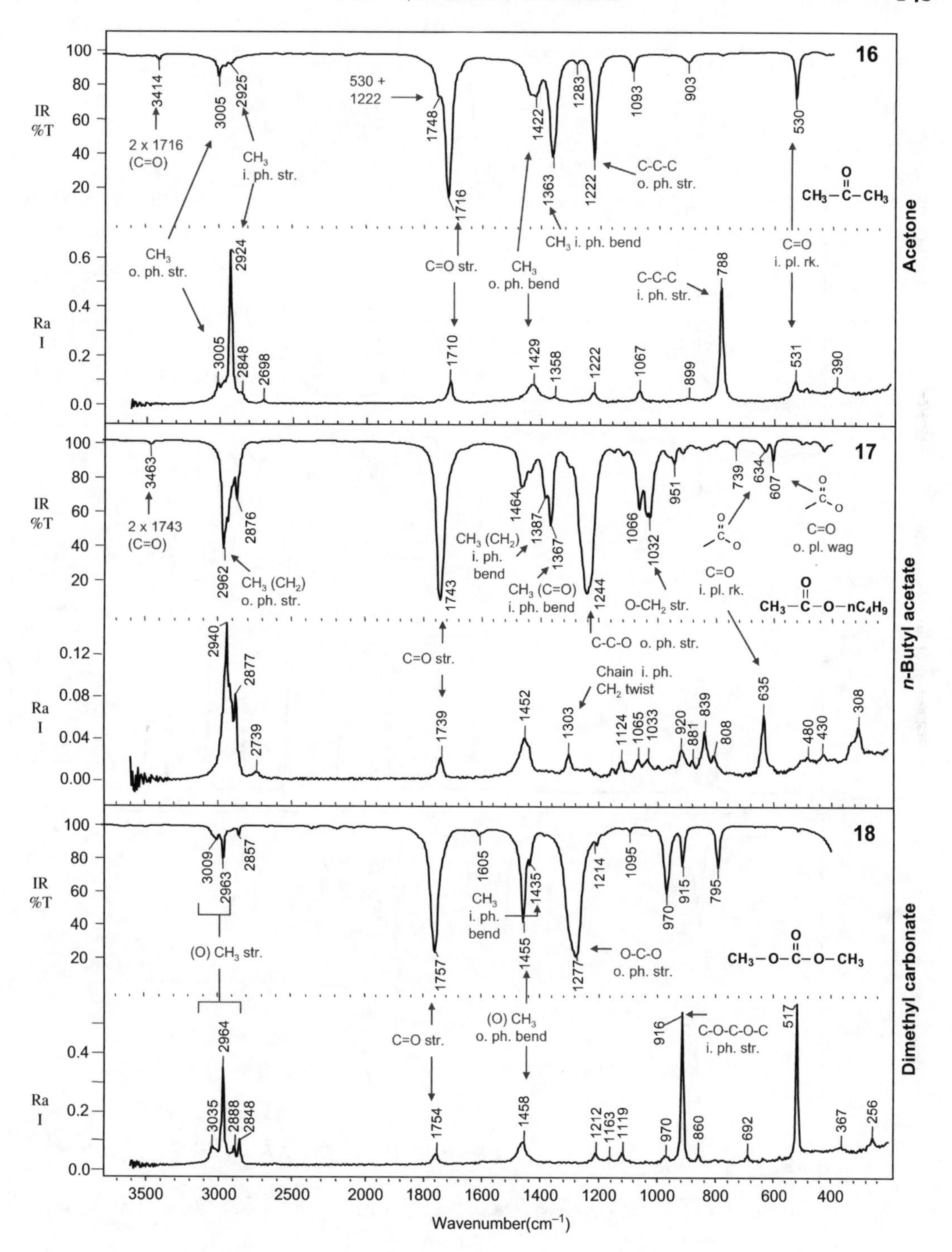
16
Acetone
17
n-Butyl acetate
18
Dimethyl carbonate
IR %T
Ra I
Wavenumber(cm⁻¹)
2 x 1716 (C=O)
CH3 i. ph. str.
530 + 1222
C-C-C o. ph. str.
CH3 i. ph. bend
CH3 o. ph. str.
C=O str.
CH3 o. ph. bend
C-C-C i. ph. str.
C=O i. pl. rk.
2 x 1743 (C=O)
CH3 (CH2) o. ph. str.
CH3 (CH2) i. ph. bend
CH3 (C=O) i. ph. bend
O-CH2 str.
C=O i. pl. rk.
C=O o. pl. wag
C-C-O o. ph. str.
Chain i. ph. CH2 twist
(O) CH3 str.
CH3 i. ph. bend
O-C-O o. ph. str.
(O) CH3 o. ph. bend
C-O-C-O-C i. ph. str.

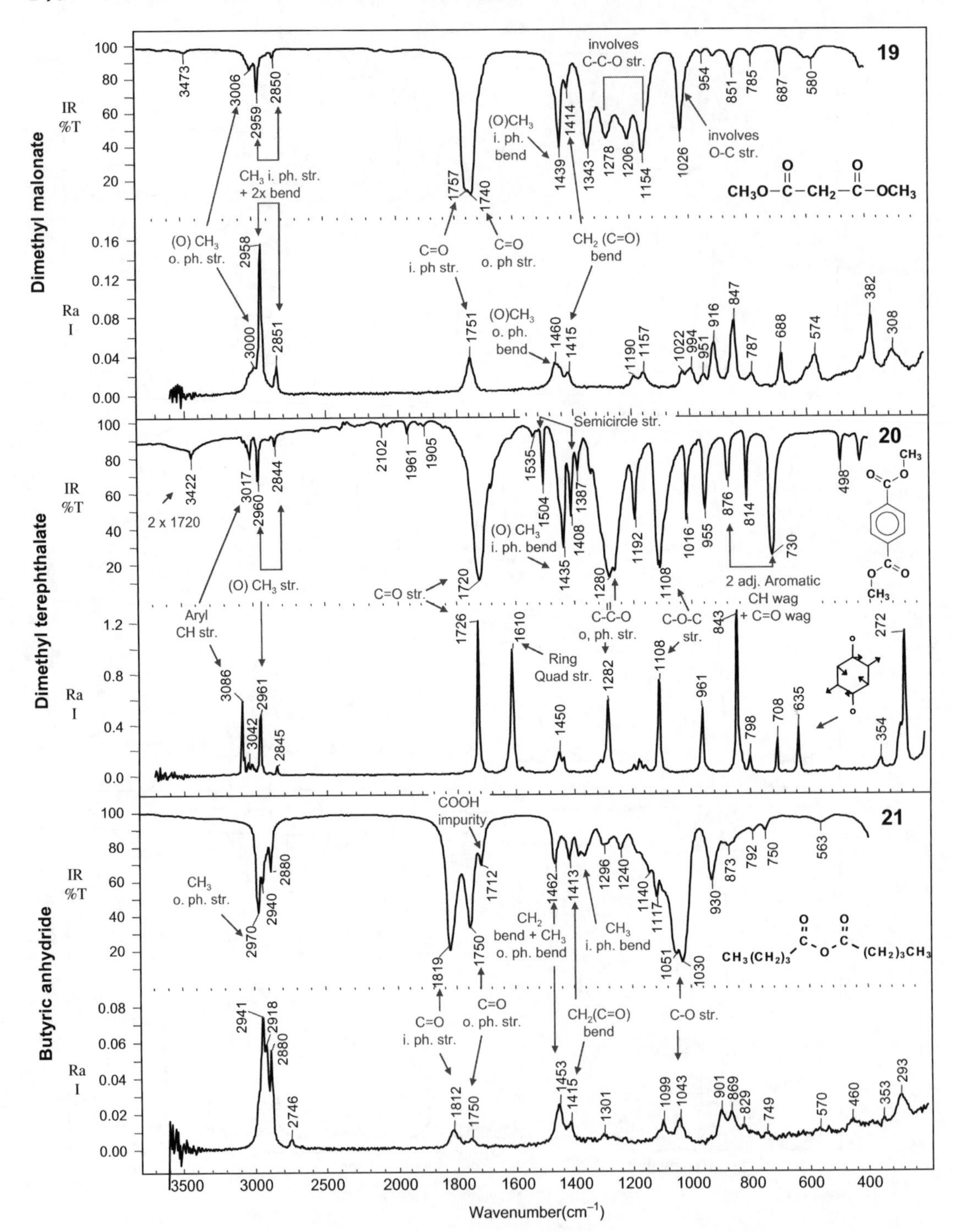

Dimethyl malonate
19
IR %T
Ra I
involves C-C-O str.
involves O-C str.
(O)CH3 i. ph. bend
CH3 i. ph. str. + 2x bend
(O) CH3 o. ph. str.
C=O i. ph str.
C=O o. ph str.
CH2 (C=O) bend
(O)CH3 o. ph. bend
Dimethyl terephthalate
20
Semicircle str.
2 x 1720
(O) CH3 str.
(O) CH3 i. ph. bend
C=O str.
Aryl CH str.
Ring Quad str.
C-C-O o. ph. str.
C-O-C str.
2 adj. Aromatic CH wag + C=O wag
Butyric anhydride
21
COOH impurity
CH3 o. ph. str.
CH2 bend + CH3 o. ph. bend
CH3 i. ph. bend
C=O i. ph. str.
C=O o. ph. str.
CH2(C=O) bend
C-O str.
Wavenumber(cm−1)

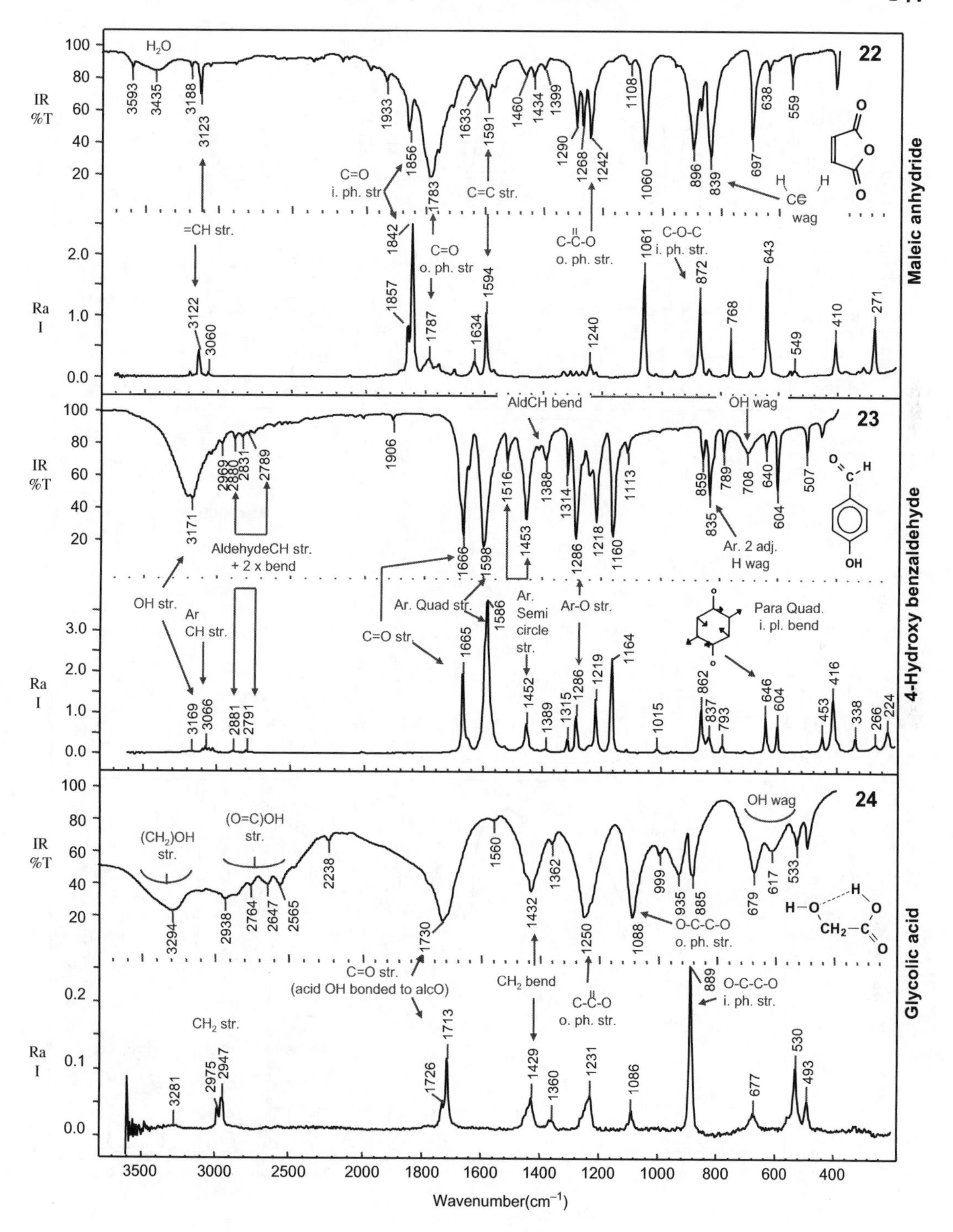
22
Maleic anhydride
23
4-Hydroxy benzaldehyde
24
Glycolic acid
IR %T
Ra I
Wavenumber(cm−1)
C=O i. ph. str
C=O o. ph. str
C=C str.
=CH str.
C-C-O o. ph. str.
C-O-C i. ph. str.
CH wag
OH str.
Ar CH str.
AldehydeCH str. + 2 x bend
Ar. Quad str.
C=O str.
Ar. Semi circle str.
AldCH bend
Ar-O str.
OH wag
Ar. 2 adj. H wag
Para Quad. i. pl. bend
(CH2)OH str.
(O=C)OH str.
CH2 str.
C=O str. (acid OH bonded to alcO)
CH2 bend
O-C-C-O o. ph. str.
O-C-C-O i. ph. str.

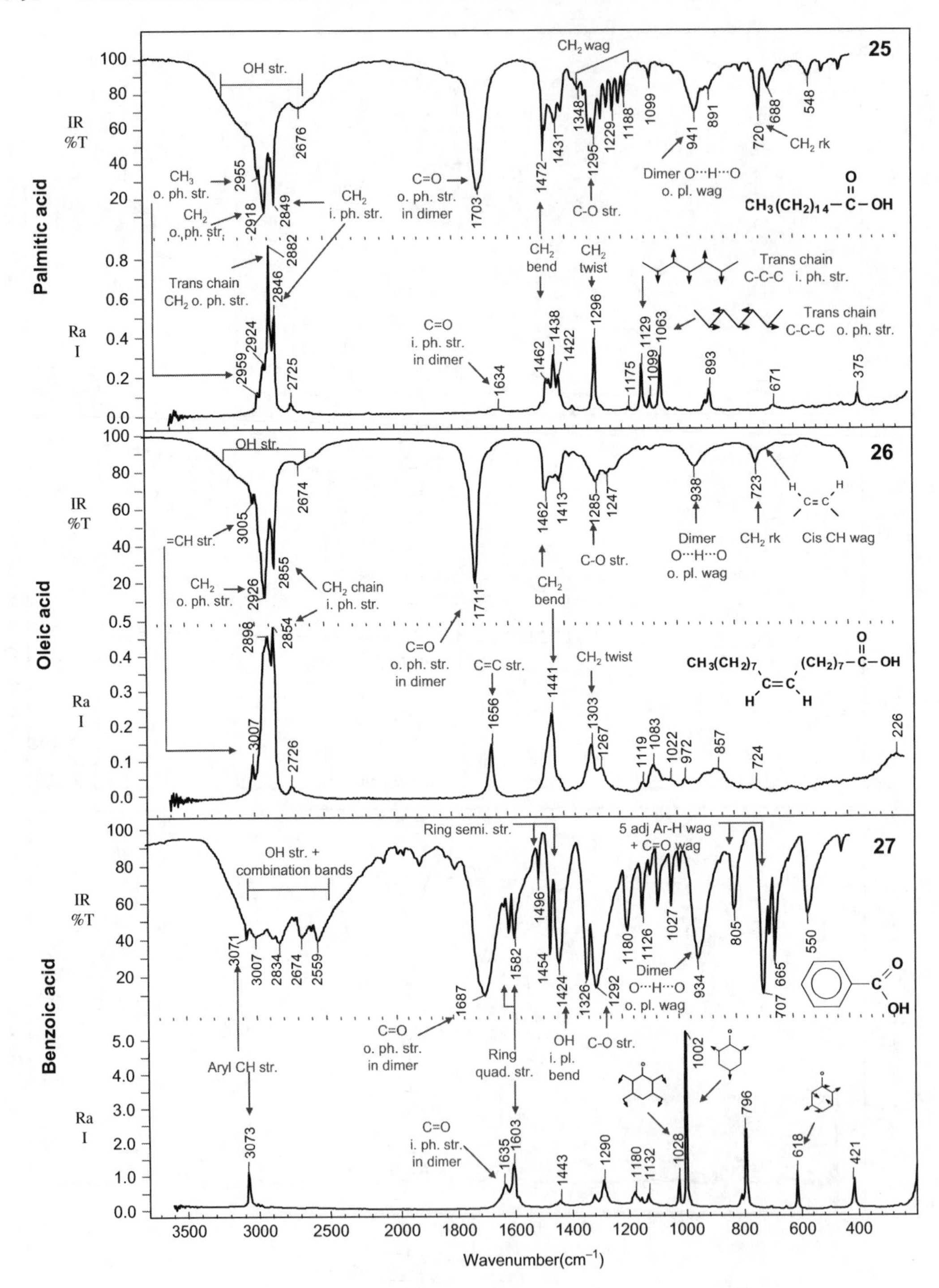

Palmitic acid
25
OH str.
CH_2 wag
CH_3 o. ph. str.
CH_2 o. ph. str.
CH_2 i. ph. str.
C=O o. ph. str. in dimer
C-O str.
Dimer O···H···O o. pl. wag
CH_2 rk
$CH_3(CH_2)_{14}$–C(=O)–OH
CH_2 bend
CH_2 twist
Trans chain C-C-C i. ph. str.
Trans chain C-C-C o. ph. str.
Trans chain CH_2 o. ph. str.
C=O i. ph. str. in dimer
Oleic acid
26
=CH str.
CH_2 chain i. ph. str.
C=O o. ph. str. in dimer
C=C str.
CH_2 bend
CH_2 twist
C-O str.
Dimer O···H···O o. pl. wag
CH_2 rk
Cis CH wag
$CH_3(CH_2)_7$CH=CH$(CH_2)_7$–C(=O)–OH
Benzoic acid
27
OH str. + combination bands
Ring semi. str.
5 adj Ar-H wag + C=O wag
Aryl CH str.
Ring quad. str.
OH i. pl. bend
C-O str.
Dimer O···H···O o. pl. wag
C=O i. ph. str. in dimer
IR %T
Ra I
Wavenumber(cm^{-1})

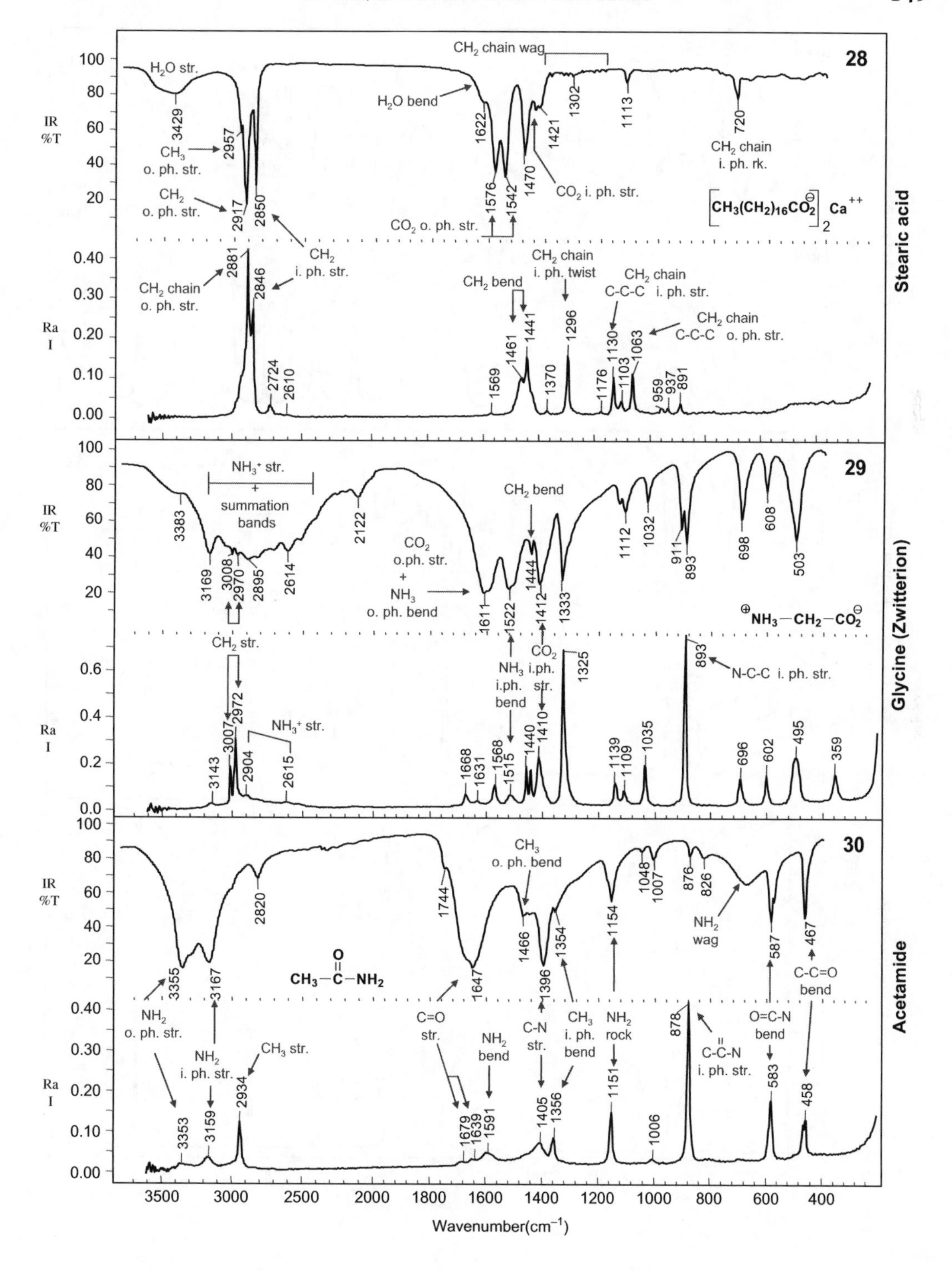
28
Stearic acid
29
Glycine (Zwitterion)
30
Acetamide
Wavenumber(cm−1)

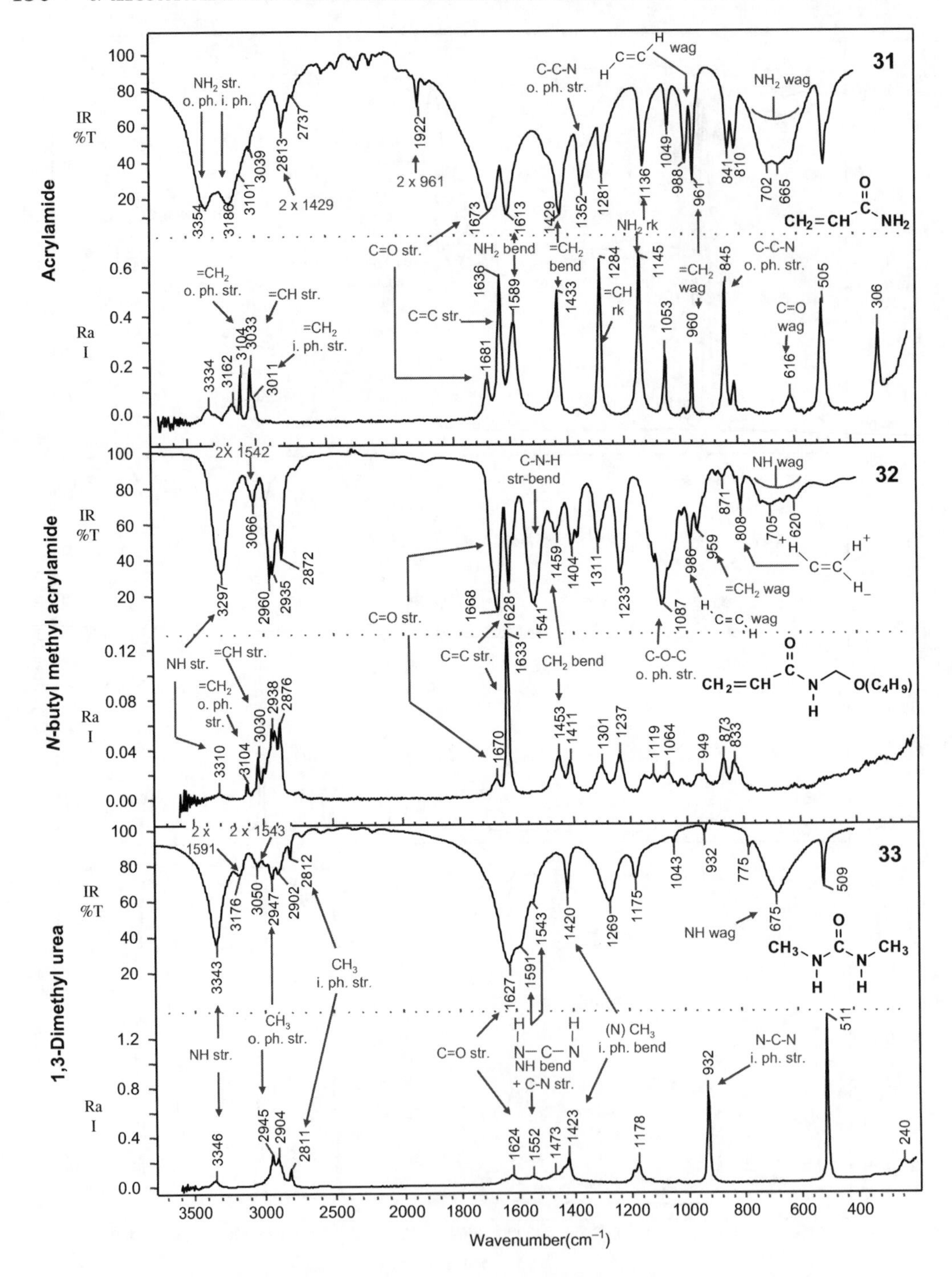
Acrylamide
N-butyl methyl acrylamide
1,3-Dimethyl urea
IR %T
Ra I
31
32
33
NH2 str.
o. ph. i. ph.
2 x 1429
2 x 961
C-C-N
o. ph. str.
wag
NH2 wag
C=O str.
NH2 bend
=CH2 bend
NH2 rk
=CH2 o. ph. str.
=CH str.
=CH2 i. ph. str.
C=C str.
=CH rk
=CH2 wag
C-C-N o. ph. str.
C=O wag
2X 1542
C-N-H str-bend
NH wag
=CH2 wag
C-O-C o. ph. str.
NH str.
CH2 bend
2 x 1591
2 x 1543
CH3 i. ph. str.
CH3 o. ph. str.
NH bend + C-N str.
(N) CH3 i. ph. bend
N-C-N i. ph. str.
Wavenumber(cm−1)

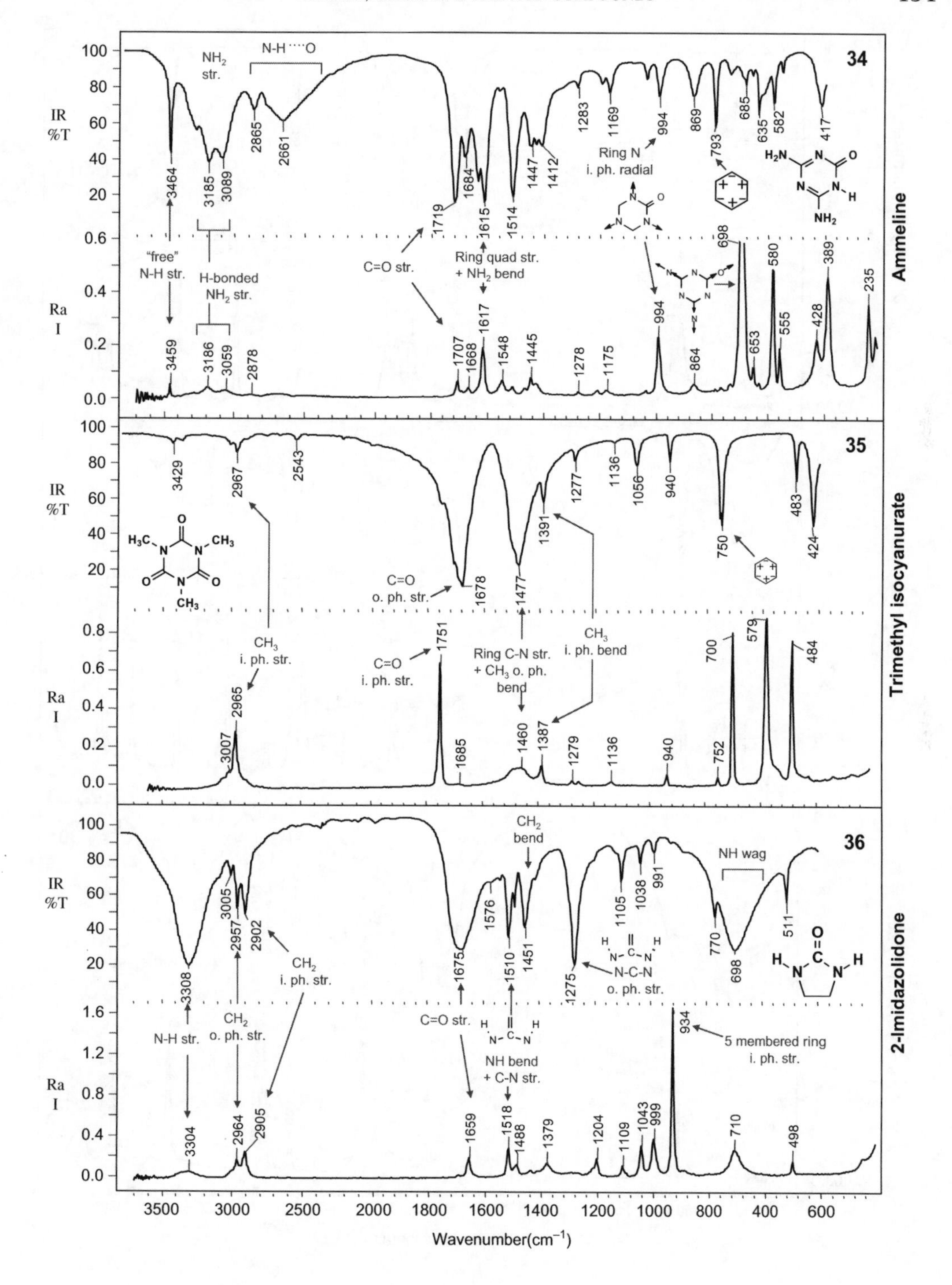
34
Ammeline
35
Trimethyl isocyanurate
36
2-Imidazolidone
IR %T
Ra I
Wavenumber(cm^{-1})
"free" N-H str.
H-bonded NH_2 str.
NH_2 str.
N-H····O
C=O str.
Ring quad str. + NH_2 bend
Ring N i. ph. radial
C=O o. ph. str.
C=O i. ph. str.
CH_3 i. ph. str.
Ring C-N str. + CH_3 o. ph. bend
CH_3 i. ph. bend
CH_2 bend
NH wag
CH_2 i. ph. str.
CH_2 o. ph. str.
N-H str.
N-C-N o. ph. str.
NH bend + C-N str.
5 membered ring i. ph. str.

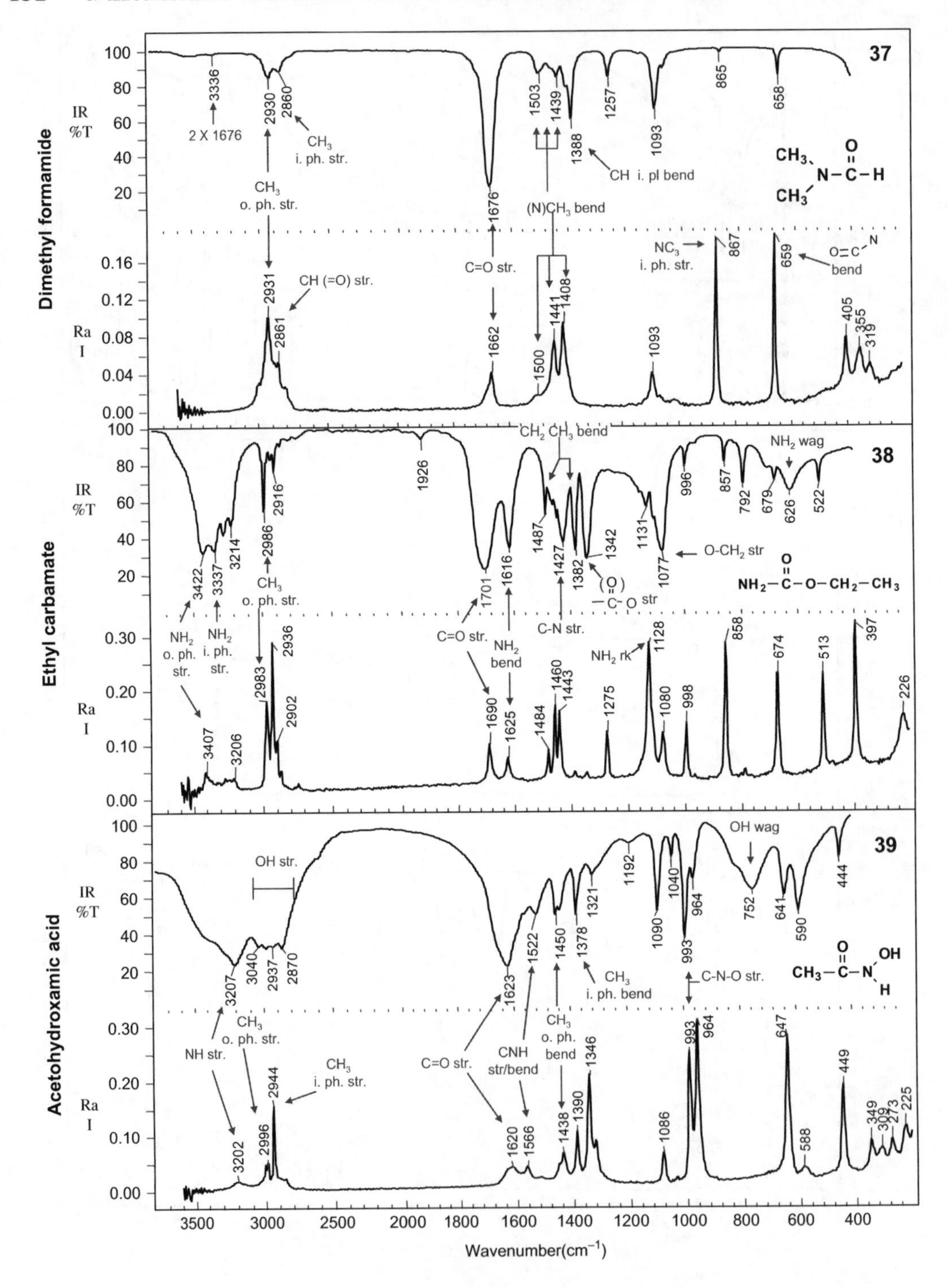
37
Dimethyl formamide
IR %T
Ra I
3336
2 X 1676
2930
2860
CH3 o. ph. str.
CH3 i. ph. str.
1676
1503
1439
1388
(N)CH3 bend
CH i. pl bend
1257
1093
865
658
C=O str.
2931
2861
CH (=O) str.
1662
1500
1441
1408
1093
NC3 i. ph. str.
867
659
O=C-N bend
405
355
319
38
Ethyl carbamate
3422
3337
3214
2986
2916
1926
1701
1616
1487
1427
1382
1342
1131
1077
996
857
792
679
626
522
CH2 CH3 bend
NH2 wag
O-CH2 str
C-O str
C-N str.
NH2 o. ph. str.
NH2 i. ph. str.
CH3 o. ph. str.
C=O str.
NH2 bend
NH2 rk
3407
3206
2983
2936
2902
1690
1625
1484
1460
1443
1275
1128
1080
998
858
674
513
397
226
39
Acetohydroxamic acid
OH str.
3207
3040
2937
2870
1623
1522
1450
1378
1321
1192
1090
1040
993
964
752
641
590
444
OH wag
CH3 i. ph. bend
C-N-O str.
NH str.
CH3 o. ph. str.
CH3 i. ph. str.
C=O str.
CNH str/bend
CH3 o. ph. bend
3202
2996
2944
1620
1566
1438
1390
1346
1086
993
964
647
588
449
349
309
273
225
Wavenumber(cm−1)

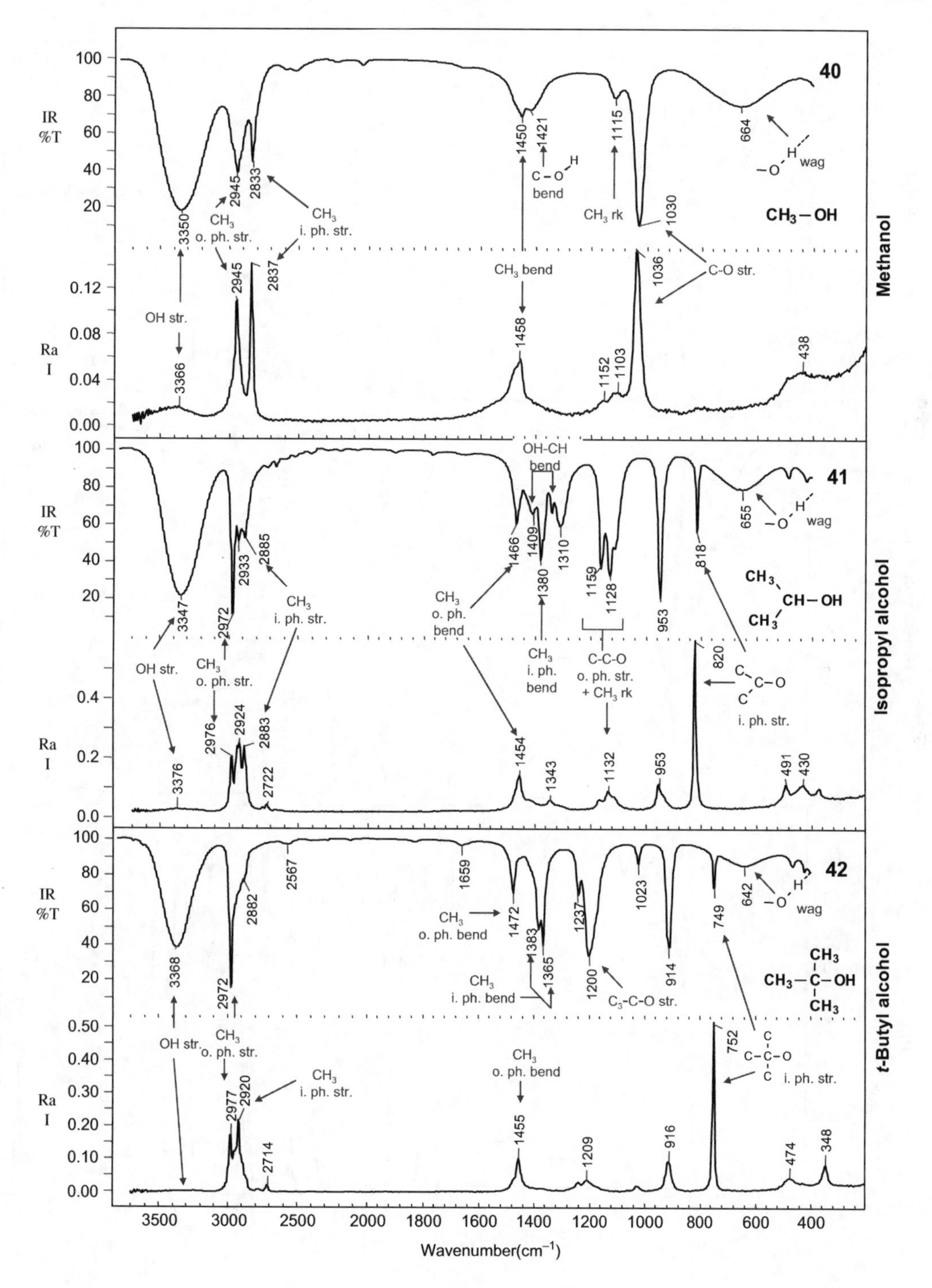
40
CH3 – OH
Methanol
3350
2945
2833
1450
1421
1115
1030
664
CH3 o. ph. str.
CH3 i. ph. str.
C–O–H bend
CH3 rk
H wag
OH str.
3366
2945
2837
CH3 bend
1458
1152
1103
1036
C-O str.
438
41
Isopropyl alcohol
OH-CH bend
3347
2972
2933
2885
1466
1409
1380
1310
1159
1128
953
818
655
CH3 o. ph. bend
CH3 i. ph. bend
C-C-O o. ph. str. + CH3 rk
i. ph. str.
3376
2976
2924
2883
2722
1454
1343
1132
953
820
491
430
42
t-Butyl alcohol
3368
2972
2882
2567
1659
1472
1383
1365
1237
1200
1023
914
749
642
C3-C-O str.
2977
2920
2714
1455
1209
916
752
474
348
IR %T
Ra I
Wavenumber(cm−1)

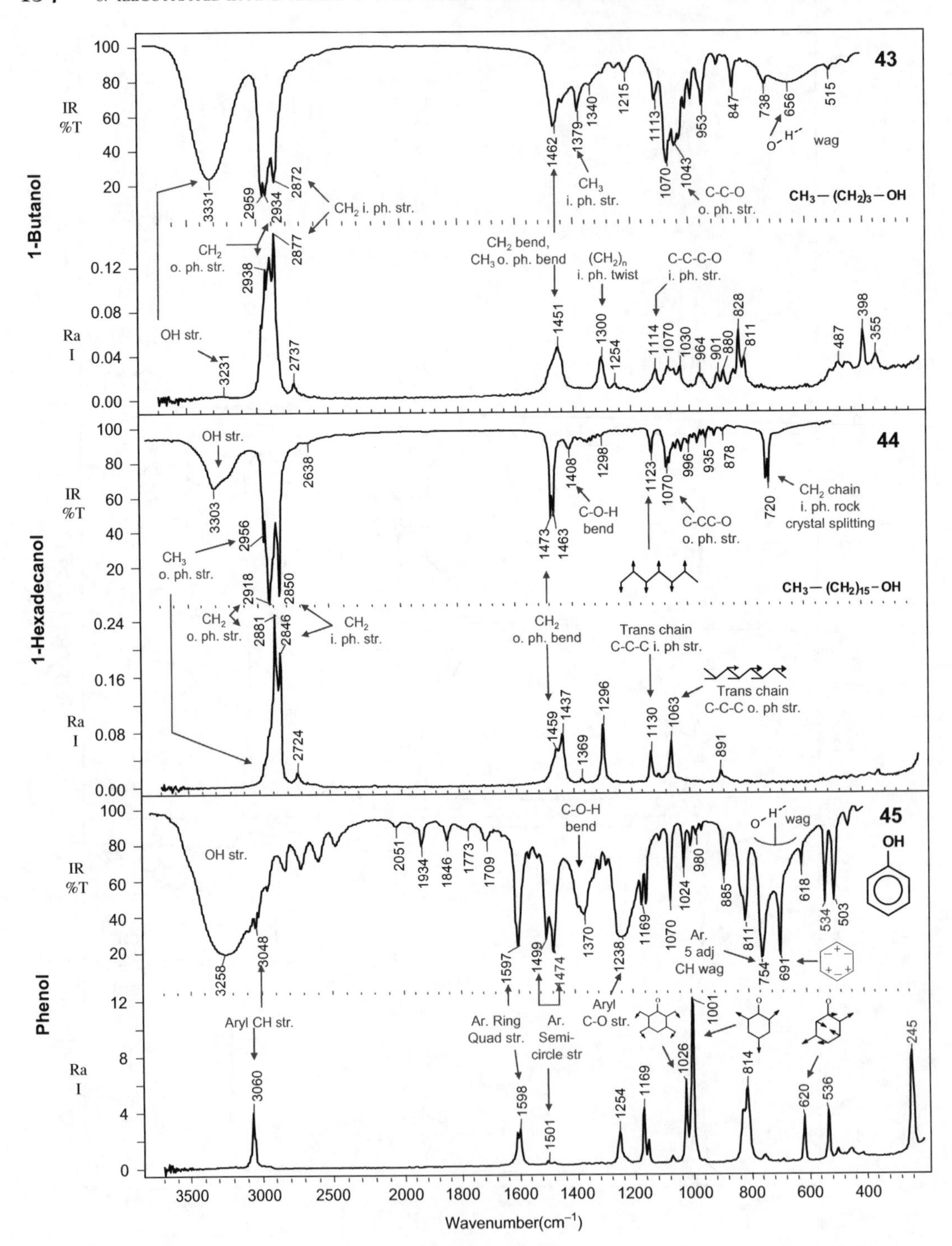
1-Butanol
43
$CH_3-(CH_2)_3-OH$
OH str.
CH_2 i. ph. str.
CH_2 o. ph. str.
CH_2 bend, CH_3 o. ph. bend
CH_3 i. ph. str.
$(CH_2)_n$ i. ph. twist
C-C-C-O i. ph. str.
C-C-O o. ph. str.
O-H wag
1-Hexadecanol
44
$CH_3-(CH_2)_{15}-OH$
OH str.
CH_3 o. ph. str.
CH_2 o. ph. str.
CH_2 i. ph. str.
C-O-H bend
C-CC-O o. ph. str.
CH_2 chain i. ph. rock crystal splitting
CH_2 o. ph. bend
Trans chain C-C-C i. ph str.
Trans chain C-C-C o. ph str.
Phenol
45
OH
OH str.
C-O-H bend
O-H wag
Ar. 5 adj CH wag
Aryl CH str.
Ar. Ring Quad str.
Ar. Semi-circle str
Aryl C-O str.
IR %T
Ra I
Wavenumber(cm⁻¹)

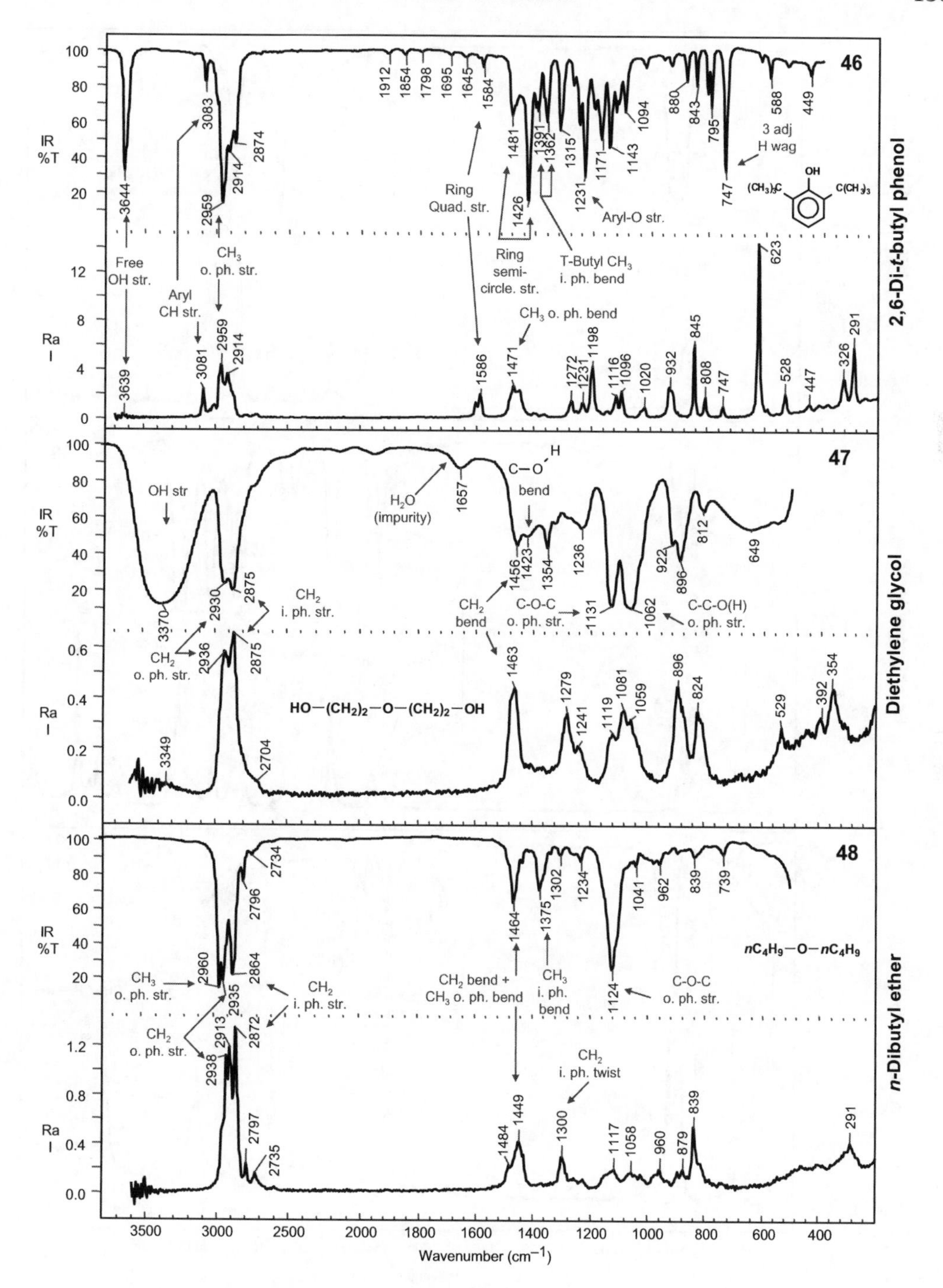

46
2,6-Di-t-butyl phenol
Free OH str.
Aryl CH str.
CH3 o. ph. str.
Ring Quad. str.
Ring semi-circle. str.
T-Butyl CH3 i. ph. bend
CH3 o. ph. bend
Aryl-O str.
3 adj H wag
47
Diethylene glycol
OH str
H2O (impurity)
C-O-H bend
CH2 i. ph. str.
CH2 bend
C-O-C o. ph. str.
C-C-O(H) o. ph. str.
CH2 o. ph. str.
HO—(CH2)2—O—(CH2)2—OH
48
n-Dibutyl ether
nC4H9—O—nC4H9
CH3 o. ph. str.
CH2 i. ph. str.
CH2 bend + CH3 o. ph. bend
CH3 i. ph. bend
C-O-C o. ph. str.
CH2 o. ph. str.
CH2 i. ph. twist
IR %T
Ra I
Wavenumber (cm−1)

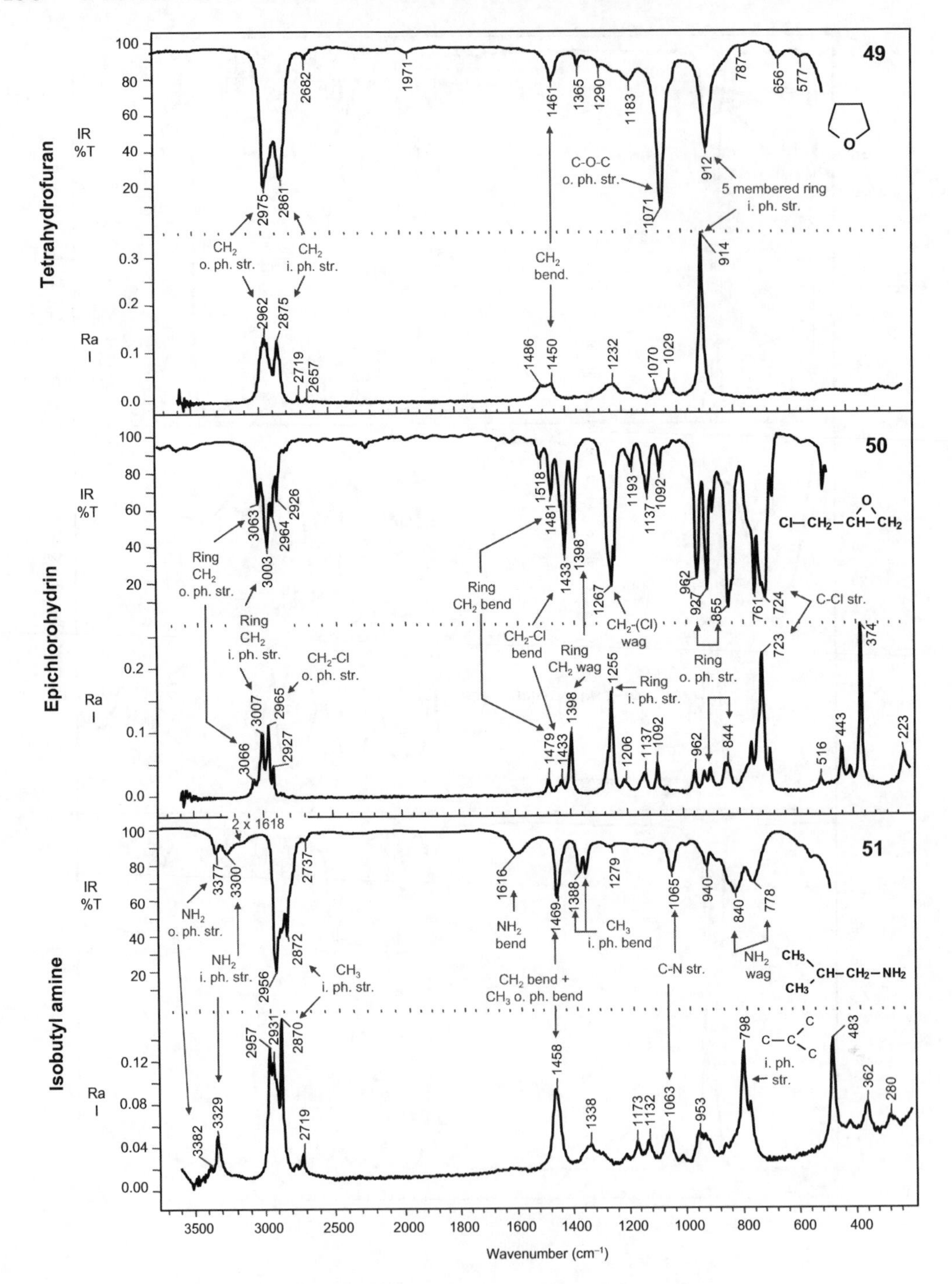

Tetrahydrofuran
49
IR %T
Ra I
CH2 o. ph. str.
CH2 i. ph. str.
CH2 bend.
C-O-C o. ph. str.
5 membered ring i. ph. str.
Epichlorohydrin
50
Ring CH2 o. ph. str.
Ring CH2 i. ph. str.
CH2-Cl o. ph. str.
Ring CH2 bend
CH2-Cl bend
Ring CH2 wag
CH2-(Cl) wag
Ring i. ph. str.
Ring o. ph. str.
C-Cl str.
Isobutyl amine
51
2 x 1618
NH2 o. ph. str.
NH2 i. ph. str.
CH3 i. ph. str.
NH2 bend
CH2 bend + CH3 o. ph. bend
CH3 i. ph. bend
C-N str.
NH2 wag
i. ph. str.
Wavenumber (cm−1)

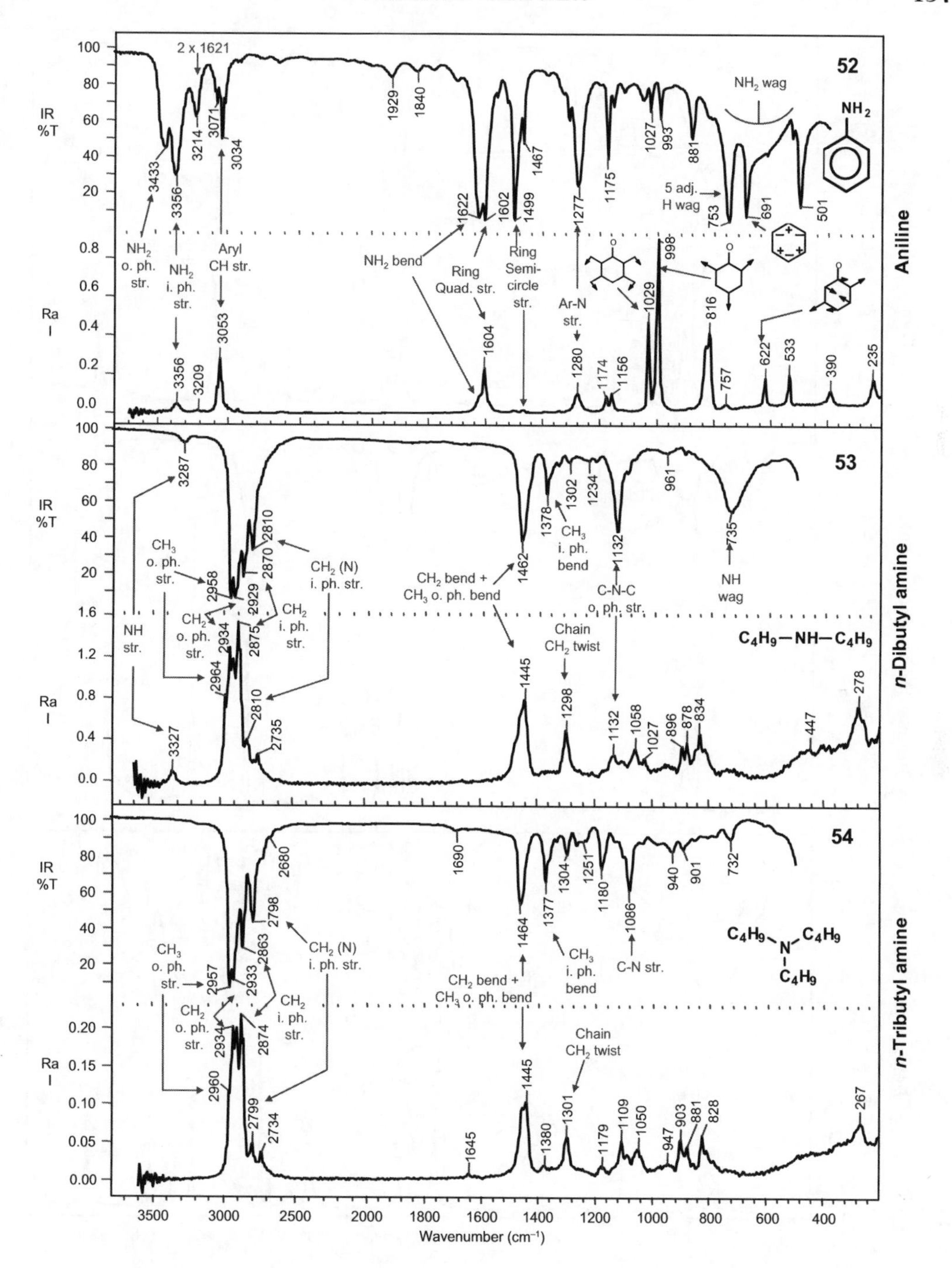
52
Aniline
NH2 o. ph. str.
NH2 i. ph. str.
Aryl CH str.
NH2 bend
Ring Quad. str.
Ring Semi-circle str.
Ar-N str.
NH2 wag
5 adj. H wag
53
n-Dibutyl amine
C4H9—NH—C4H9
NH str.
CH3 o. ph. str.
CH2 o. ph. str.
CH2 (N) i. ph. str.
CH2 i. ph. str.
CH2 bend + CH3 o. ph. bend
CH3 i. ph. bend
C-N-C o. ph. str.
Chain CH2 twist
NH wag
54
n-Tributyl amine
C-N str.
Wavenumber (cm−1)
IR %T
Ra I

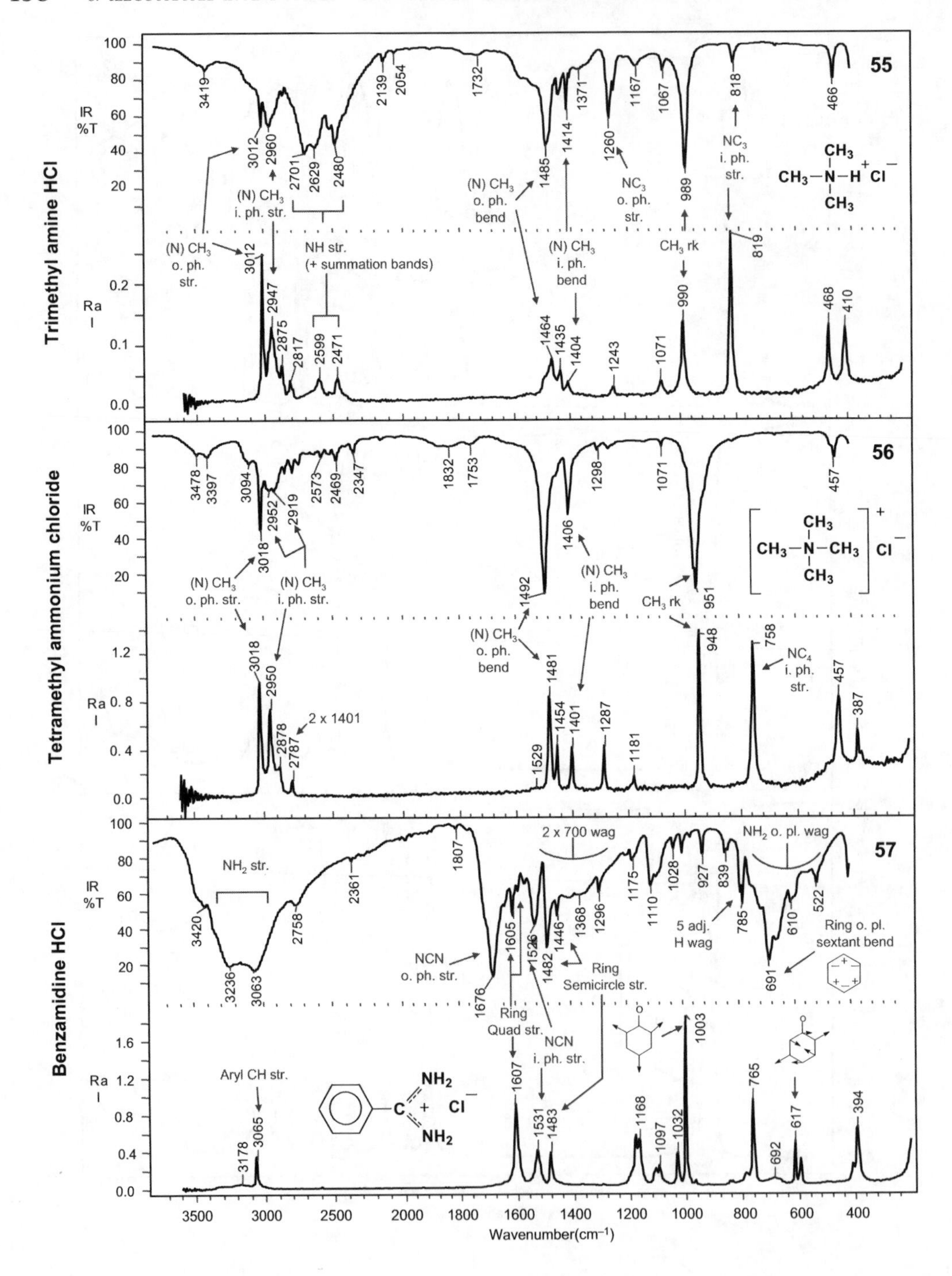

55
Trimethyl amine HCl
IR %T
Ra I
(N) CH3 o. ph. str.
(N) CH3 i. ph. str.
NH str. (+ summation bands)
(N) CH3 o. ph. bend
(N) CH3 i. ph. bend
NC3 o. ph. str.
CH3 rk
NC3 i. ph. str.
56
Tetramethyl ammonium chloride
2 x 1401
NC4 i. ph. str.
57
Benzamidine HCl
NH2 str.
Aryl CH str.
NCN o. ph. str.
Ring Quad str.
NCN i. ph. str.
Ring Semicircle str.
2 x 700 wag
5 adj. H wag
NH2 o. pl. wag
Ring o. pl. sextant bend
Wavenumber(cm−1)

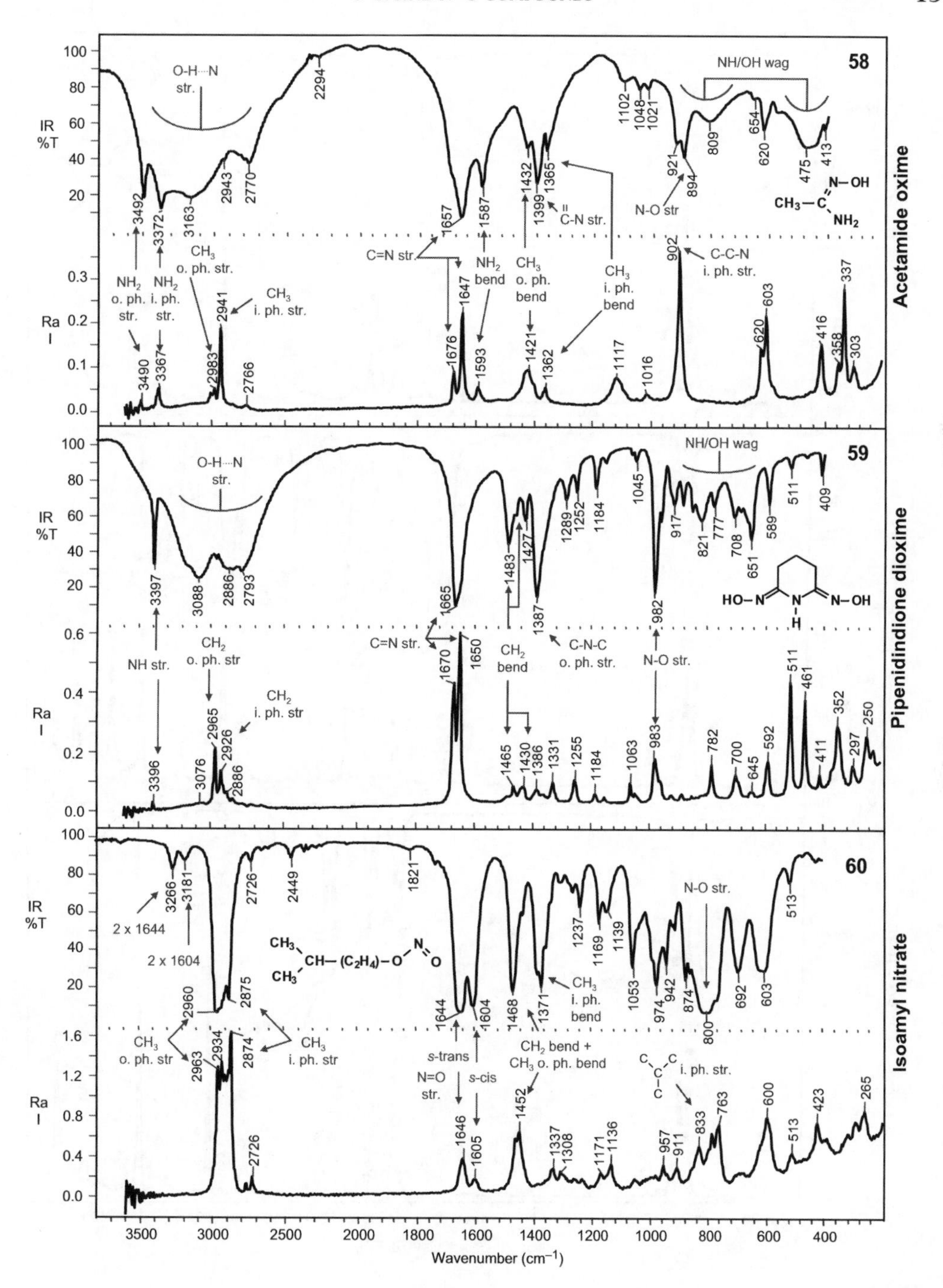
58
Acetamide oxime
NH/OH wag
O-H···N str.
N-O str
C-N str.
C=N str.
NH2 bend
CH3 o. ph. bend
CH3 i. ph. bend
C-C-N i. ph. str.
NH2 o. ph. str.
NH2 i. ph. str.
CH3 o. ph. str.
CH3 i. ph. str.
59
Pipenidindione dioxime
NH str.
CH2 o. ph. str
CH2 i. ph. str
CH2 bend
C-N-C o. ph. str.
N-O str.
60
Isoamyl nitrate
2 x 1644
2 x 1604
N-O str.
CH3 i. ph. bend
CH3 o. ph. str
CH3 i. ph. str
s-trans
s-cis
N=O str.
CH2 bend + CH3 o. ph. bend
C-C-C i. ph. str.
IR %T
Ra I
Wavenumber (cm−1)

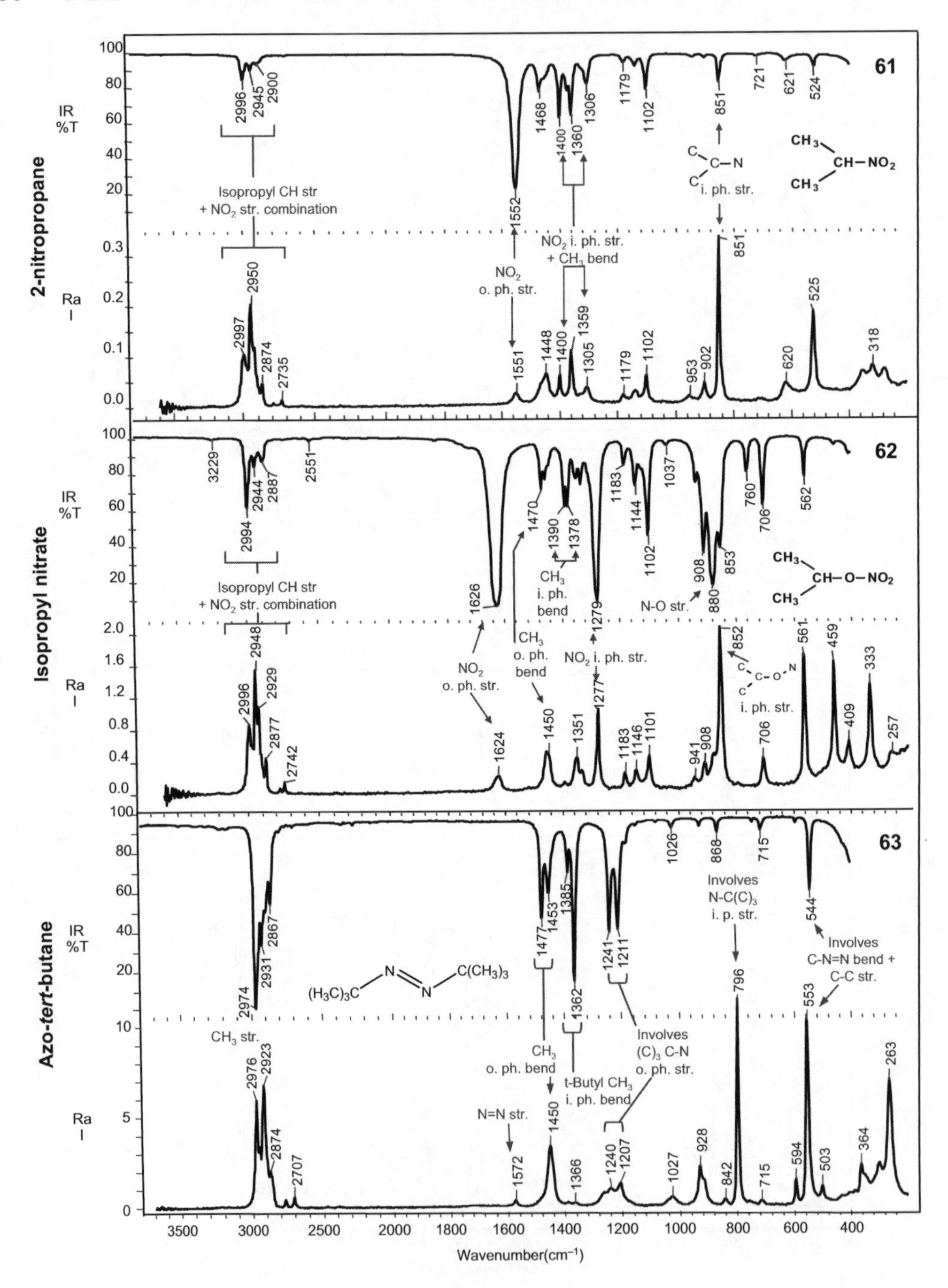
61
2-nitropropane
IR %T
Ra I
Isopropyl CH str + NO2 str. combination
NO2 o. ph. str.
NO2 i. ph. str. + CH3 bend
C-N i. ph. str.
62
Isopropyl nitrate
Isopropyl CH str + NO2 str. combination
NO2 o. ph. str.
CH3 o. ph. bend
CH3 i. ph. bend
NO2 i. ph. str.
N-O str.
i. ph. str.
63
Azo-tert-butane
CH3 str.
CH3 o. ph. bend
N=N str.
t-Butyl CH3 i. ph. bend
Involves (C)3 C-N o. ph. str.
Involves N-C(C)3 i. p. str.
Involves C-N=N bend + C-C str.
Wavenumber(cm−1)

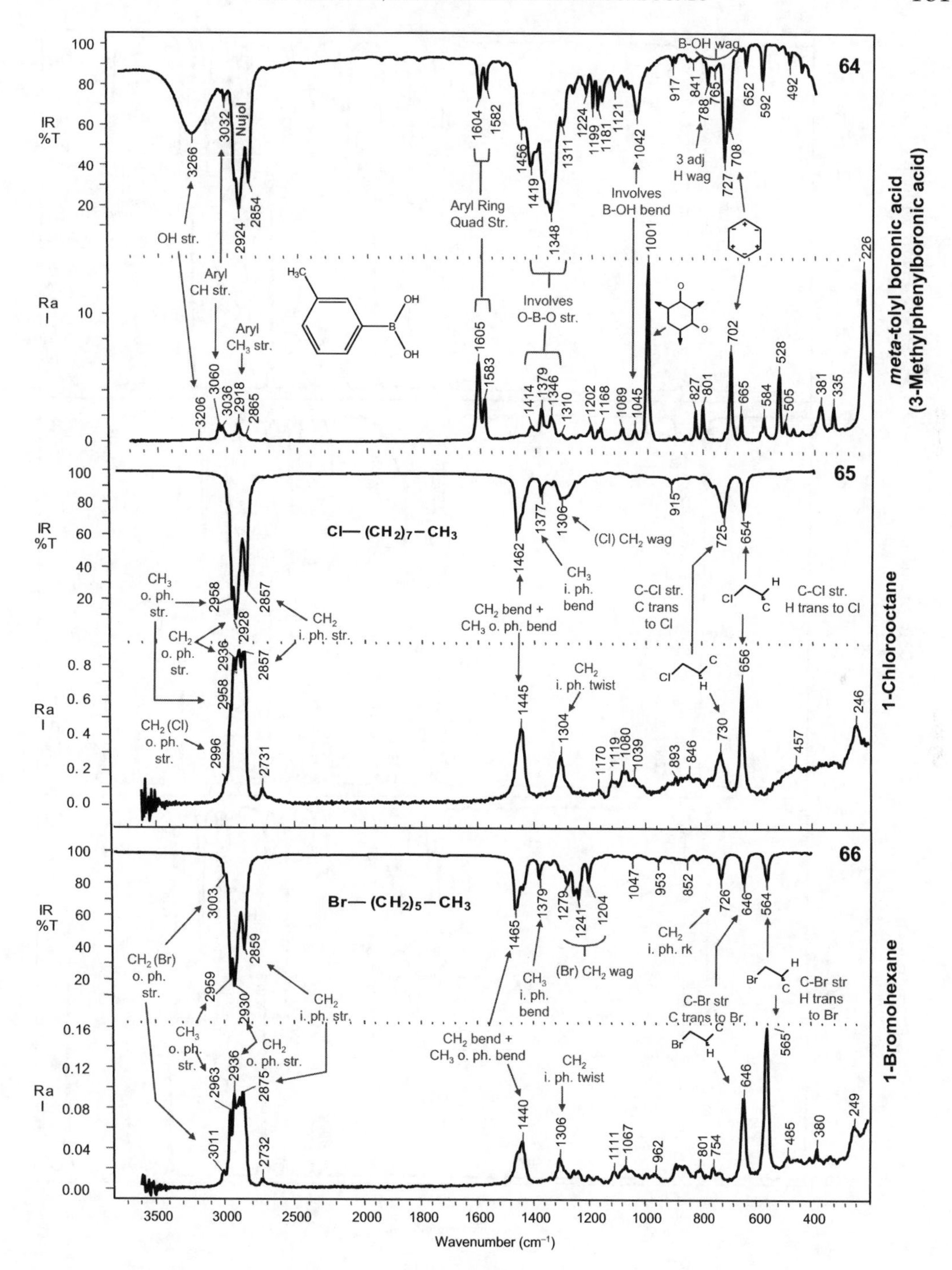
64
meta-tolyl boronic acid
(3-Methylphenylboronic acid)
OH str.
Aryl CH str.
Aryl CH3 str.
Nujol
Aryl Ring Quad Str.
Involves O-B-O str.
Involves B-OH bend
B-OH wag
3 adj H wag
65
1-Chlorooctane
Cl—(CH2)7—CH3
(Cl) CH2 wag
CH3 i. ph. bend
CH2 bend + CH3 o. ph. bend
C-Cl str. C trans to Cl
C-Cl str. H trans to Cl
CH2 i. ph. twist
CH2 (Cl) o. ph. str.
66
1-Bromohexane
Br—(CH2)5—CH3
(Br) CH2 wag
CH2 i. ph. rk
C-Br str C trans to Br
C-Br str H trans to Br
CH2 (Br) o. ph. str.
CH3 o. ph. str.
CH2 o. ph. str.
CH2 i. ph. str.
IR %T
Ra I
Wavenumber (cm−1)

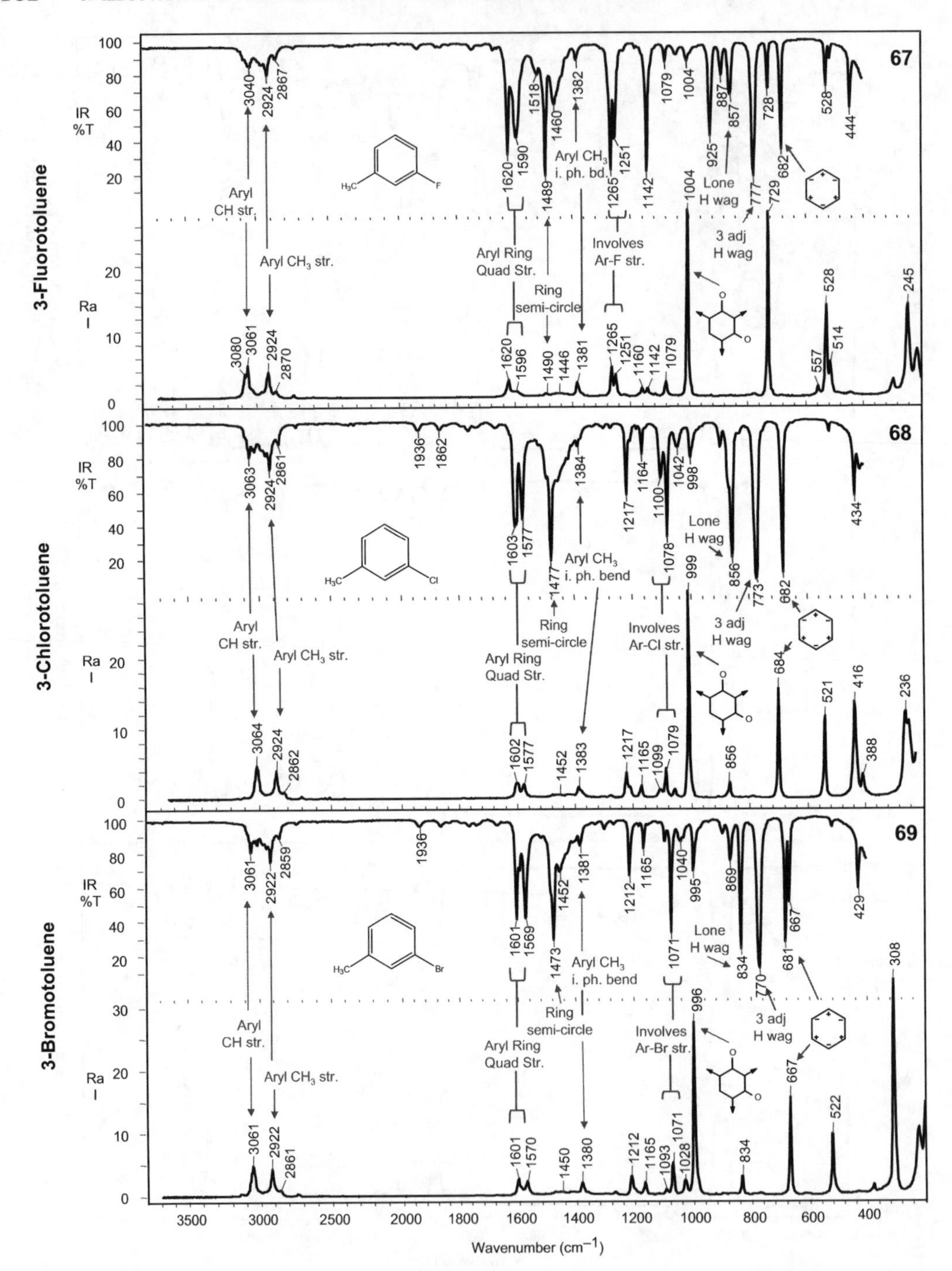
67
3-Fluorotoluene
IR %T
Ra I
Aryl CH str.
Aryl CH_3 str.
Aryl Ring Quad Str.
Ring semi-circle
Aryl CH_3 i. ph. bd.
Involves Ar-F str.
Lone H wag
3 adj H wag
3040
2924
2867
1620
1590
1518
1489
1460
1382
1265
1251
1142
1079
1004
925
887
857
777
728
729
682
528
444
3080
3061
2924
2870
1620
1596
1490
1446
1381
1265
1251
1160
1142
1079
1004
557
528
514
245
68
3-Chlorotoluene
Aryl CH str.
Aryl CH_3 str.
Aryl Ring Quad Str.
Ring semi-circle
Aryl CH_3 i. ph. bend
Involves Ar-Cl str.
Lone H wag
3 adj H wag
3063
2924
2861
1936
1862
1603
1577
1477
1384
1217
1164
1100
1078
1042
998
999
856
773
682
434
3064
2924
2862
1602
1577
1452
1383
1217
1165
1099
1079
999
856
684
521
416
388
236
69
3-Bromotoluene
Aryl CH str.
Aryl CH_3 str.
Aryl Ring Quad Str.
Ring semi-circle
Aryl CH_3 i. ph. bend
Involves Ar-Br str.
Lone H wag
3 adj H wag
3061
2922
2859
1936
1601
1569
1473
1452
1381
1212
1165
1071
1040
995
996
869
834
770
681
667
429
308
3061
2922
2861
1601
1570
1450
1380
1212
1165
1093
1071
1028
834
667
522
Wavenumber (cm^{-1})

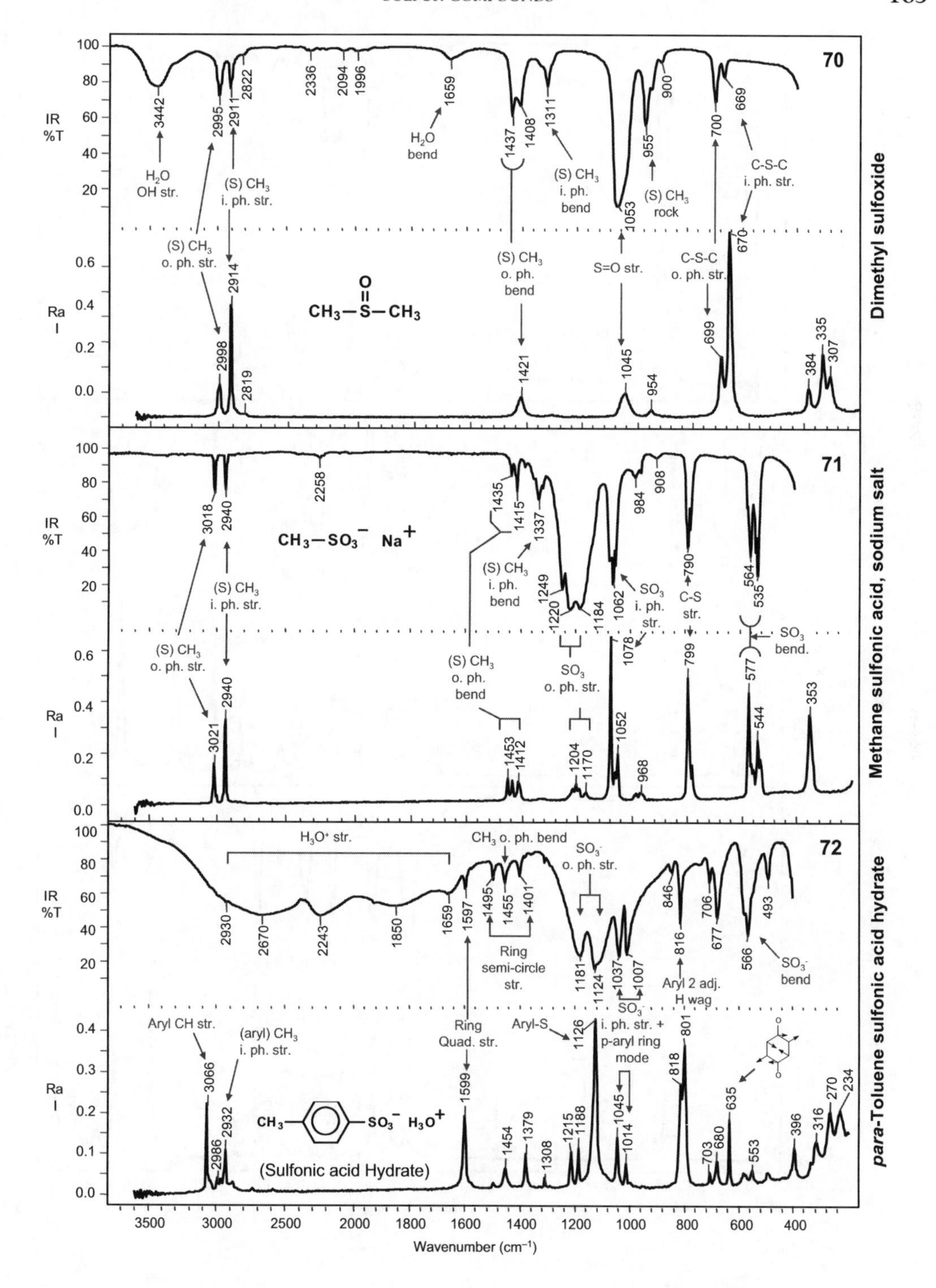
70
Dimethyl sulfoxide
$CH_3-S(=O)-CH_3$
H2O OH str.
(S) CH3 i. ph. str.
(S) CH3 o. ph. str.
H2O bend
(S) CH3 o. ph. bend
(S) CH3 i. ph. bend
S=O str.
(S) CH3 rock
C-S-C o. ph. str.
C-S-C i. ph. str.
71
Methane sulfonic acid, sodium salt
$CH_3-SO_3^-$ Na^+
(S) CH3 i. ph. str.
(S) CH3 o. ph. str.
(S) CH3 i. ph. bend
(S) CH3 o. ph. bend
SO3 o. ph. str.
SO3 i. ph. str.
C-S str.
SO3 bend.
72
para-Toluene sulfonic acid hydrate
$CH_3-C_6H_4-SO_3^-$ H_3O^+
(Sulfonic acid Hydrate)
H3O+ str.
CH3 o. ph. bend
SO3- o. ph. str.
Ring semi-circle str.
Aryl 2 adj. H wag
SO3- bend
Aryl CH str.
(aryl) CH3 i. ph. str.
Ring Quad. str.
Aryl-S
SO3- i. ph. str. + p-aryl ring mode
IR %T
Ra I
Wavenumber (cm−1)

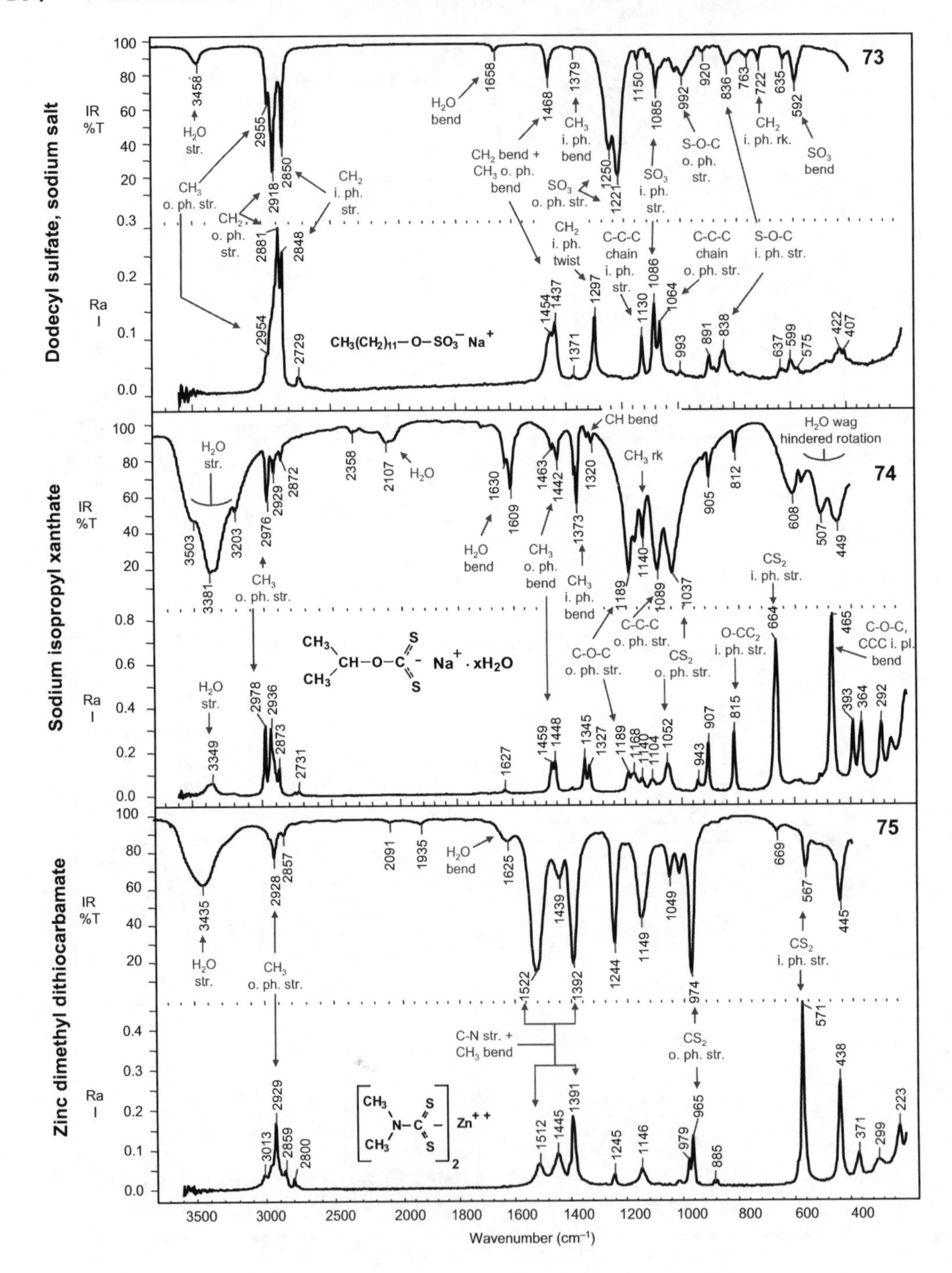

73
Dodecyl sulfate, sodium salt
IR %T
Ra I
$CH_3(CH_2)_{11}—O—SO_3^-\ Na^+$
3458
H_2O str.
2955
CH_3 o. ph. str.
2918
2850
CH_2 i. ph. str.
CH_2 o. ph. str.
2881
2848
2954
2729
1658
H_2O bend
1468
CH_2 bend + CH_3 o. ph. bend
1379
CH_3 i. ph. bend
1250
1221
SO_3 o. ph. str.
1150
1085
SO_3 i. ph. str.
992
S-O-C o. ph. str.
920
836
763
722
CH_2 i. ph. rk.
635
592
SO_3 bend
CH_2 i. ph. twist
C-C-C chain i. ph. str.
C-C-C chain o. ph. str.
S-O-C i. ph. str.
1454
1437
1371
1297
1130
1086
1064
993
891
838
637
599
575
422
407
74
Sodium isopropyl xanthate
H_2O str.
3503
3381
3203
2976
2929
2872
CH_3 o. ph. str.
2358
2107
H_2O
1630
1609
H_2O bend
1463
1442
CH_3 o. ph. bend
1373
CH_3 i. ph. bend
1320
CH bend
CH_3 rk
1189
1140
1089
1037
C-C-C o. ph. str.
C-O-C o. ph. str.
CS_2 o. ph. str.
905
812
O-CC_2 i. ph. str.
CS_2 i. ph. str.
664
608
H_2O wag hindered rotation
507
449
465
C-O-C, CCC i. pl. bend
3349
2978
2936
2873
2731
1627
1459
1448
1345
1327
1189
1168
1140
1104
1052
943
907
815
393
364
292
CH_3 CH-O-C(=S)-S$^-$ $Na^+ \cdot xH_2O$
75
Zinc dimethyl dithiocarbamate
3435
H_2O str.
2928
2857
CH_3 o. ph. str.
2091
1935
H_2O bend
1625
1522
1439
1392
1244
1149
1049
974
669
567
445
CS_2 i. ph. str.
C-N str. + CH_3 bend
CS_2 o. ph. str.
571
438
3013
2929
2859
2800
1512
1445
1391
1245
1146
979
965
885
371
299
223
$[(CH_3)_2N-C(=S)-S^-]_2\ Zn^{++}$
Wavenumber (cm^{-1})

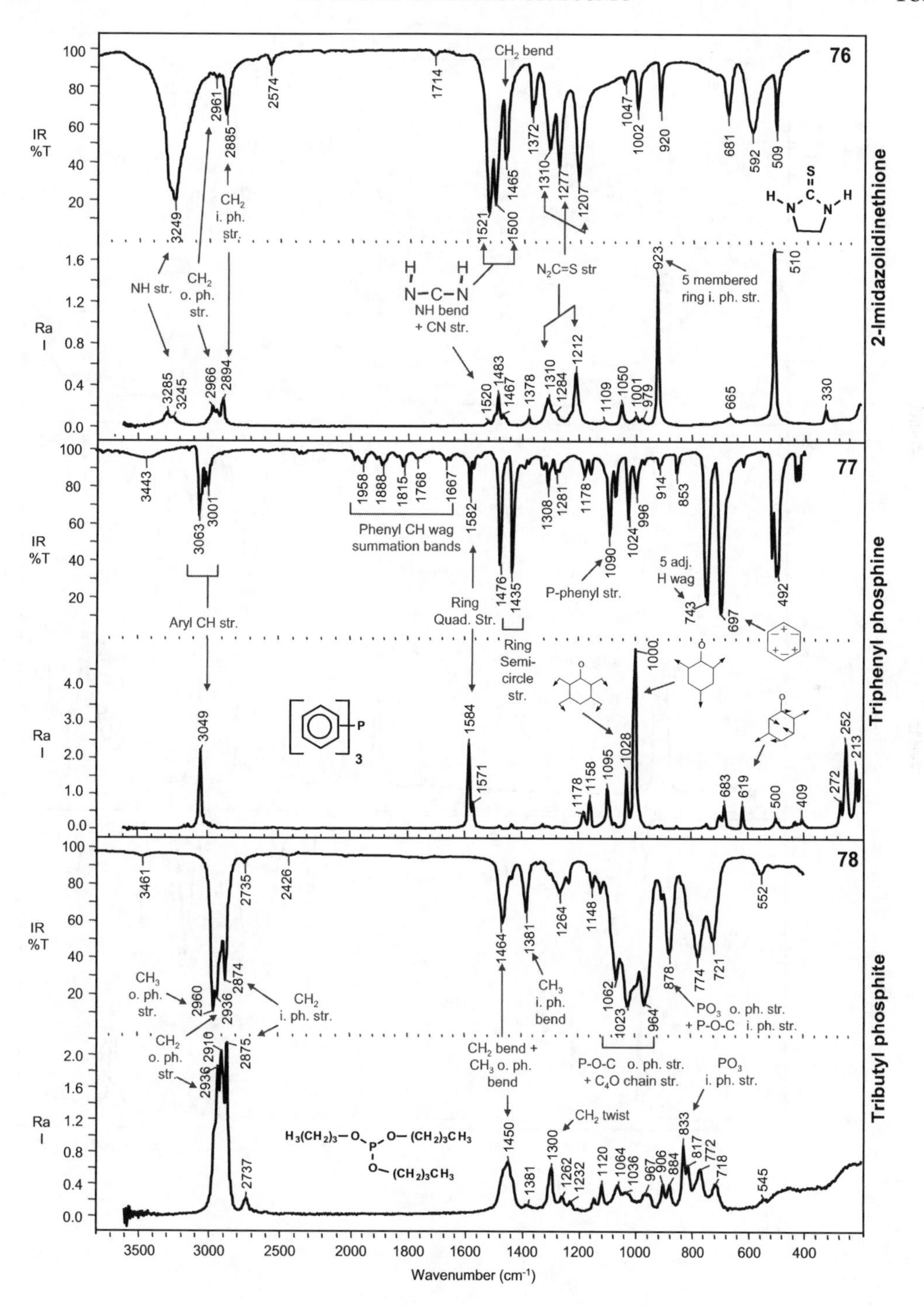
76
2-Imidazolidinethione
77
Triphenyl phosphine
78
Tributyl phosphite
IR %T
Ra I
Wavenumber (cm-1)
CH2 bend
NH str.
CH2 o. ph. str.
CH2 i. ph. str.
NH bend + CN str.
N2C=S str
5 membered ring i. ph. str.
Phenyl CH wag summation bands
Aryl CH str.
Ring Quad. Str.
Ring Semi-circle str.
P-phenyl str.
5 adj. H wag
CH3 o. ph. str.
CH3 i. ph. bend
CH2 bend + CH3 o. ph. bend
P-O-C o. ph. str. + C4O chain str.
PO3 o. ph. str. + P-O-C i. ph. str.
PO3 i. ph. str.
CH2 twist
H3(CH2)3—O P O—(CH2)3CH3 O—(CH2)3CH3

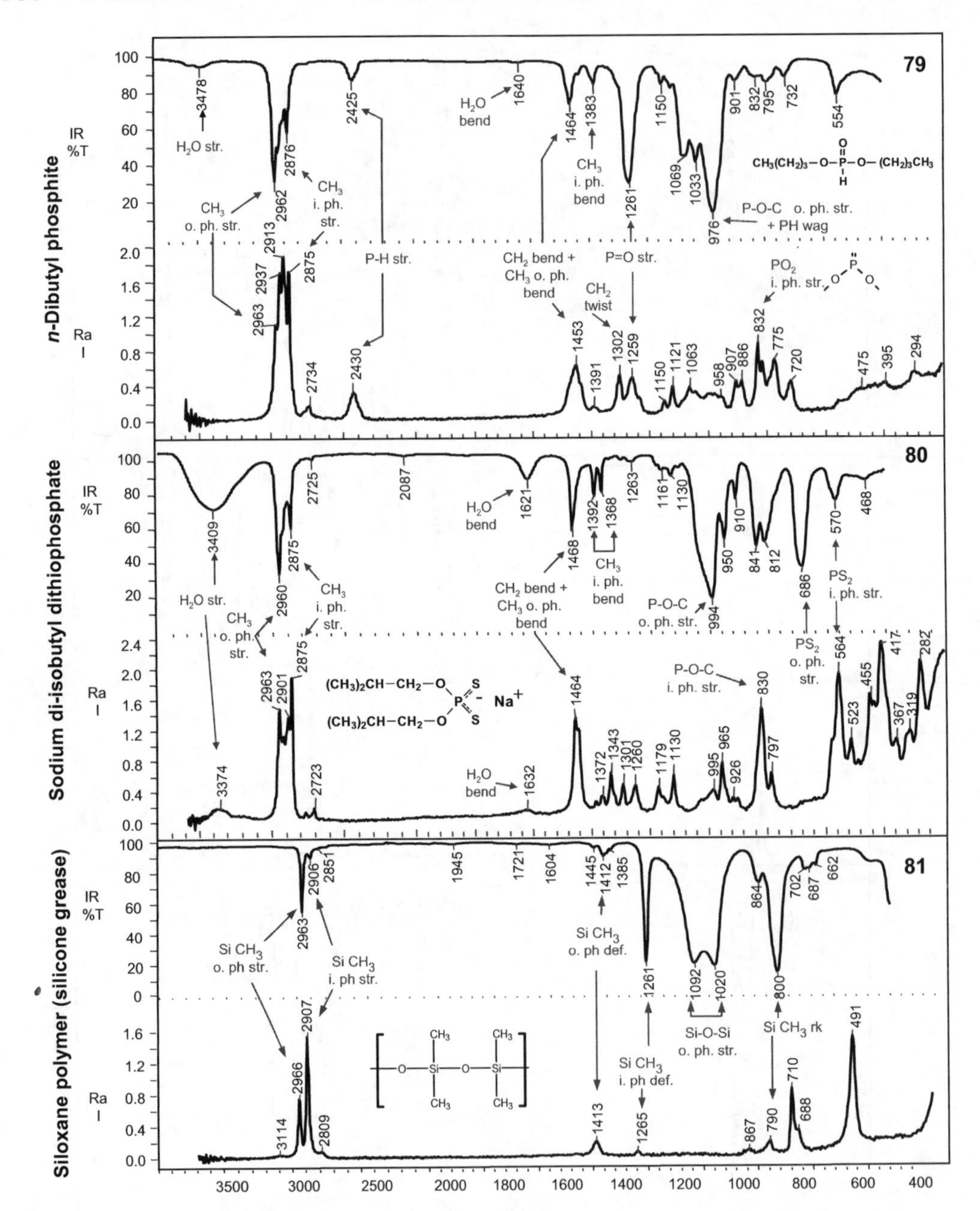
n-Dibutyl phosphite
79
IR %T
Ra I
H2O str.
CH3 o. ph. str.
CH3 i. ph. str.
P-H str.
H2O bend
CH2 bend + CH3 o. ph. bend
CH3 i. ph. bend
CH2 twist
P=O str.
P-O-C o. ph. str. + PH wag
PO2 i. ph. str.
CH3(CH2)3—O—P—O—(CH2)3CH3
Sodium di-isobutyl dithiophosphate
80
H2O str.
CH3 o. ph. str.
CH3 i. ph. str.
H2O bend
CH2 bend + CH3 o. ph. bend
CH3 i. ph. bend
P-O-C o. ph. str.
P-O-C i. ph. str.
PS2 i. ph. str.
PS2 o. ph. str.
(CH3)2CH—CH2—O
Na+
Siloxane polymer (silicone grease)
81
Si CH3 o. ph str.
Si CH3 i. ph str.
Si CH3 o. ph def.
Si CH3 i. ph def.
Si-O-Si o. ph. str.
Si CH3 rk
3500 3000 2500 2000 1800 1600 1400 1200 1000 800 600 400

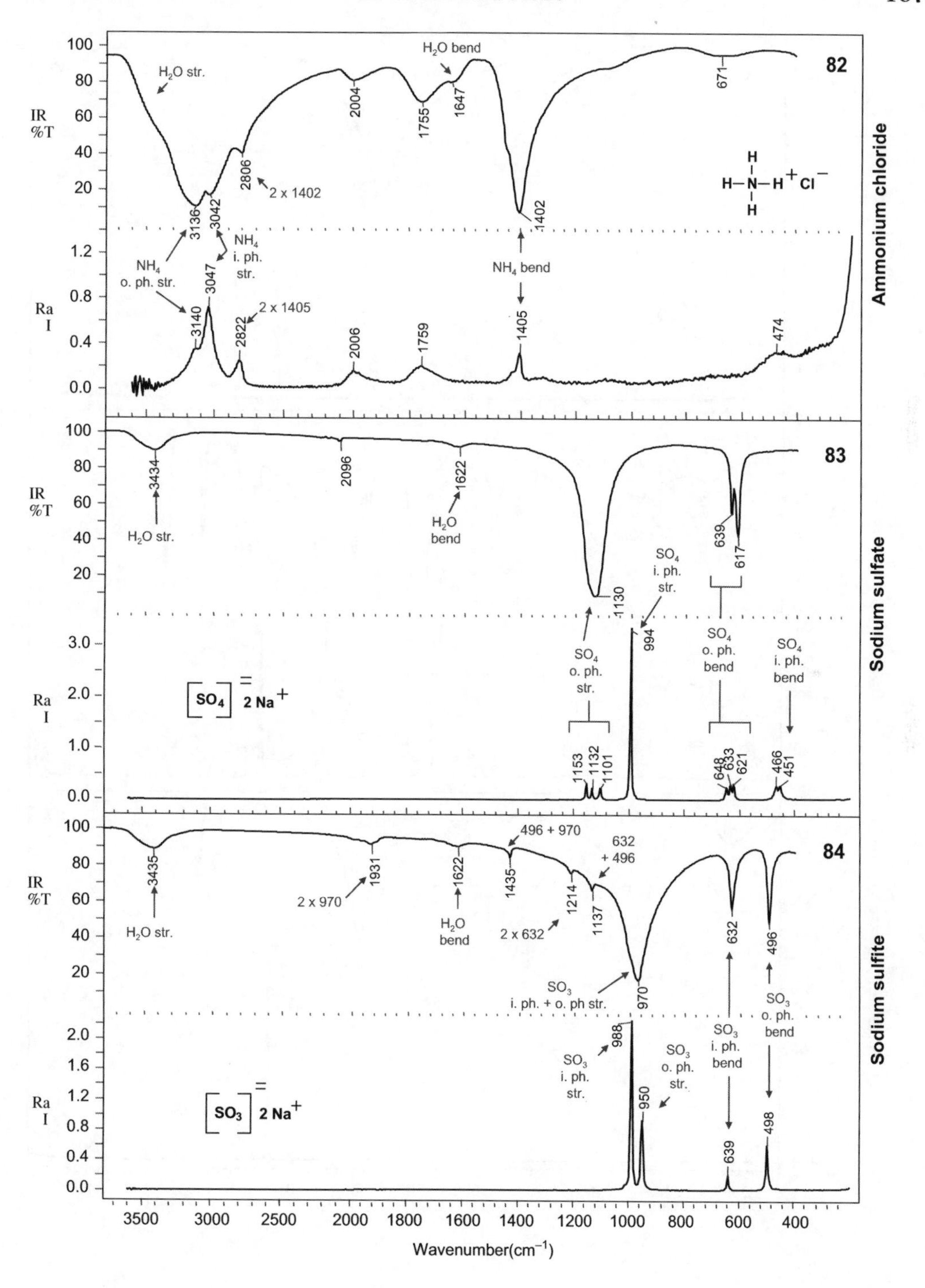
82
Ammonium chloride
IR %T
Ra I
H$_2$O str.
H$_2$O bend
2 x 1402
NH$_4$ o. ph. str.
NH$_4$ i. ph. str.
NH$_4$ bend
2 x 1405
3136
3042
2806
2004
1755
1647
1402
671
3140
3047
2822
2006
1759
1405
474
83
Sodium sulfate
3434
2096
1622
1130
639
617
H$_2$O str.
H$_2$O bend
SO$_4$ i. ph. str.
SO$_4$ o. ph. str.
SO$_4$ o. ph. bend
SO$_4$ i. ph. bend
994
1153
1132
1101
648
633
621
466
451
[SO$_4$]= 2 Na+
84
Sodium sulfite
3435
1931
1622
1435
1214
1137
970
632
496
H$_2$O str.
2 x 970
H$_2$O bend
2 x 632
496 + 970
632 + 496
SO$_3$ i. ph. + o. ph str.
SO$_3$ o. ph. bend
SO$_3$ i. ph. bend
SO$_3$ i. ph. str.
SO$_3$ o. ph. str.
988
950
639
498
[SO$_3$]= 2 Na+
Wavenumber(cm^{-1})

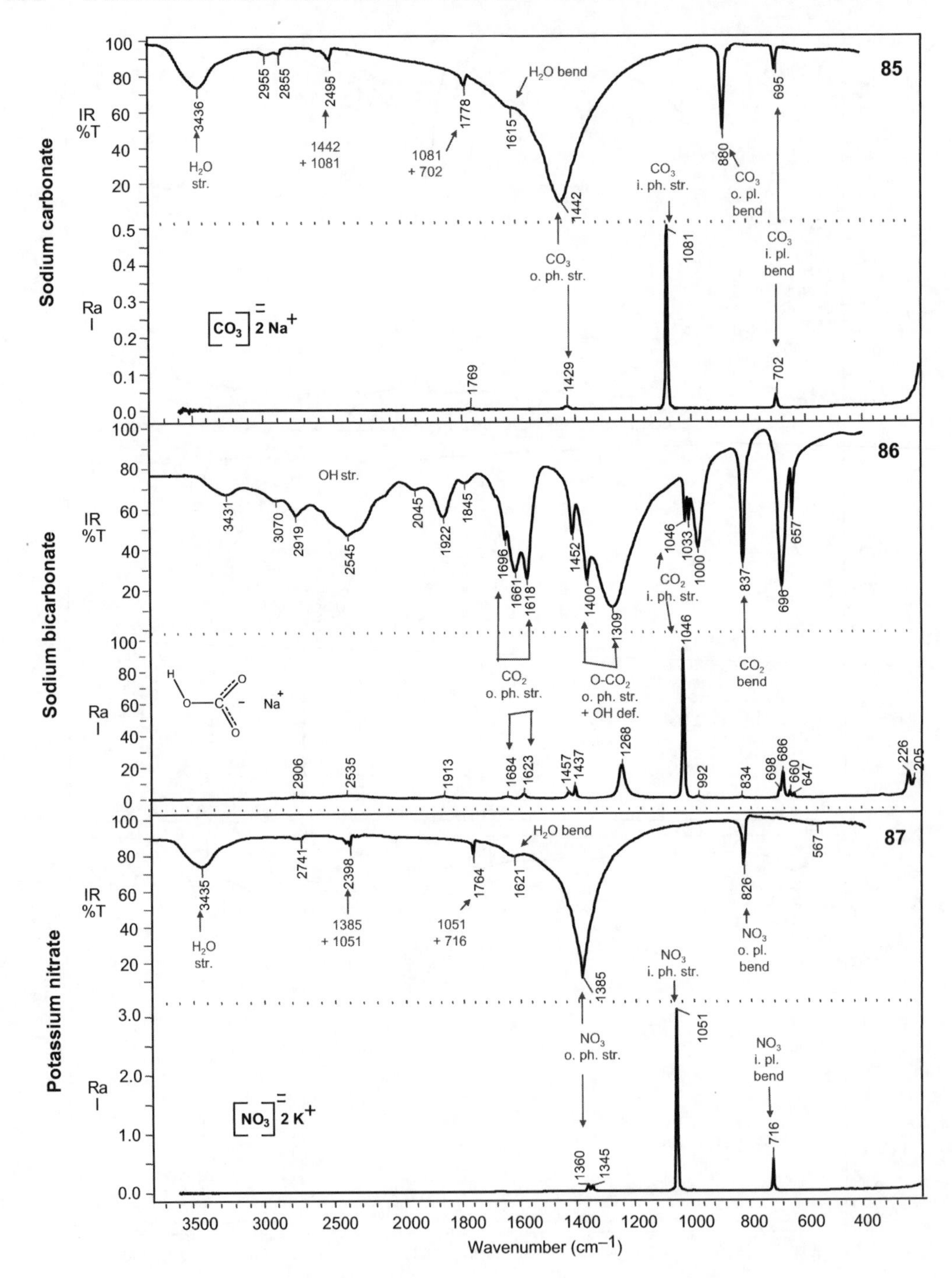

Sodium carbonate
85
IR %T
Ra I
H2O str.
3436
2955
2855
2495
1442 + 1081
1778
1081 + 702
1615
H2O bend
1442
CO3 o. ph. str.
CO3 i. ph. str.
880
CO3 o. pl. bend
695
CO3 i. pl. bend
1081
1769
1429
702
[CO3]= 2 Na+
Sodium bicarbonate
86
OH str.
3431
3070
2919
2545
2045
1922
1845
1696
1661
1618
1452
1400
1309
1046
1033
1000
CO2 i. ph. str.
837
698
657
CO2 o. ph. str.
O-CO2 o. ph. str. + OH def.
CO2 bend
Na+
2906
2535
1913
1684
1623
1457
1437
1268
992
834
698
686
660
647
226
205
Potassium nitrate
87
H2O str.
3435
2741
2398
1385 + 1051
1764
1051 + 716
1621
H2O bend
1385
NO3 o. ph. str.
NO3 i. ph. str.
826
NO3 o. pl. bend
567
1051
NO3 i. pl. bend
716
1360
1345
[NO3]= 2 K+
Wavenumber (cm−1)

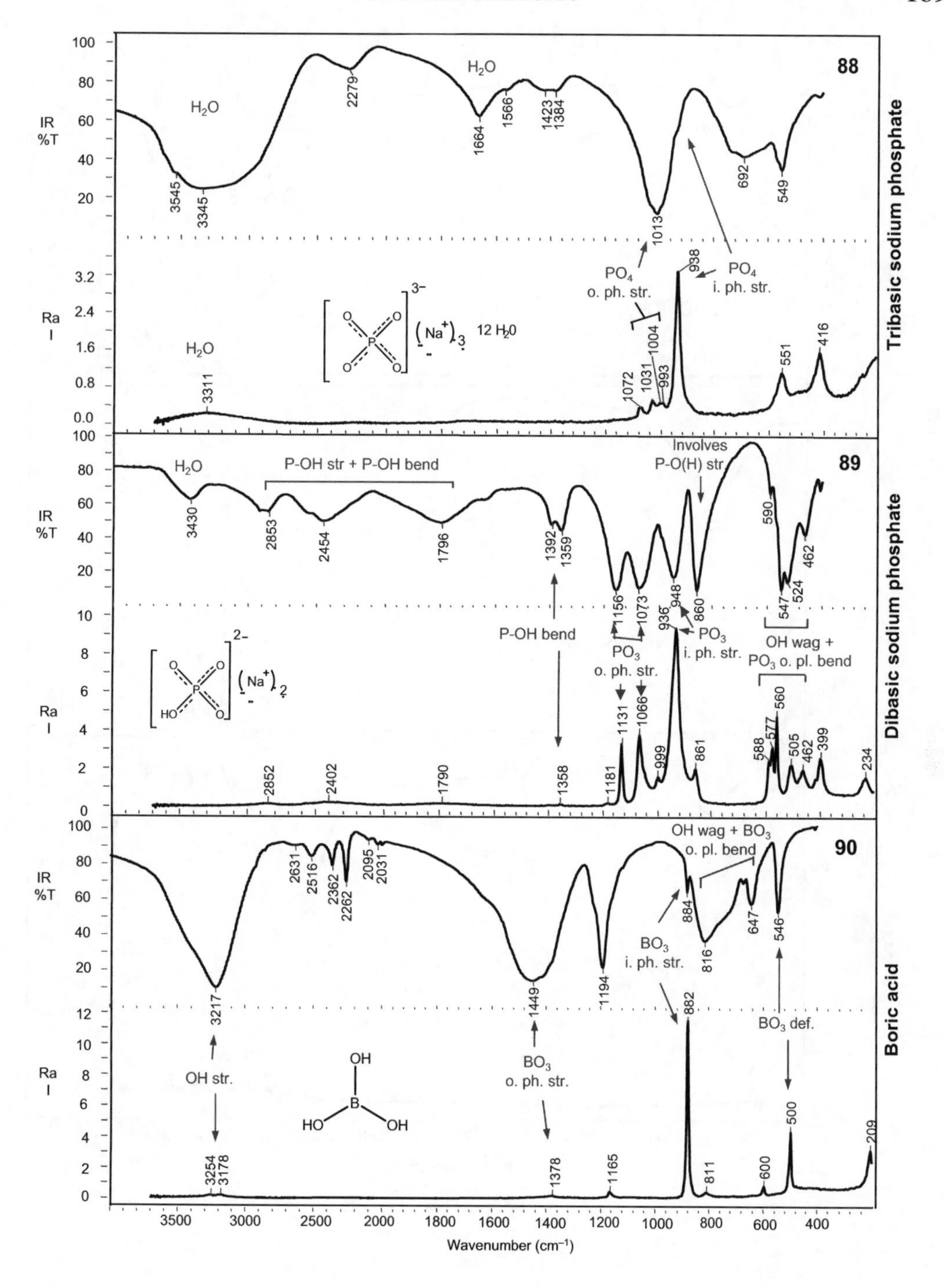
88
Tribasic sodium phosphate
IR %T
Ra I
H2O
PO4 o. ph. str.
PO4 i. ph. str.
12 H2O
89
Dibasic sodium phosphate
P-OH str + P-OH bend
Involves P-O(H) str.
P-OH bend
PO3 o. ph. str.
PO3 i. ph. str.
OH wag + PO3 o. pl. bend
90
Boric acid
OH wag + BO3 o. pl. bend
BO3 i. ph. str.
OH str.
BO3 o. ph. str.
BO3 def.
Wavenumber (cm−1)

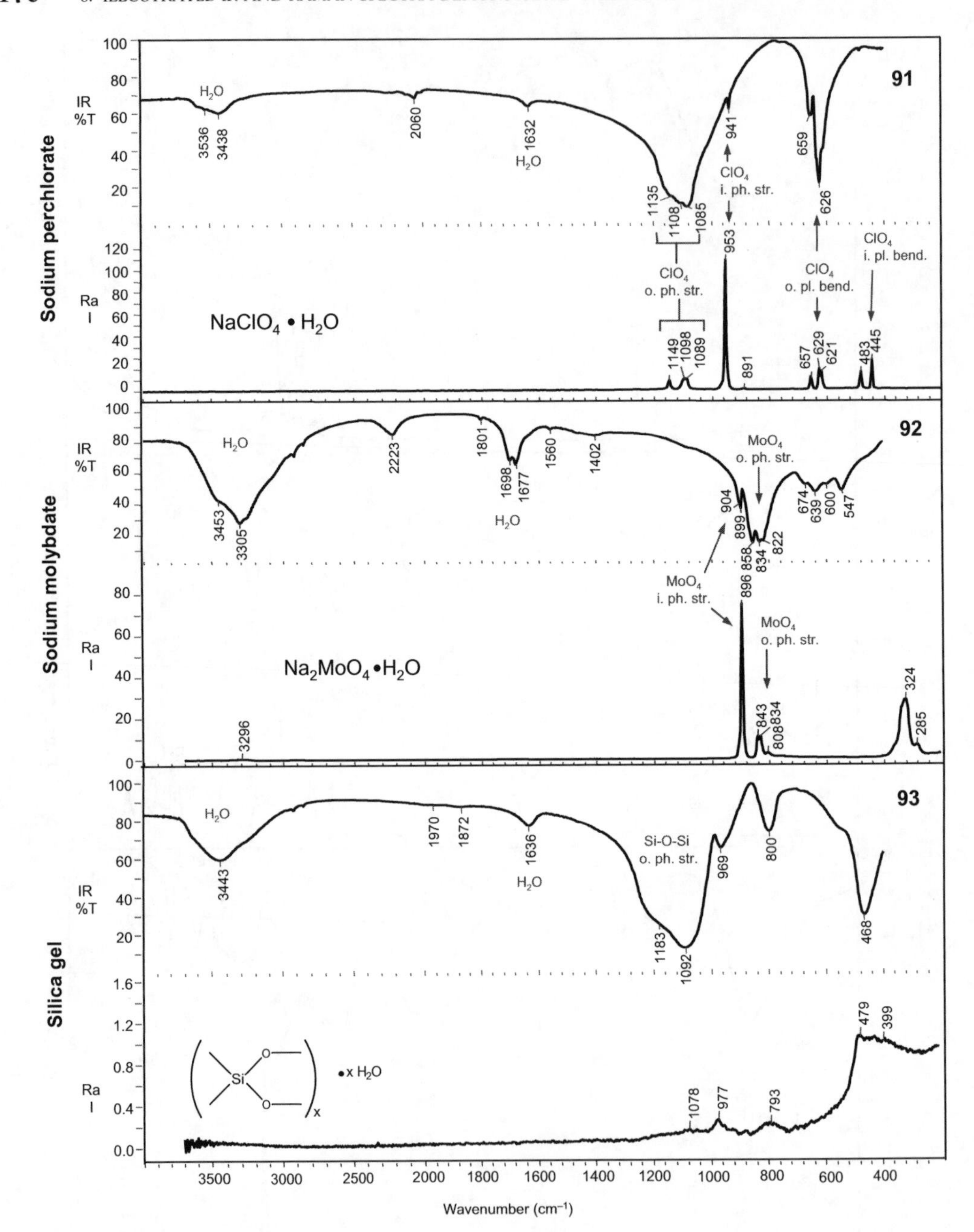
Sodium perchlorate
91
$NaClO_4 \cdot H_2O$
ClO4 i. ph. str.
ClO4 o. ph. str.
ClO4 o. pl. bend.
ClO4 i. pl. bend.
Sodium molybdate
92
$Na_2MoO_4 \cdot H_2O$
MoO4 o. ph. str.
MoO4 i. ph. str.
Silica gel
93
Si-O-Si o. ph. str.
Wavenumber (cm−1)

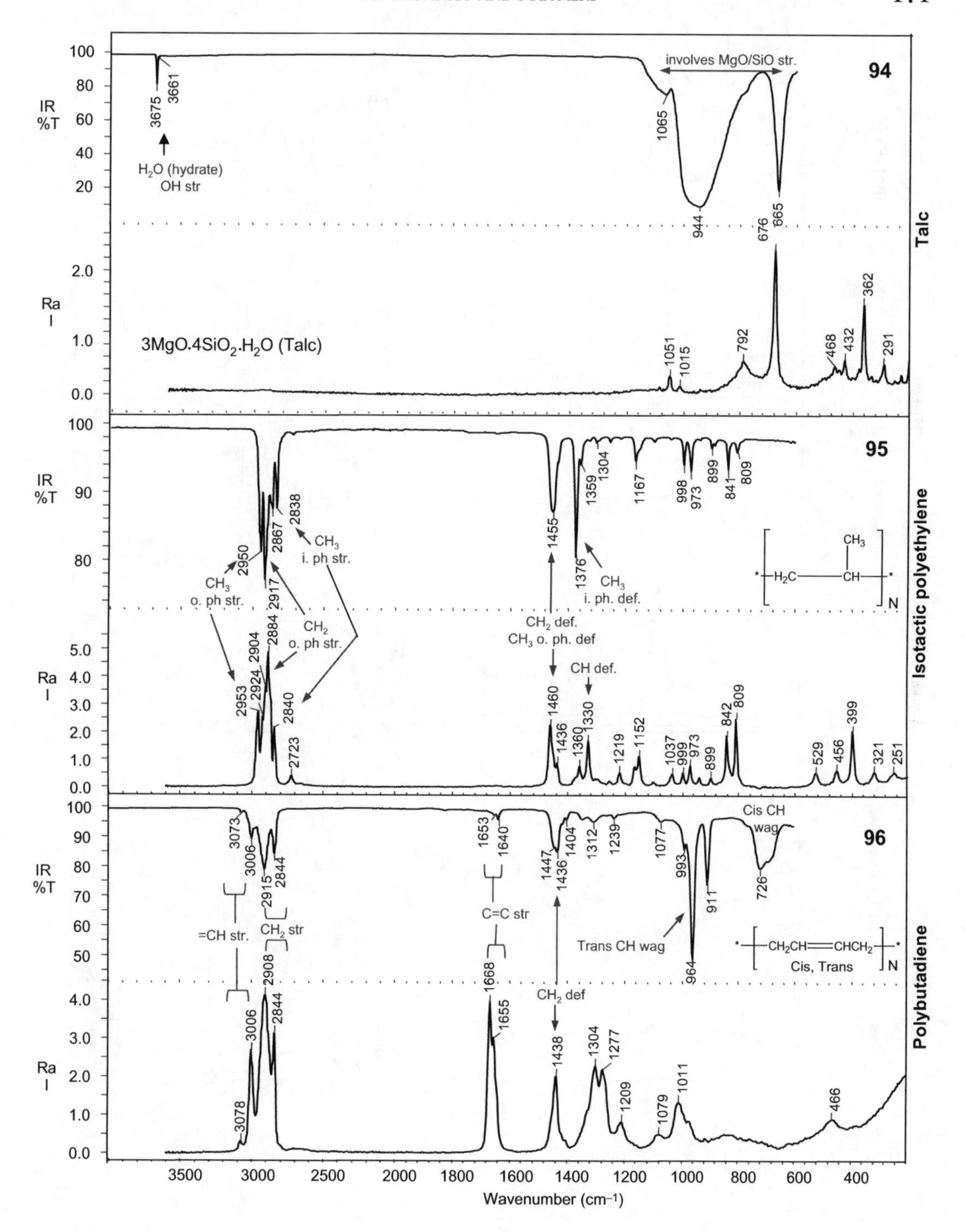

94
Talc
IR %T
Ra I
3675
3661
H2O (hydrate) OH str
involves MgO/SiO str.
1065
944
676
665
3MgO.4SiO2.H2O (Talc)
1051
1015
792
468
432
362
291
95
Isotactic polyethylene
2950
2917
2867
2838
CH3 i. ph str.
CH3 o. ph str.
CH2 o. ph str.
1455
1376
1359
1304
1167
998
973
899
841
809
CH3 i. ph. def.
CH2 def. CH3 o. ph. def
CH def.
2953
2924
2904
2884
2840
2723
1460
1436
1360
1330
1219
1152
1037
999
973
899
842
809
529
456
399
321
251
96
Polybutadiene
3073
3006
2915
2844
=CH str.
CH2 str
1653
1640
C=C str
1447
1436
1404
1312
1239
1077
993
911
964
726
Cis CH wag
Trans CH wag
Cis, Trans
3078
3006
2908
2844
1668
1655
CH2 def
1438
1304
1277
1209
1079
1011
466
Wavenumber (cm−1)

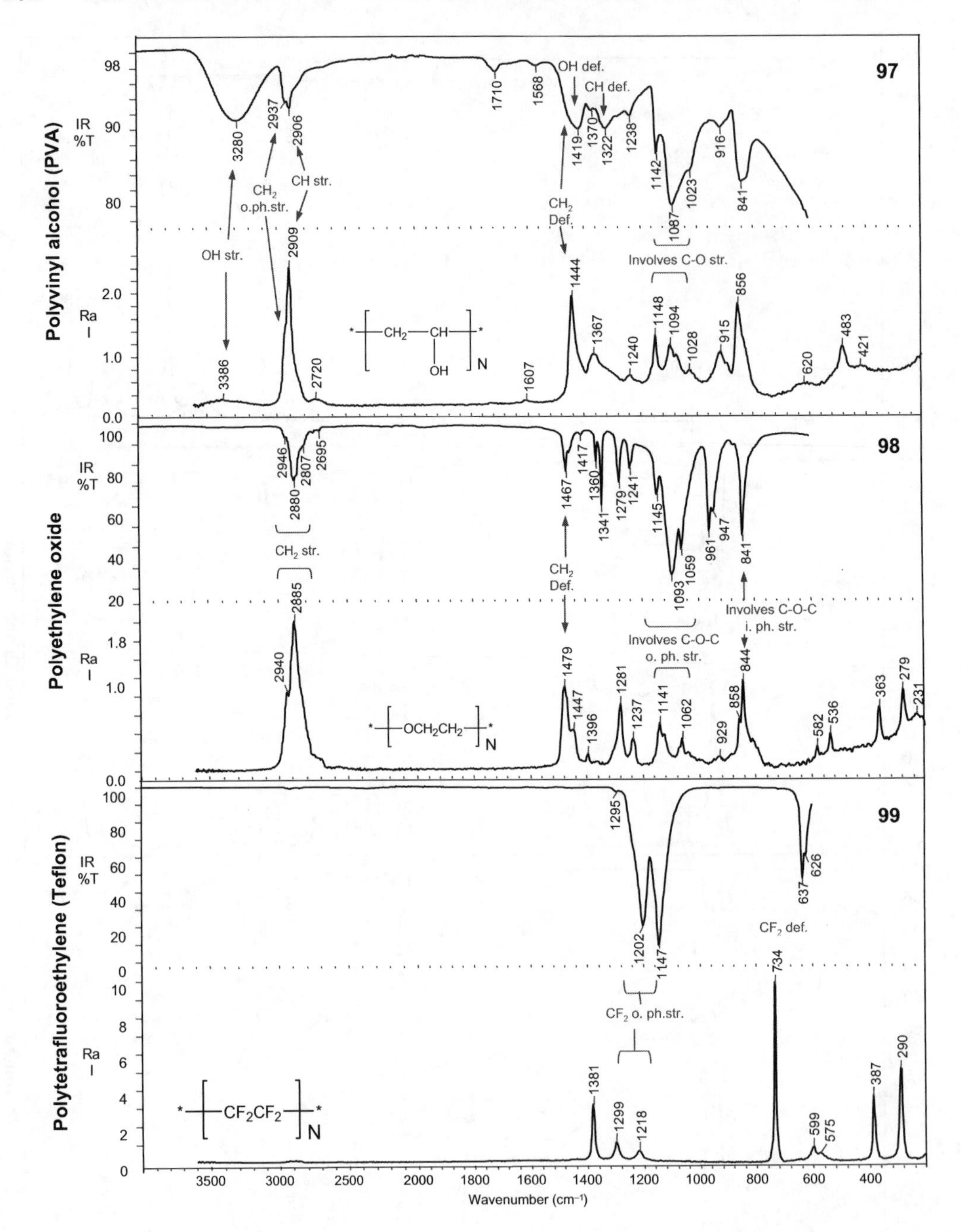
97
Polyvinyl alcohol (PVA)
IR %T
Ra I
OH str.
CH2 o.ph.str.
CH str.
OH def.
CH def.
CH2 Def.
Involves C-O str.
98
Polyethylene oxide
CH2 str.
CH2 Def.
Involves C-O-C o. ph. str.
Involves C-O-C i. ph. str.
99
Polytetrafluoroethylene (Teflon)
CF2 o. ph.str.
CF2 def.
Wavenumber (cm−1)

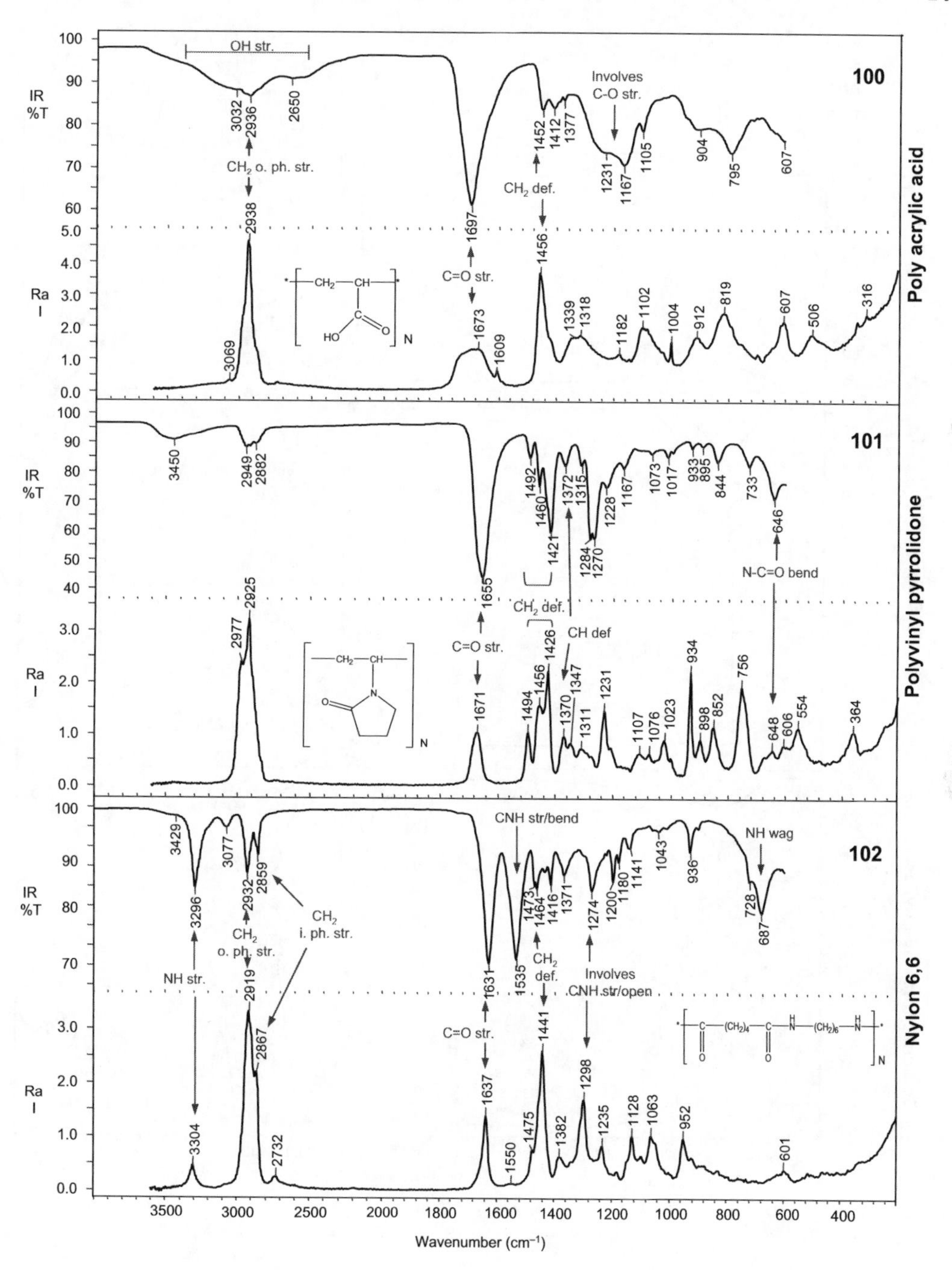
100
Poly acrylic acid
101
Polyvinyl pyrrolidone
102
Nylon 6,6
OH str.
Involves C-O str.
CH2 o. ph. str.
CH2 def.
C=O str.
N-C=O bend
CH def
CNH str/bend
NH wag
NH str.
CH2 i. ph. str.
Involves CNH str/open
IR %T
Ra I
Wavenumber (cm−1)

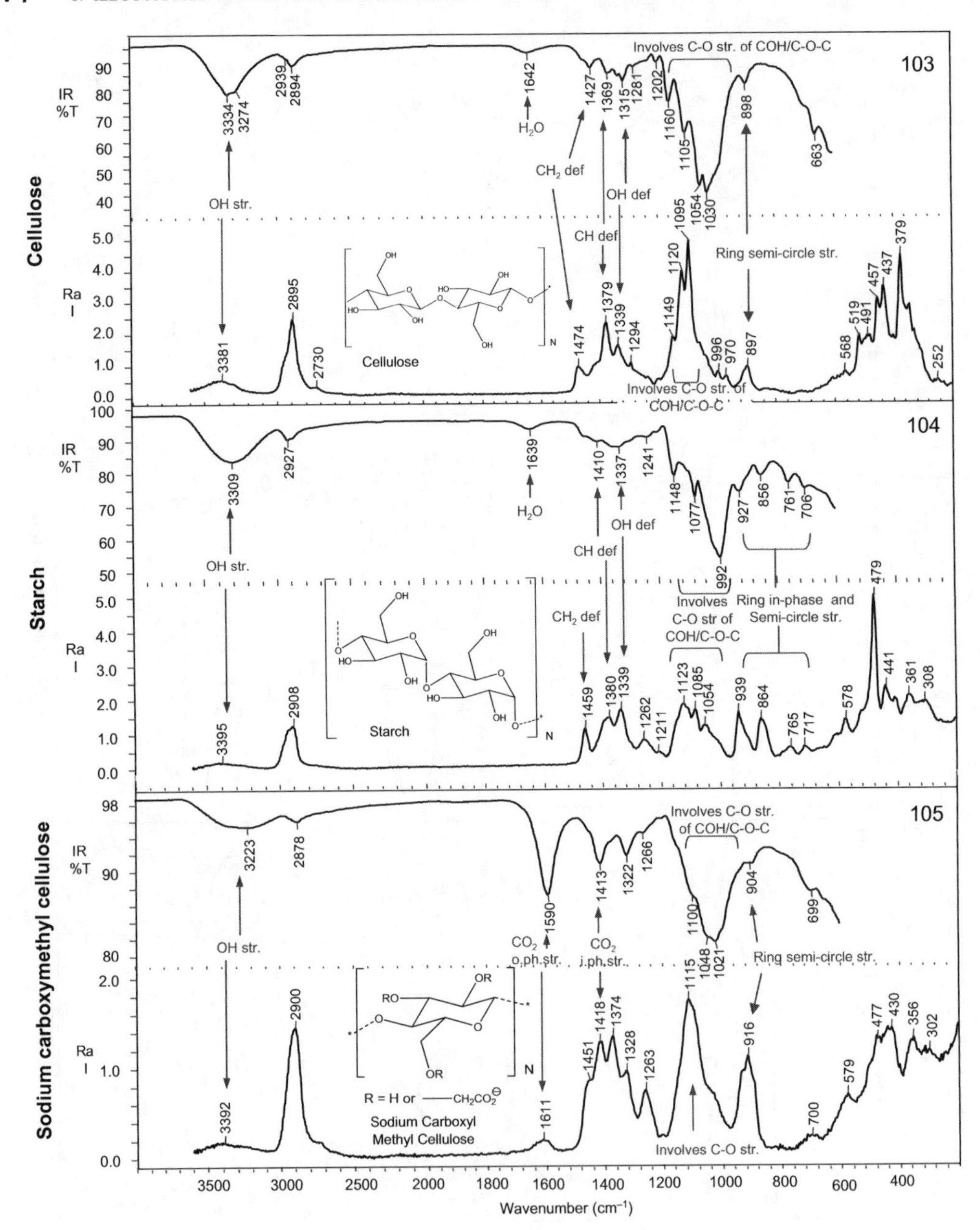
Cellulose
Starch
Sodium carboxymethyl cellulose
103
104
105
IR %T
Ra I
OH str.
CH def
OH def
CH_2 def
H_2O
Involves C-O str. of COH/C-O-C
Ring semi-circle str.
Ring in-phase and Semi-circle str.
CO_2 o.ph.str.
CO_2 i.ph.str.
Involves C-O str.
R = H or —$CH_2CO_2^⊖$
Sodium Carboxyl Methyl Cellulose
Wavenumber (cm^{-1})

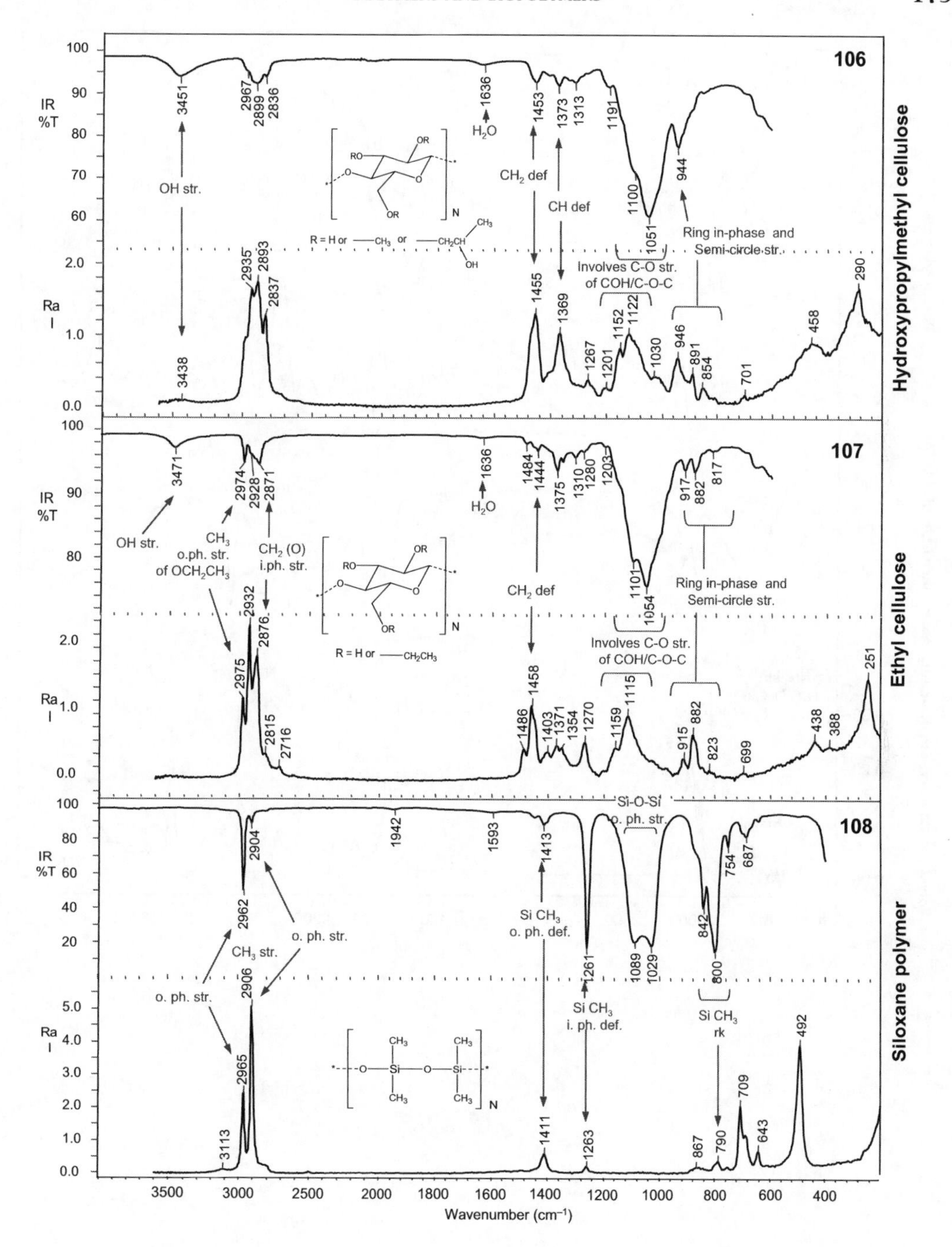
106
Hydroxypropylmethyl cellulose
OH str.
CH2 def
CH def
H2O
Ring in-phase and Semi-circle str.
Involves C-O str. of COH/C-O-C
R = H or —CH3 or —CH2CH(CH3)OH
107
Ethyl cellulose
OH str.
CH3 o.ph. str. of OCH2CH3
CH2 (O) i.ph. str.
CH2 def
Ring in-phase and Semi-circle str.
Involves C-O str. of COH/C-O-C
R = H or —CH2CH3
108
Siloxane polymer
Si-O-Si o. ph. str.
CH3 str.
o. ph. str.
Si CH3 o. ph. def.
Si CH3 i. ph. def.
Si CH3 rk
IR %T
Ra I
Wavenumber (cm−1)

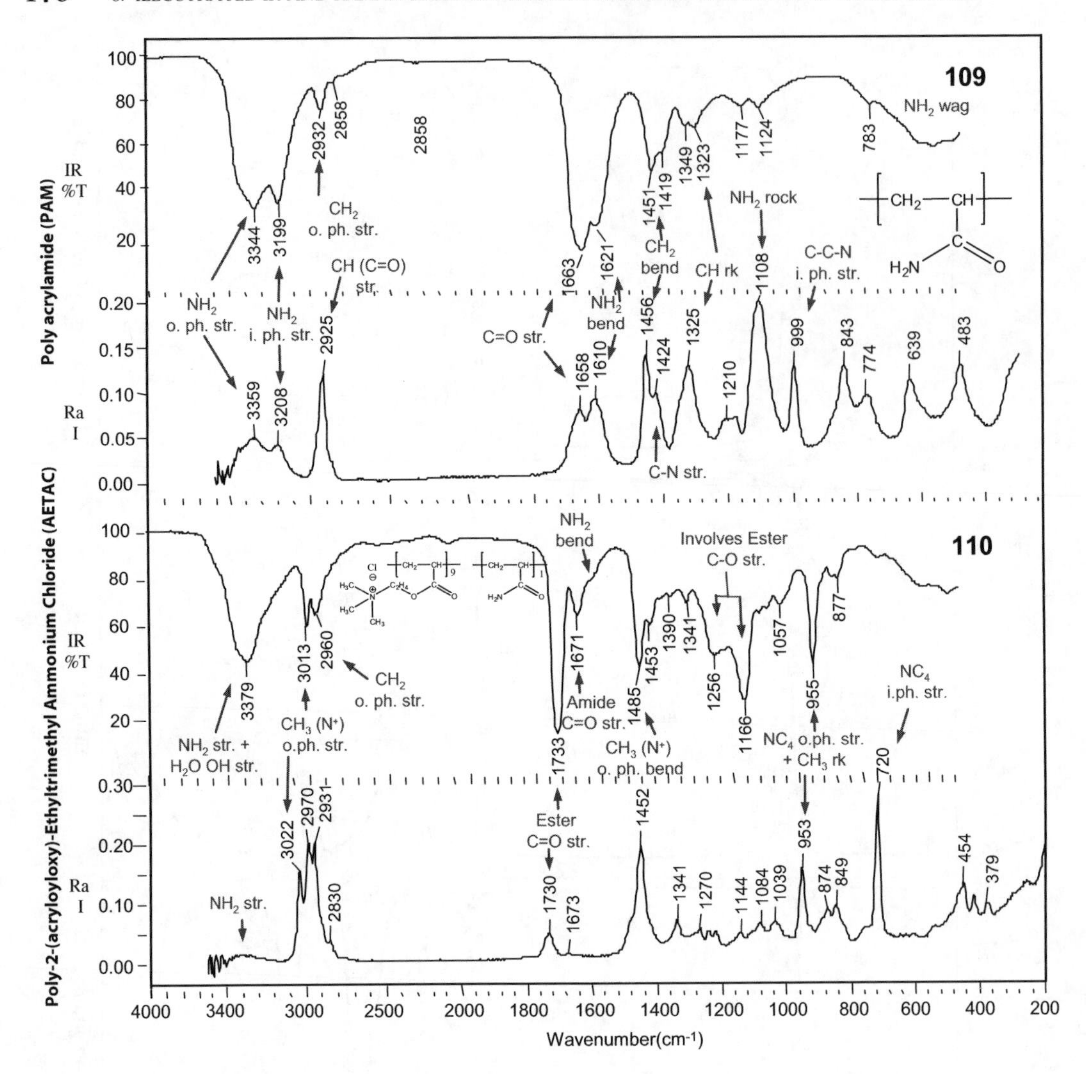
109
Poly acrylamide (PAM)
IR %T
Ra I
NH2 wag
NH2 rock
C-C-N i. ph. str.
CH2 bend
CH rk
NH2 bend
C=O str.
C-N str.
CH2 o. ph. str.
CH (C=O) str.
NH2 o. ph. str.
NH2 i. ph. str.
3344
3199
2932
2858
1663
1621
1451
1419
1349
1323
1177
1124
1108
783
3359
3208
2925
1658
1610
1456
1424
1325
1210
999
843
774
639
483
110
Poly-2-(acryloyloxy)-Ethyltrimethyl Ammonium Chloride (AETAC)
NH2 bend
Involves Ester C-O str.
NC4 i.ph. str.
NH2 str. + H2O OH str.
CH3 (N+) o.ph. str.
CH2 o. ph. str.
Amide C=O str.
CH3 (N+) o. ph. bend
NC4 o.ph. str. + CH3 rk
Ester C=O str.
NH2 str.
3379
3013
2960
1733
1671
1485
1453
1390
1341
1256
1166
1057
955
877
720
3022
2970
2931
2830
1730
1673
1452
1341
1270
1144
1084
1039
953
874
849
454
379
Wavenumber(cm-1)

CHAPTER

9

Unknown IR and Raman Spectra

On the following pages, 44 unknown spectra are presented without labels or interpretation to help test and develop the readers knowledge and understanding of functional group analysis and spectral band assignment. The spectra are found on pages 178 to 199.

The sample preparation used to acquire the spectra is provided and where appropriate hints are included to help guide the interpretation. For interpretation and an explanation of the key features useful to help identify each compound an answer key is included at the end of the unknown spectra starting on page 200.

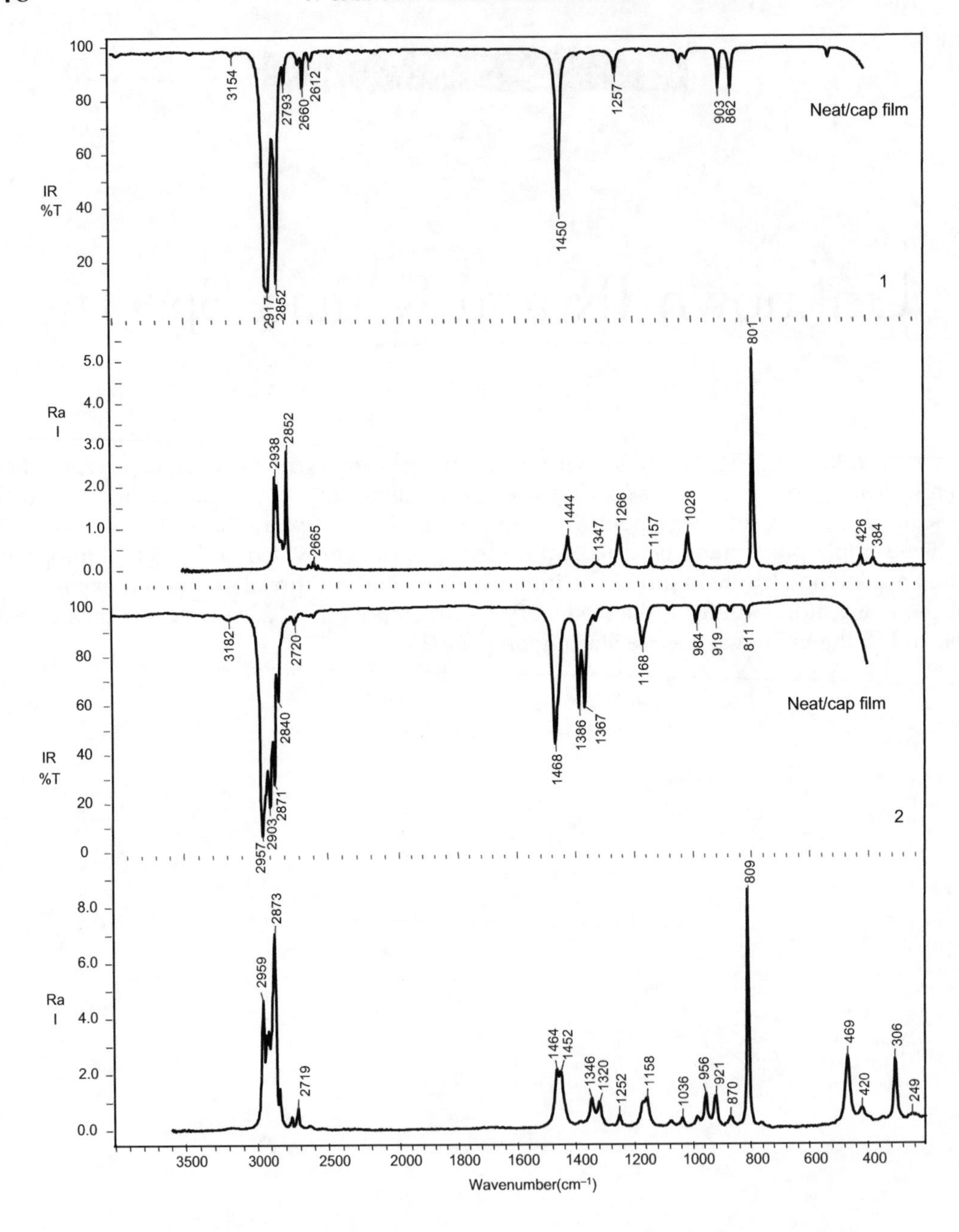

100
80
60
IR
%T
40
20
3154
2917
2852
2793
2660
2612
1450
1257
903
862
Neat/cap film
1
5.0
4.0
Ra
I
3.0
2.0
1.0
0.0
2938
2852
2665
1444
1347
1266
1157
1028
801
426
384
3182
2957
2903
2871
2840
2720
1468
1386
1367
1168
984
919
811
Neat/cap film
2
0
8.0
6.0
2959
2873
2719
1464
1452
1346
1320
1252
1158
1036
956
921
870
809
469
420
306
249
3500
3000
2500
2000
1800
1600
1400
1200
1000
800
600
400
Wavenumber(cm−1)

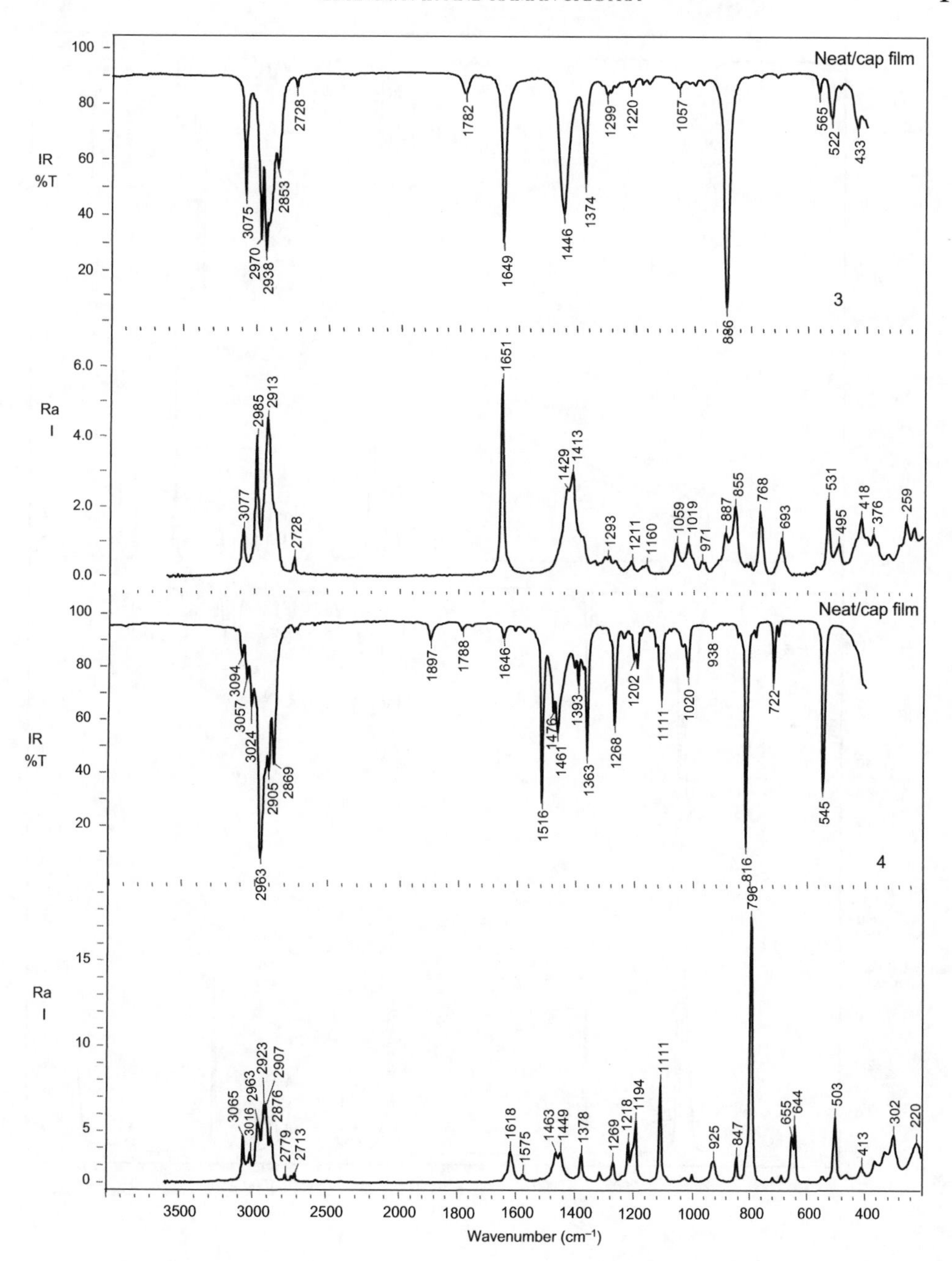
Neat/cap film
IR %T
3075
2970
2938
2853
2728
1782
1649
1446
1374
1299
1220
1057
886
565
522
433
3
Ra I
3077
2985
2913
2728
1651
1429
1413
1293
1211
1160
1059
1019
971
887
855
768
693
531
495
418
376
259
Neat/cap film
IR %T
3094
3057
3024
2963
2905
2869
1897
1788
1646
1516
1476
1461
1393
1363
1268
1202
1111
1020
938
816
722
545
4
Ra I
3065
3016
2963
2923
2907
2876
2779
2713
1618
1575
1463
1449
1378
1269
1218
1194
1111
925
847
796
655
644
503
413
302
220
Wavenumber (cm−1)

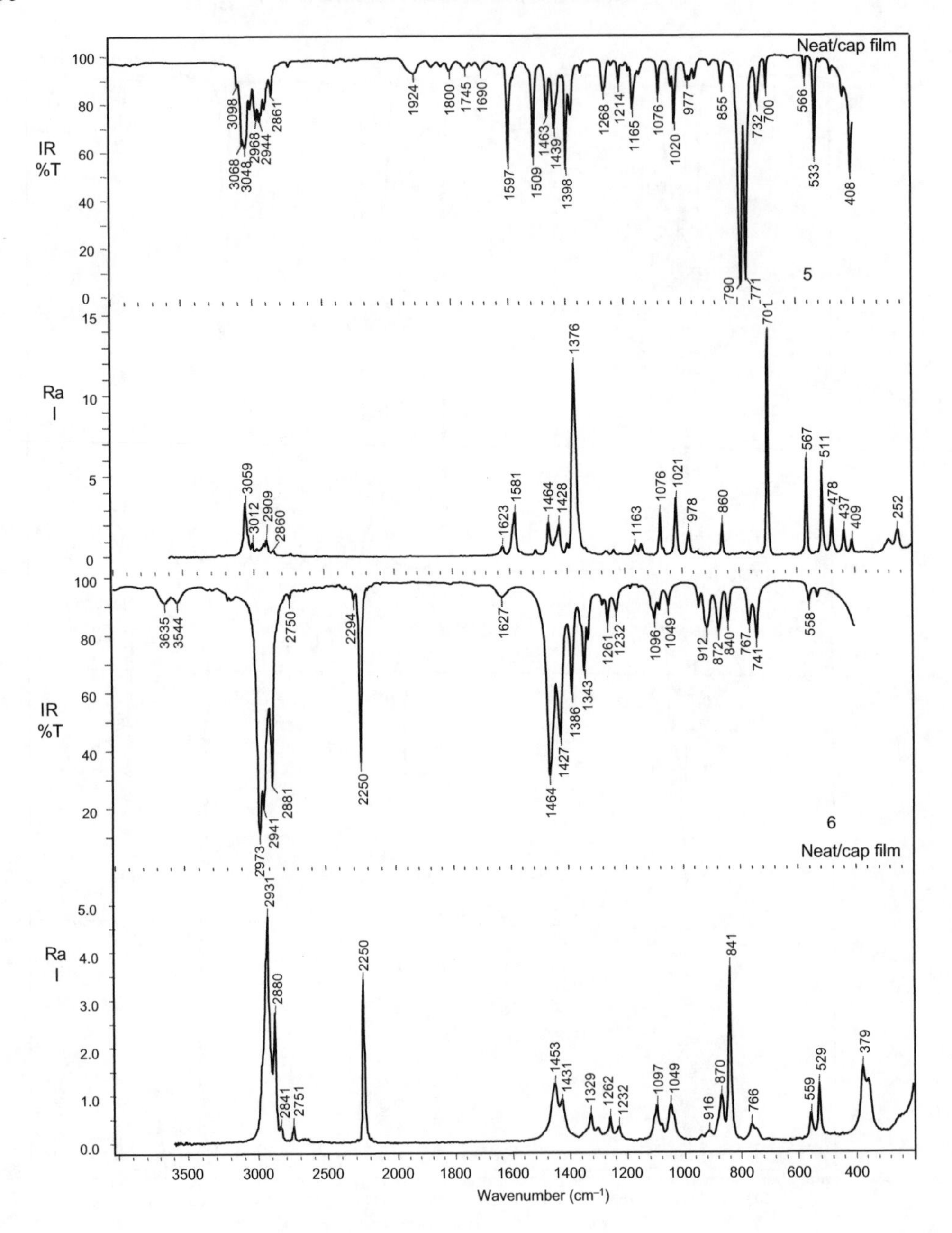

Neat/cap film
IR %T
Ra I
5
6
Neat/cap film
Wavenumber (cm⁻¹)

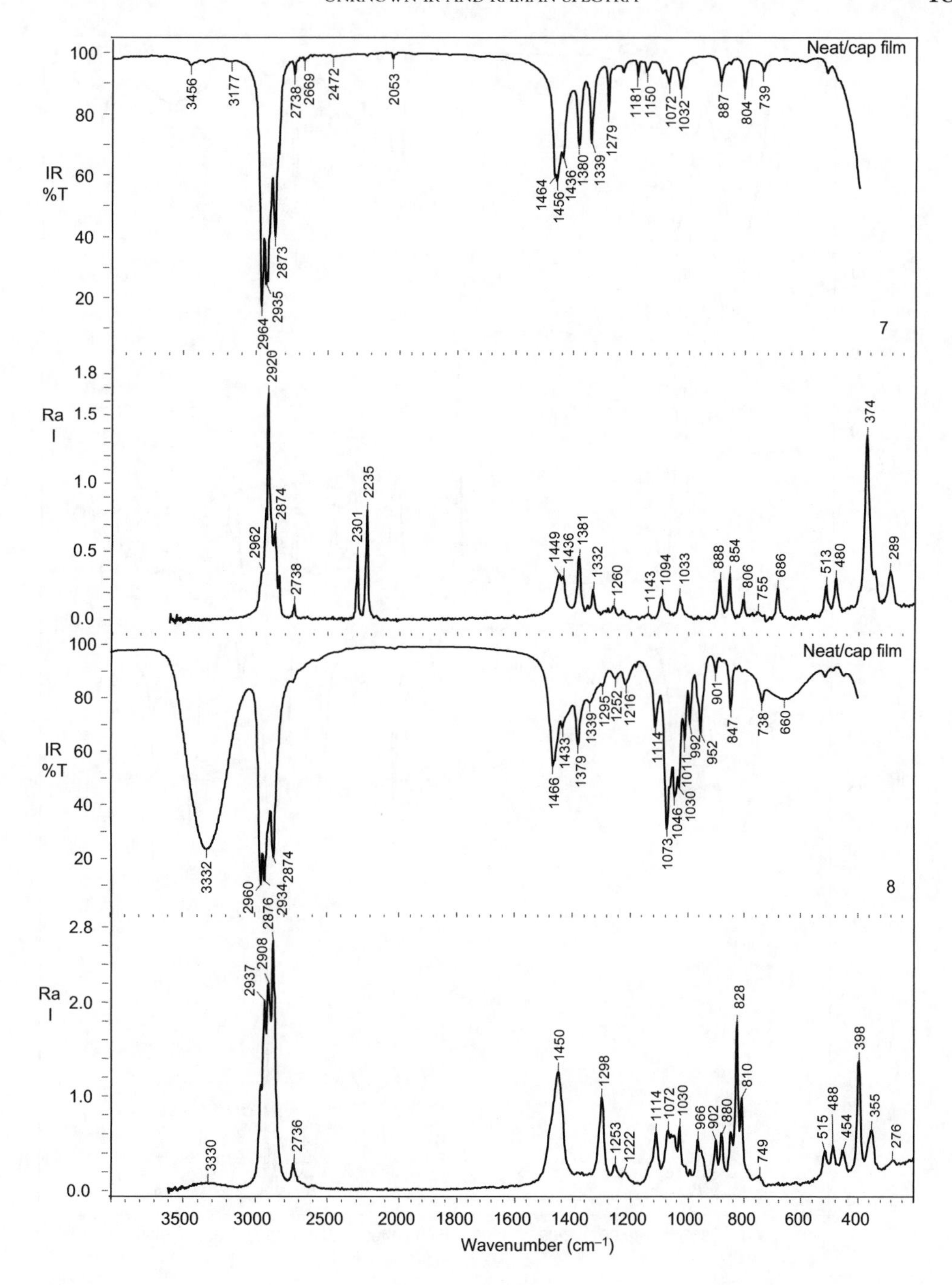
Neat/cap film
IR %T
Ra I
7
8
Wavenumber (cm−1)

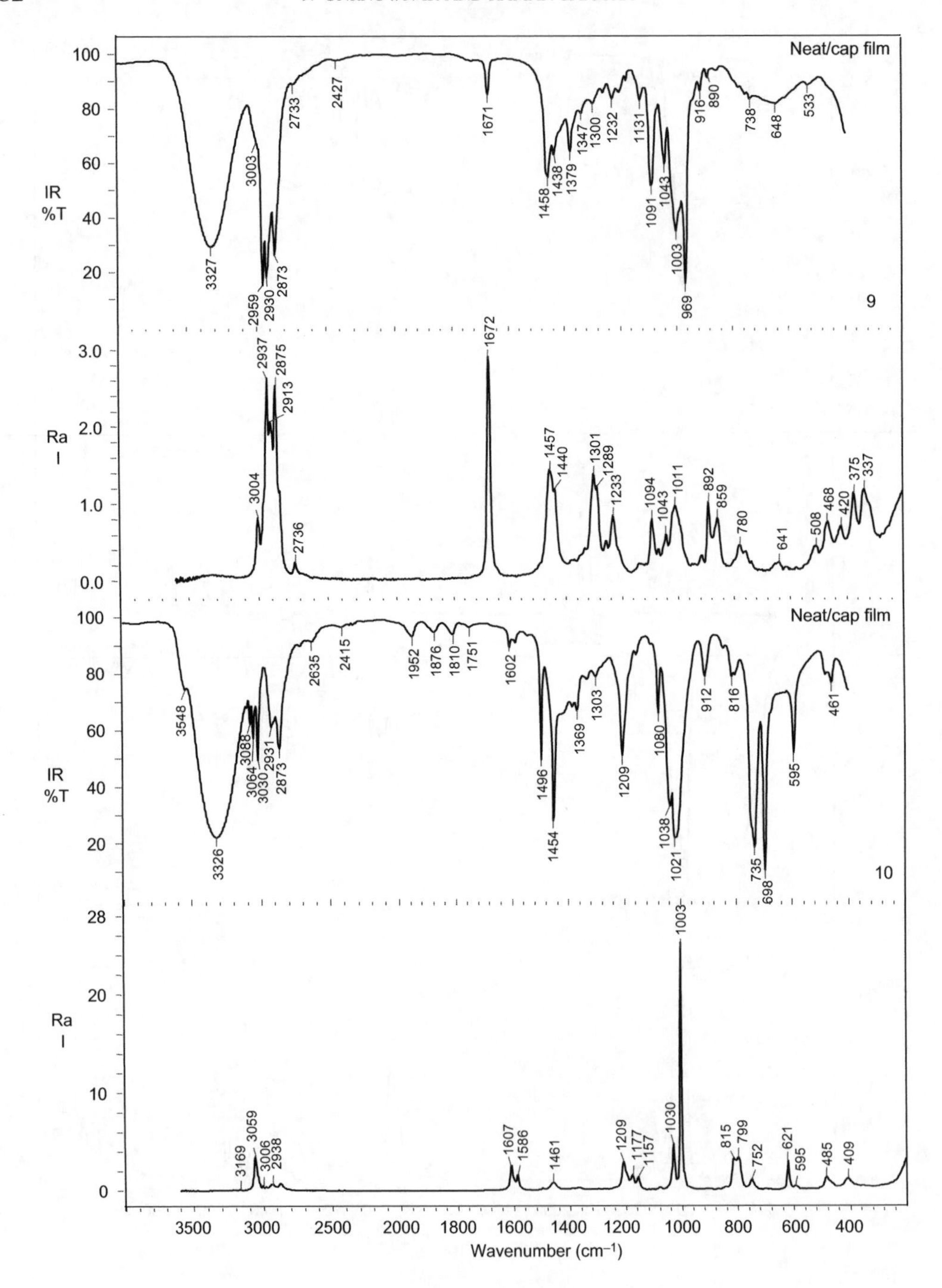

Neat/cap film
IR %T
Ra I
9
10
Wavenumber (cm−1)

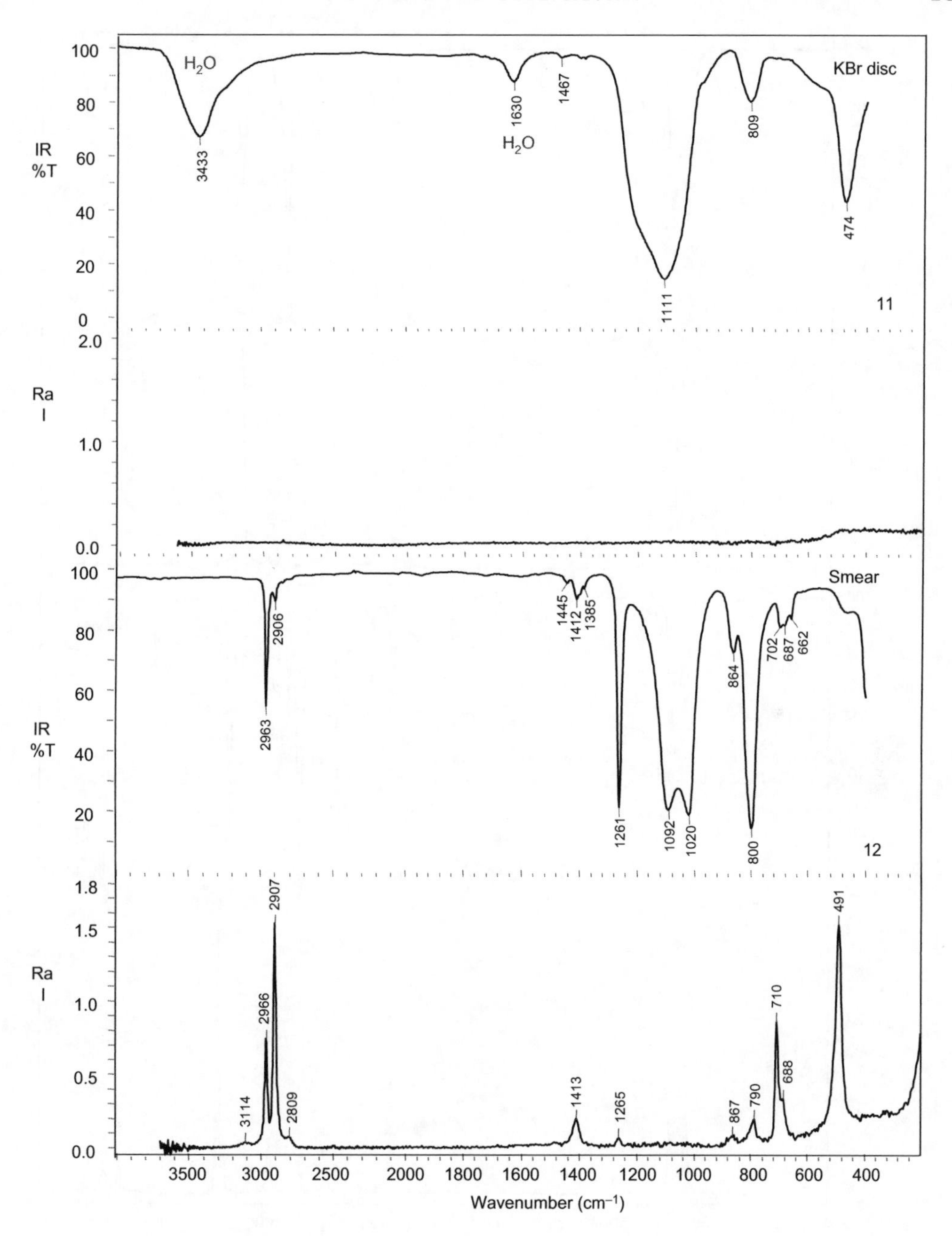
IR
%T
100
80
60
40
20
0
H_2O
3433
1630
H_2O
1467
1111
809
474
KBr disc
11
Ra
I
2.0
1.0
0.0
IR
%T
100
80
60
40
20
2963
2906
1445
1412
1385
1261
1092
1020
864
800
702
687
662
Smear
12
Ra
I
1.8
1.5
1.0
0.5
0.0
3114
2966
2907
2809
1413
1265
867
790
710
688
491
3500
3000
2500
2000
1800
1600
1400
1200
1000
800
600
400
Wavenumber (cm^{-1})

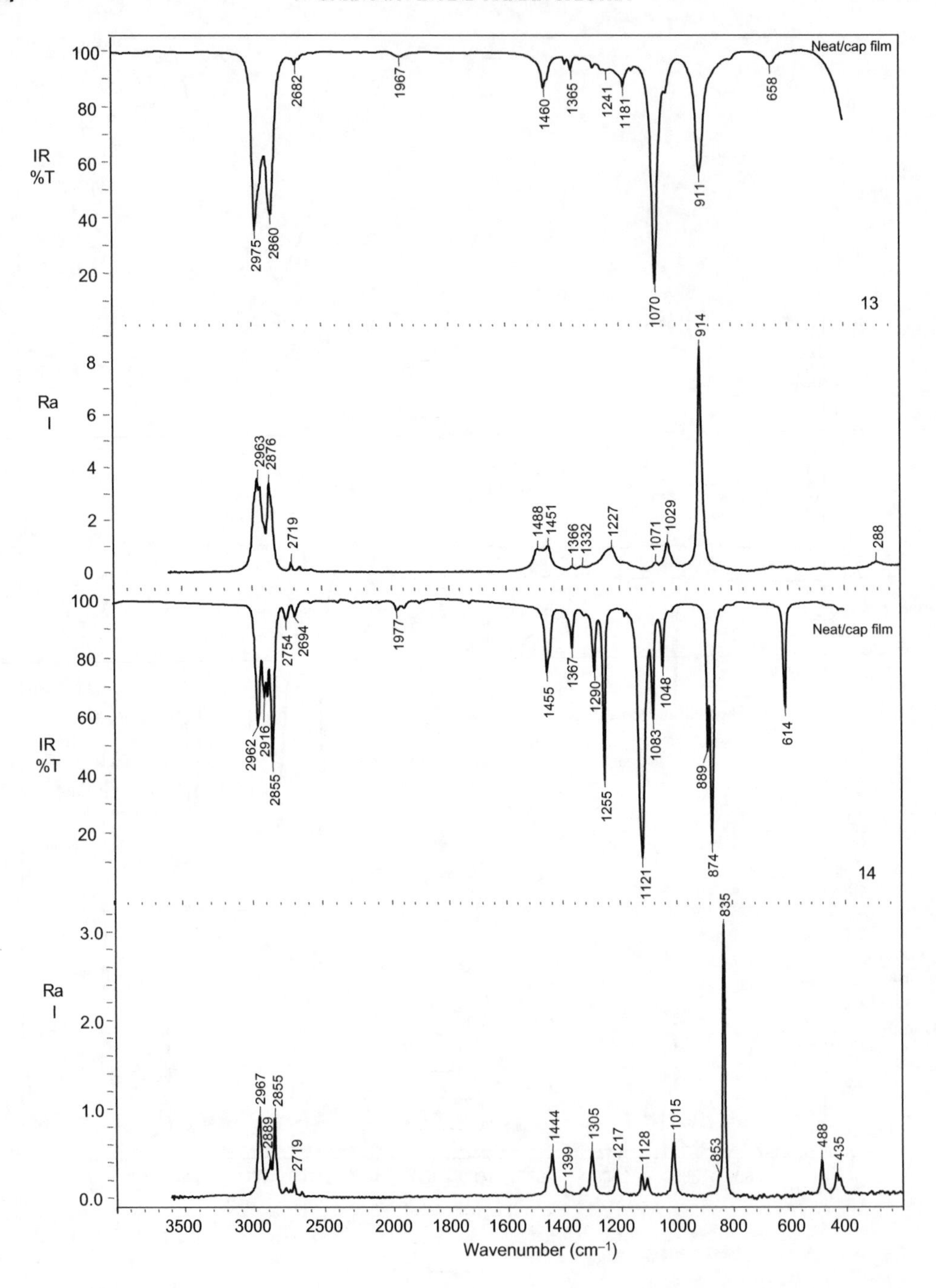

Neat/cap film
IR %T
100
80
60
40
20
2975
2860
2682
1967
1460
1365
1241
1181
1070
911
658
13
Ra I
8
6
4
2
0
2963
2876
2719
1488
1451
1366
1332
1227
1071
1029
914
288
Neat/cap film
2962
2916
2855
2754
2694
1977
1455
1367
1290
1255
1121
1083
1048
889
874
614
14
3.0
2.0
1.0
0.0
2967
2889
2855
2719
1444
1399
1305
1217
1128
1015
853
835
488
435
3500
3000
2500
2000
1800
1600
1400
1200
1000
800
600
400
Wavenumber (cm−1)

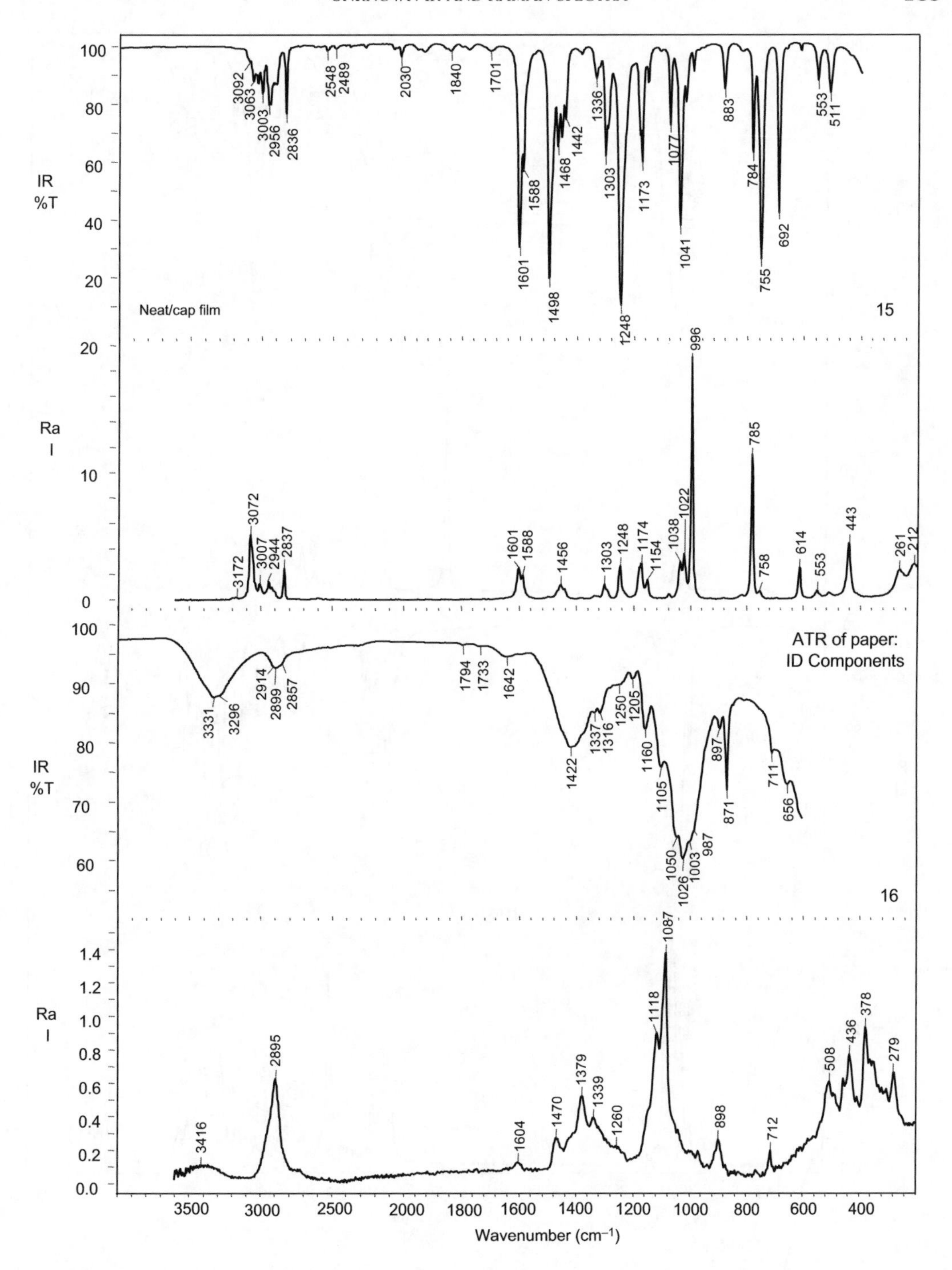
IR %T
Ra I
Neat/cap film
15
ATR of paper: ID Components
16
Wavenumber (cm−1)

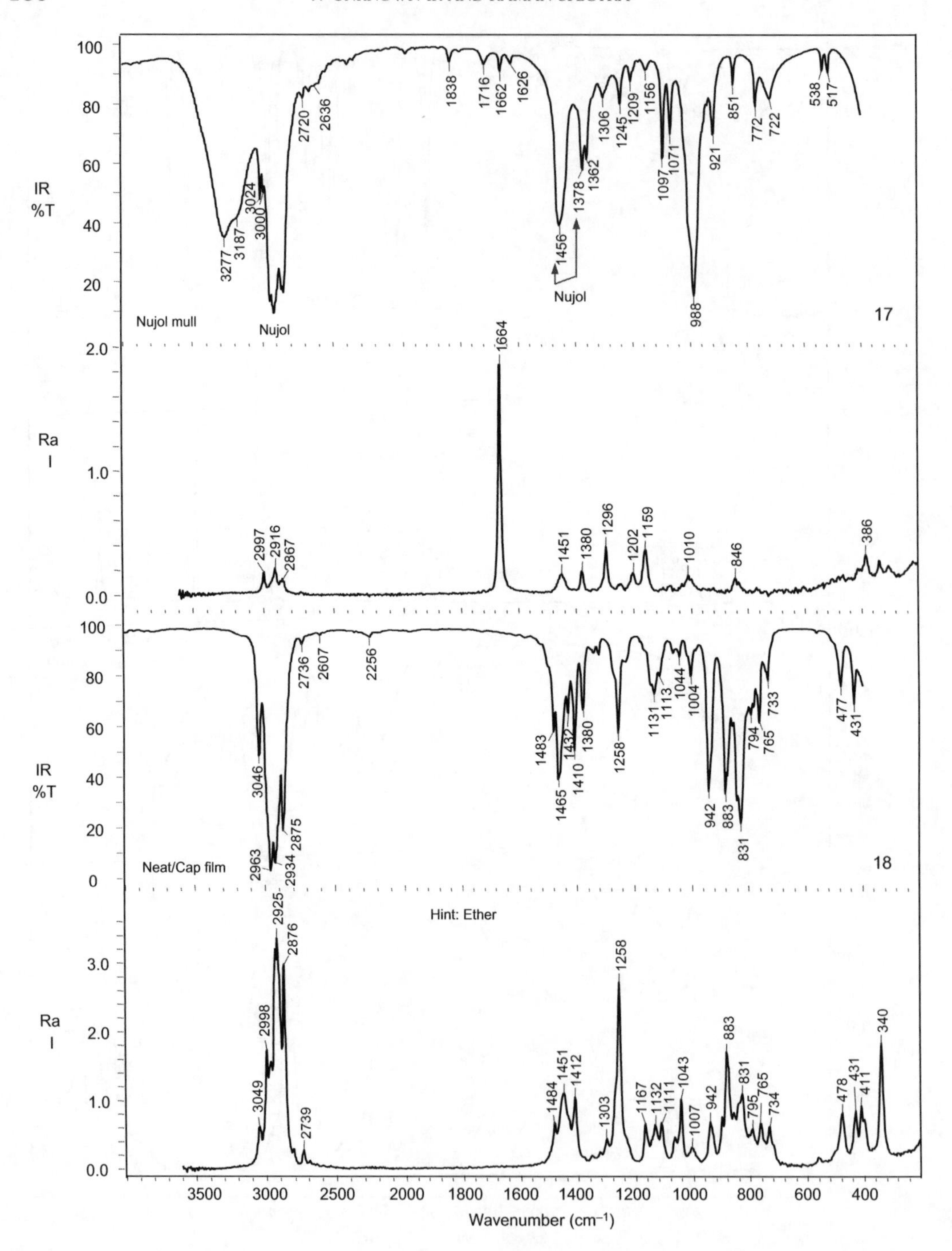
IR %T
Ra I
Nujol mull
Nujol
17
Neat/Cap film
18
Hint: Ether
Wavenumber (cm−1)

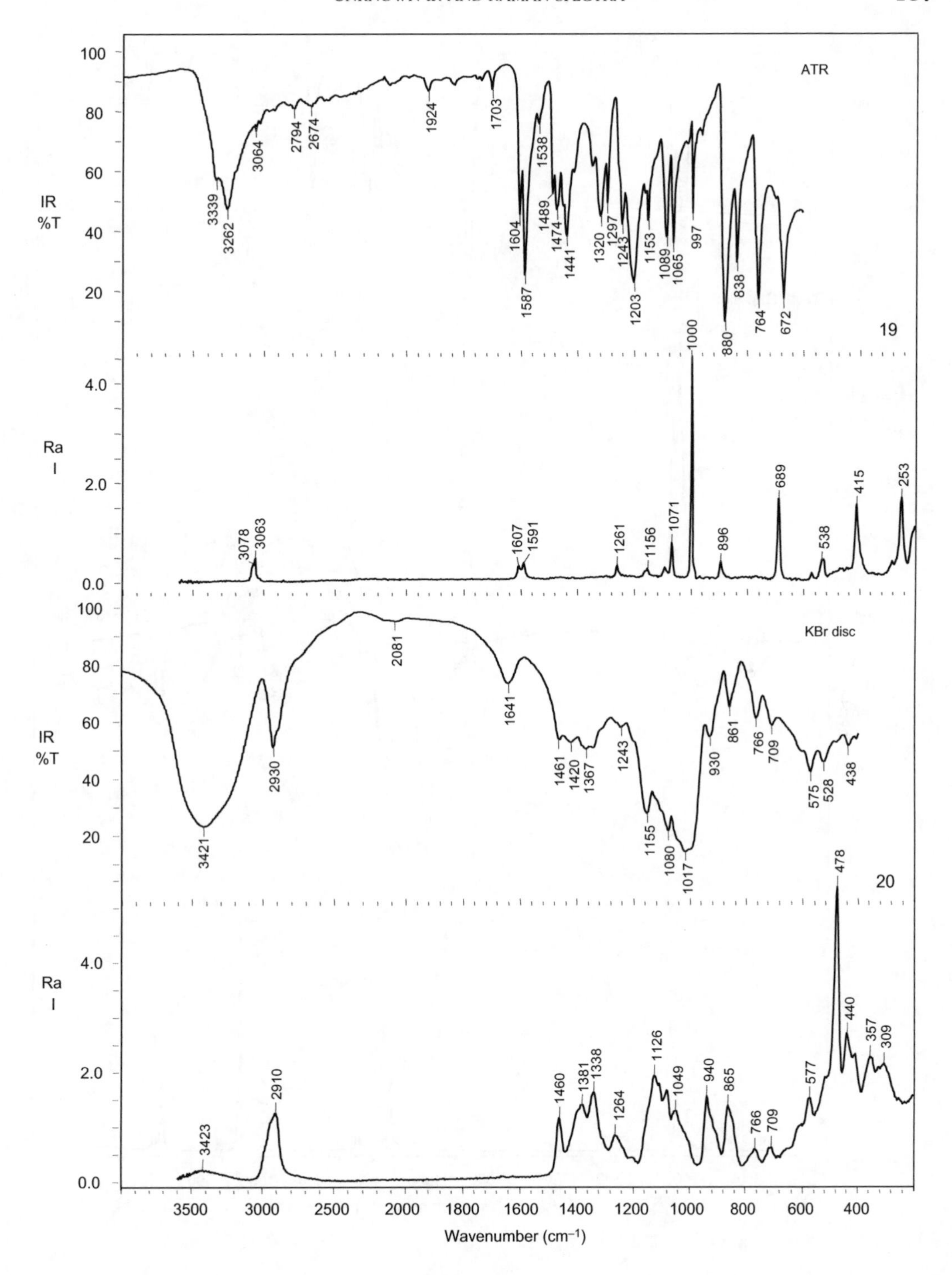
ATR
IR
%T
100
80
60
40
20
3339
3262
3064
2794
2674
1924
1703
1604
1587
1538
1489
1474
1441
1320
1297
1243
1203
1153
1089
1065
997
880
838
764
672
19
Ra
I
4.0
2.0
0.0
3078
3063
1607
1591
1261
1156
1071
1000
896
689
538
415
253
KBr disc
3421
2930
2081
1641
1461
1420
1367
1243
1155
1080
1017
930
861
766
709
575
528
438
20
3423
2910
1460
1381
1338
1264
1126
1049
940
865
766
709
577
478
440
357
309
3500
3000
2500
2000
1800
1600
1400
1200
1000
800
600
400
Wavenumber (cm−1)

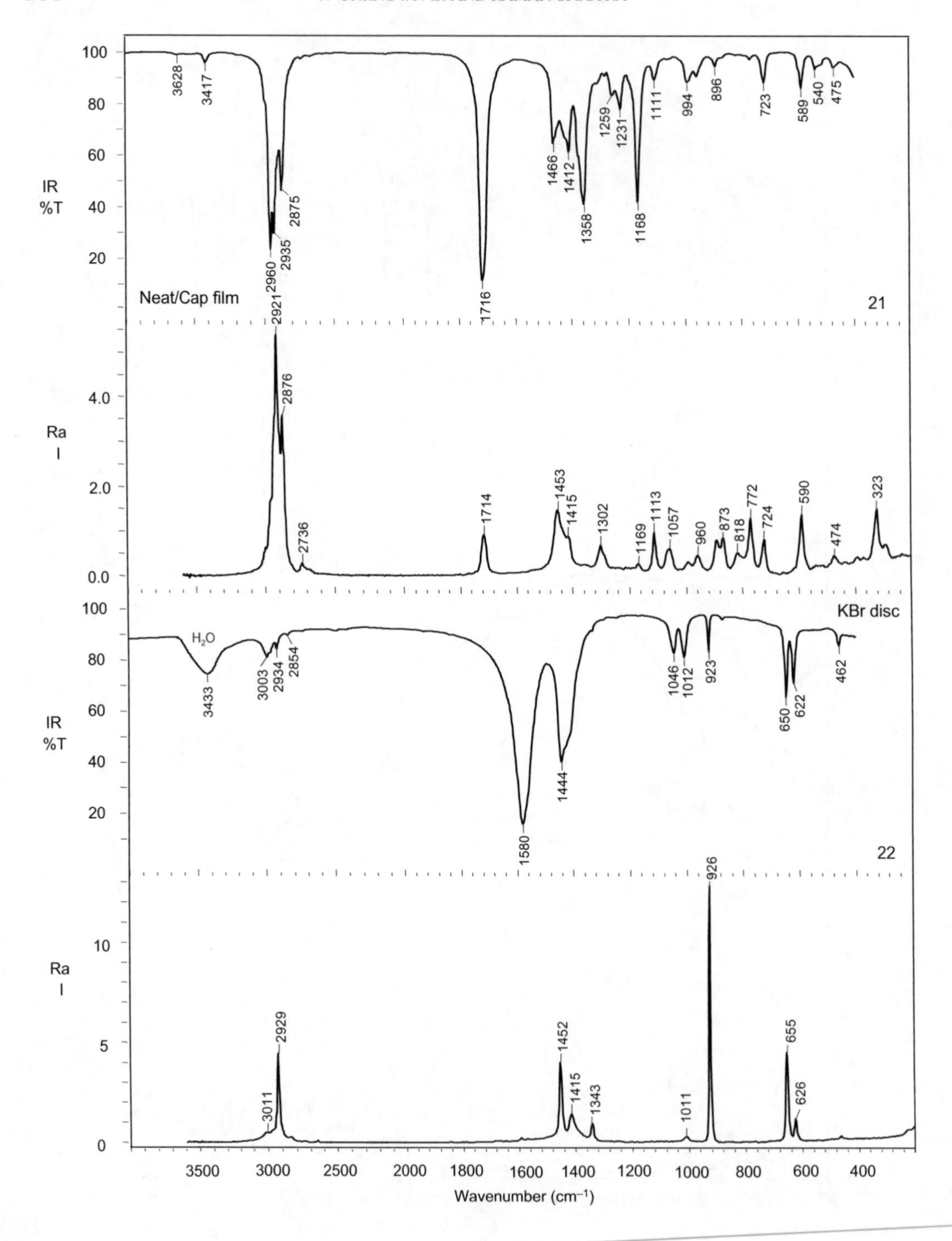

IR
%T
Ra
I
Neat/Cap film
21
KBr disc
22
H_2O
Wavenumber (cm⁻¹)
3628
3417
2960
2935
2921
2875
1716
1466
1412
1358
1259
1231
1168
1111
994
896
723
589
540
475
2876
2736
1714
1453
1415
1302
1169
1113
1057
960
873
818
772
724
590
474
323
3433
3003
2934
2854
1580
1444
1046
1012
923
650
622
462
3011
2929
1452
1343
1011
926
655
626

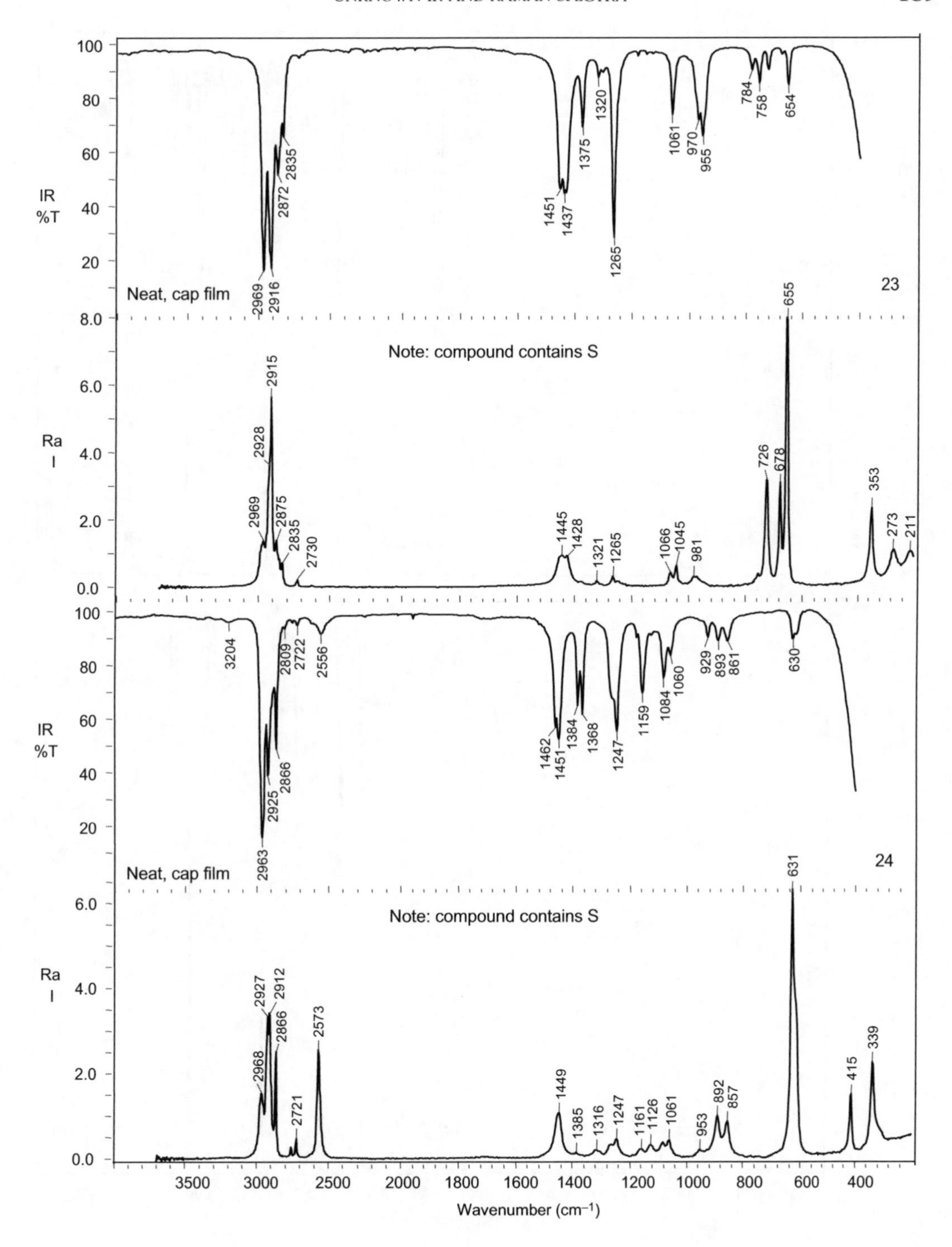

IR %T
Neat, cap film
23
Note: compound contains S
Ra I
IR %T
Neat, cap film
24
Note: compound contains S
Ra I
Wavenumber (cm−1)

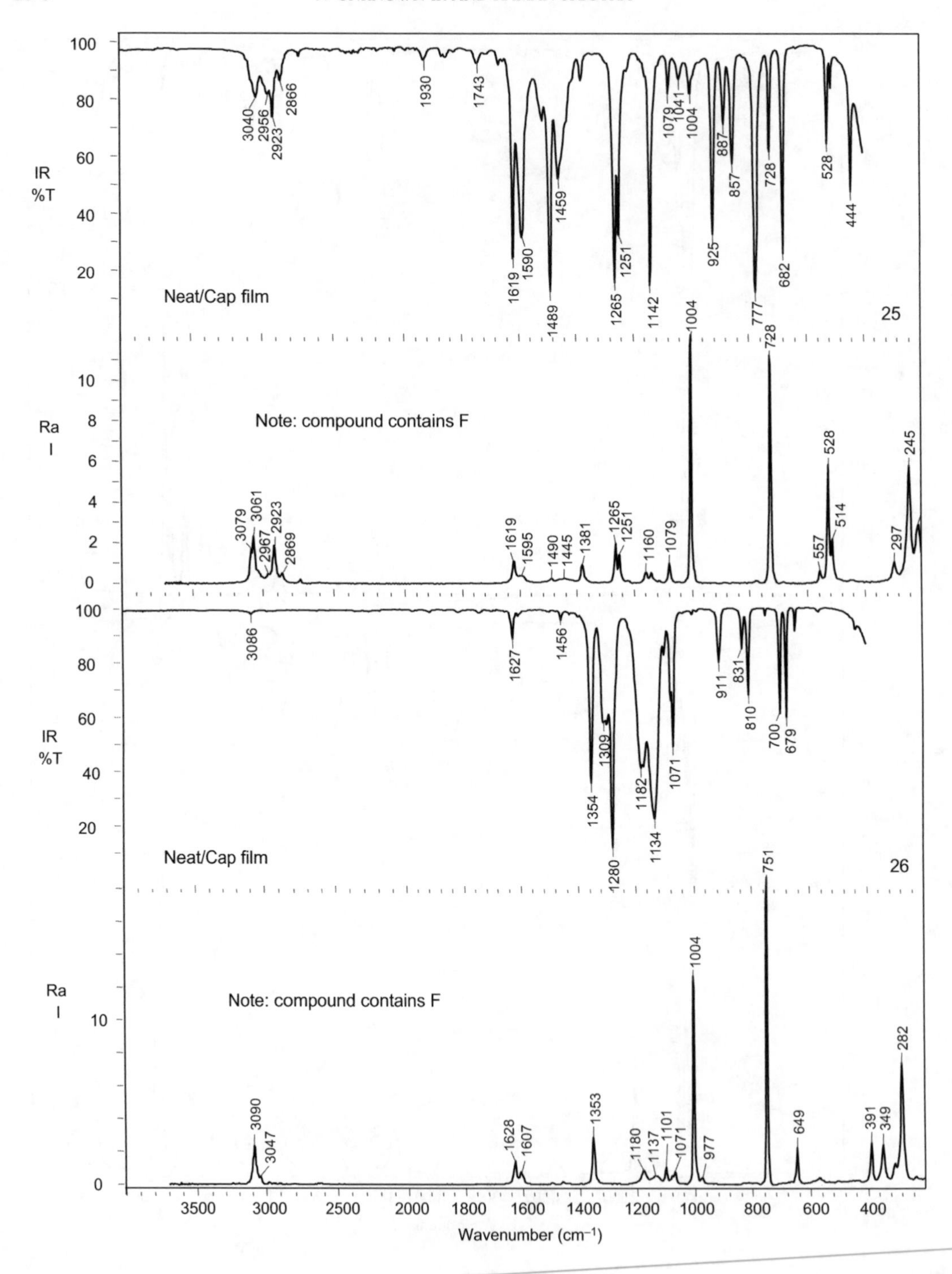

IR %T
Neat/Cap film
25
Ra I
Note: compound contains F
IR %T
Neat/Cap film
26
Ra I
Note: compound contains F
Wavenumber (cm−1)

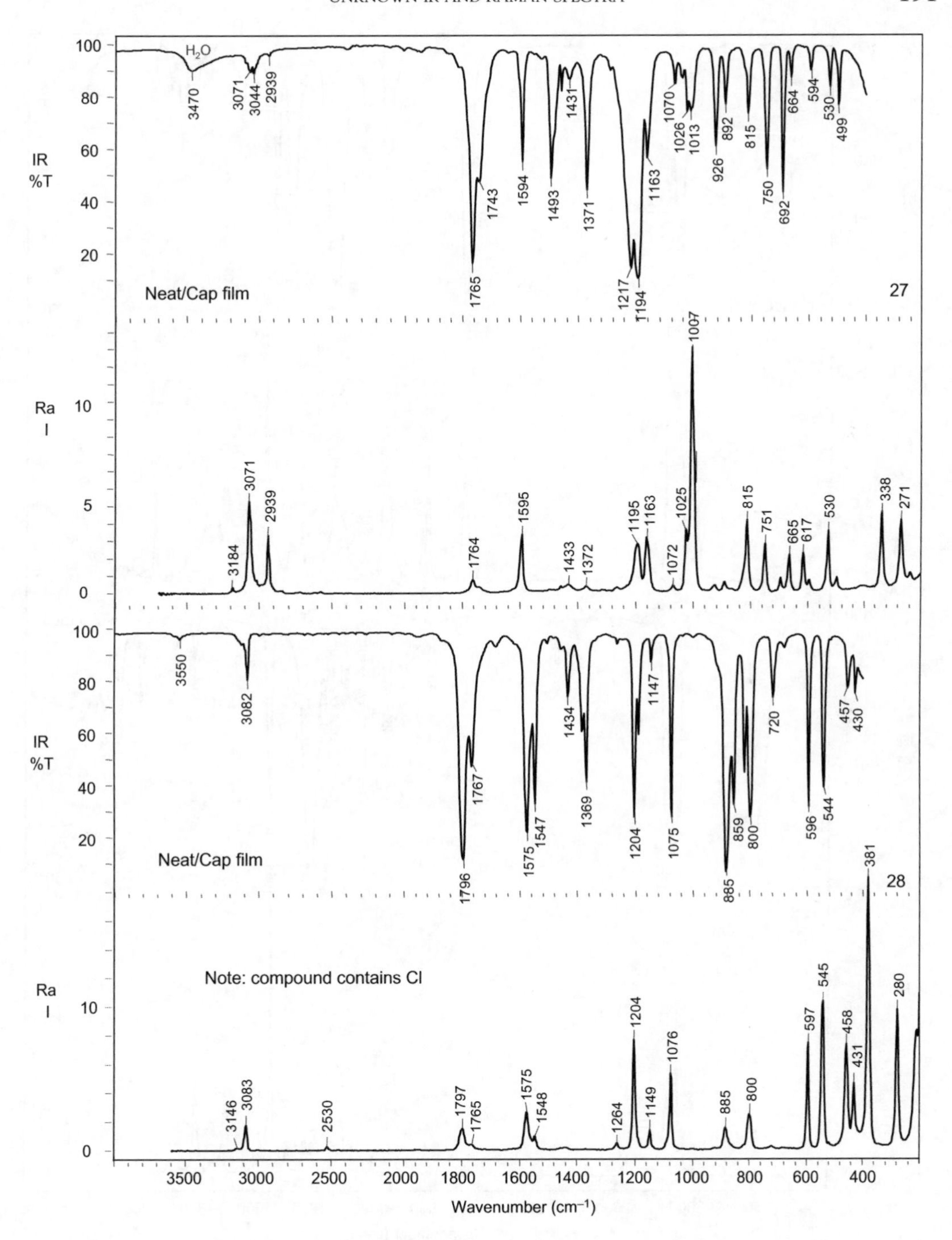
IR %T
Ra I
Neat/Cap film
27
Neat/Cap film
28
Note: compound contains Cl
Wavenumber (cm−1)

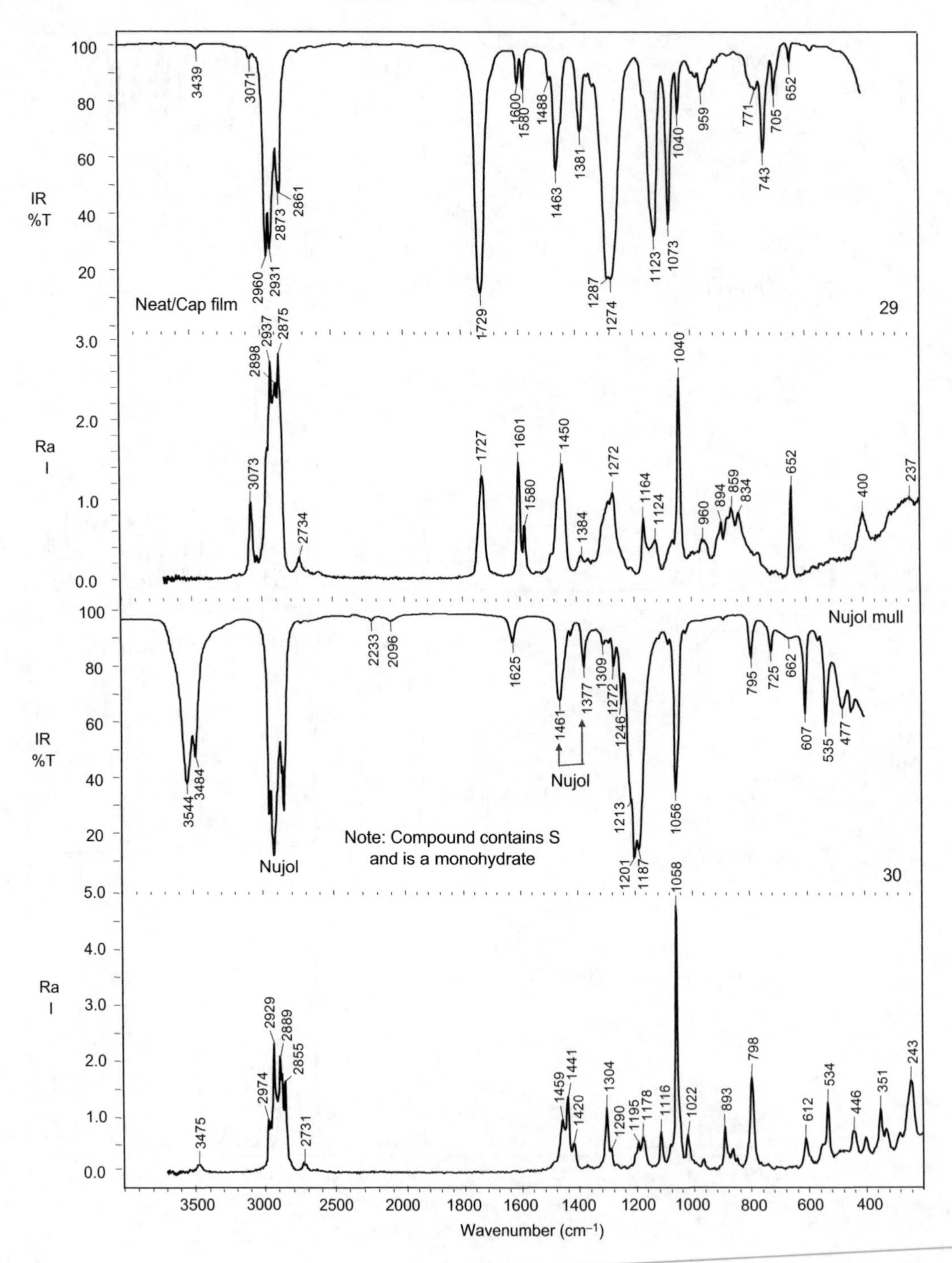
IR
%T
Ra
I
Neat/Cap film
29
3439
3071
2960
2931
2873
2861
1729
1600
1580
1488
1463
1381
1287
1274
1123
1073
1040
959
771
743
705
652
3073
2898
2937
2875
2734
1727
1601
1580
1450
1384
1272
1164
1124
1040
960
894
859
834
652
400
237
Nujol mull
3544
3484
Nujol
2233
2096
1625
1461
1377
Nujol
1309
1272
1246
1213
1201
1187
1056
795
725
662
607
535
477
Note: Compound contains S
and is a monohydrate
30
3475
2974
2929
2889
2855
2731
1459
1441
1420
1304
1290
1195
1178
1116
1058
1022
893
798
612
534
446
351
243
Wavenumber (cm−1)

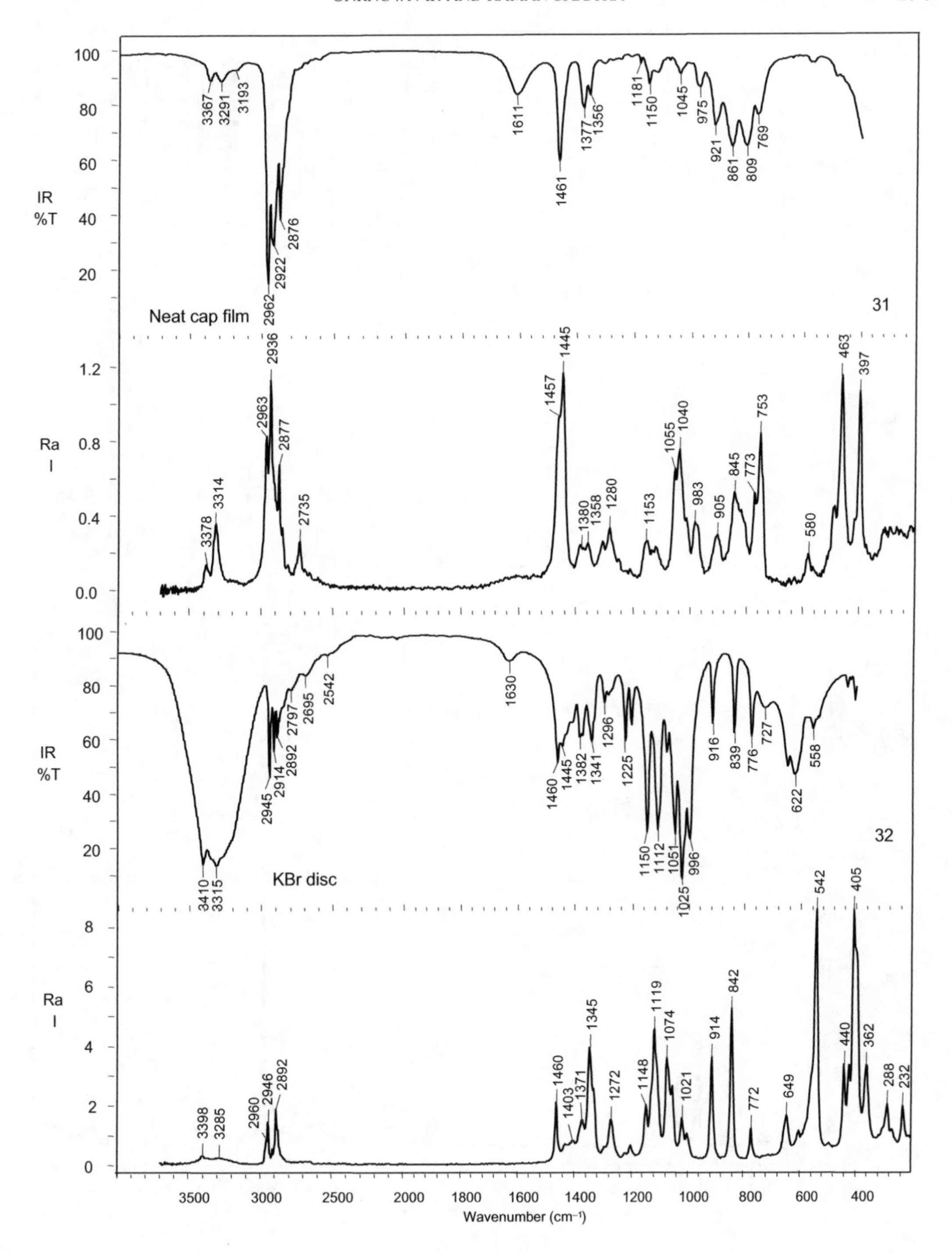
IR
%T
Neat cap film
31
Ra
I
IR
%T
KBr disc
32
Ra
I
Wavenumber (cm⁻¹)

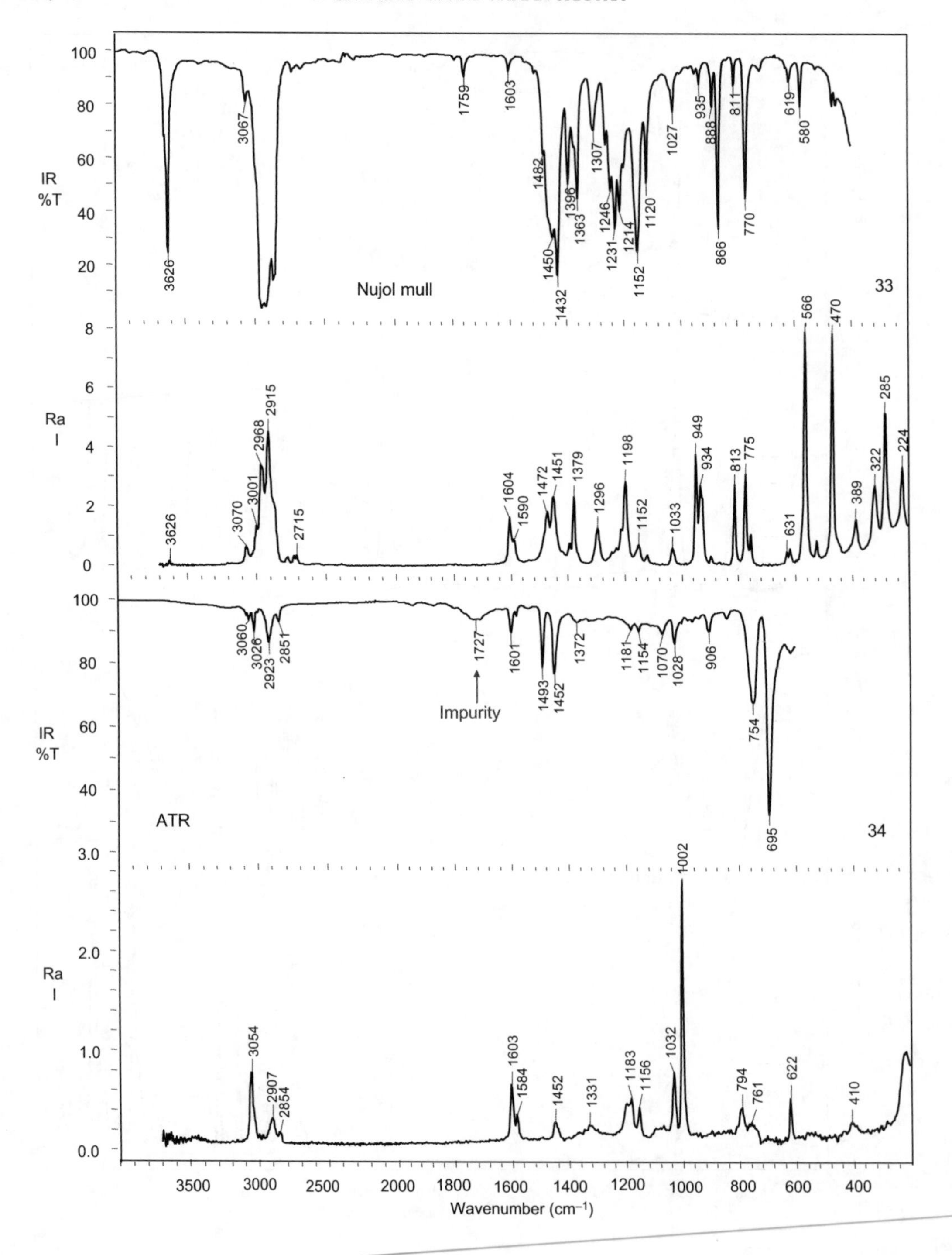

IR
%T
Nujol mull
33
Ra
I
IR
%T
Impurity
ATR
34
Ra
I
Wavenumber (cm−1)

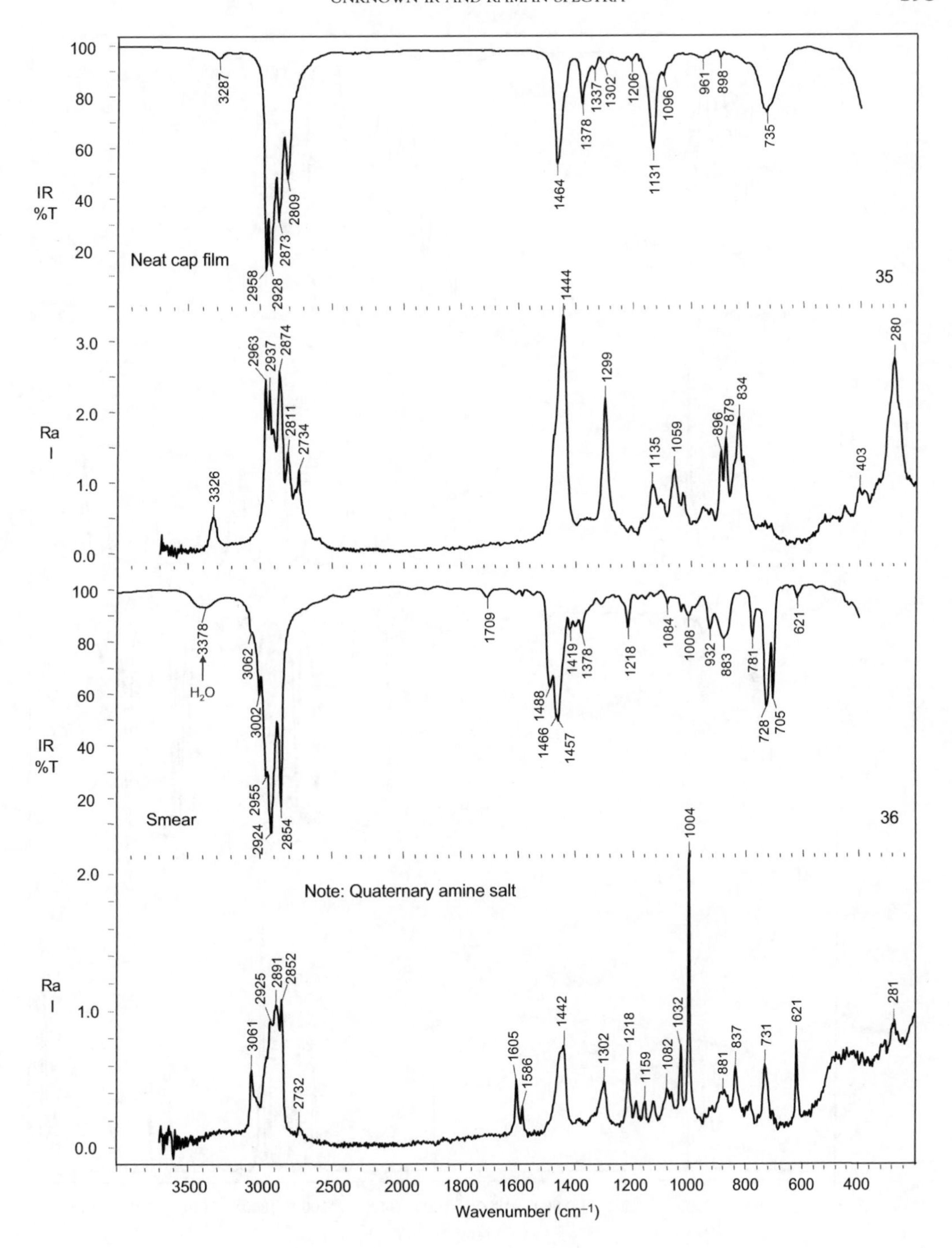
IR
%T
Ra
I
Neat cap film
35
Smear
36
H_2O
Note: Quaternary amine salt
Wavenumber (cm−1)
3287
2958
2928
2873
2809
1464
1378
1337
1302
1206
1131
1096
961
898
735
3326
2963
2937
2874
2811
2734
1444
1299
1135
1059
896
879
834
403
280
3378
3062
3002
2955
2924
2854
1709
1488
1466
1457
1419
1378
1218
1084
1008
932
883
781
728
705
621
3061
2925
2891
2852
2732
1605
1586
1442
1302
1218
1159
1082
1032
1004
881
837
731
621
281

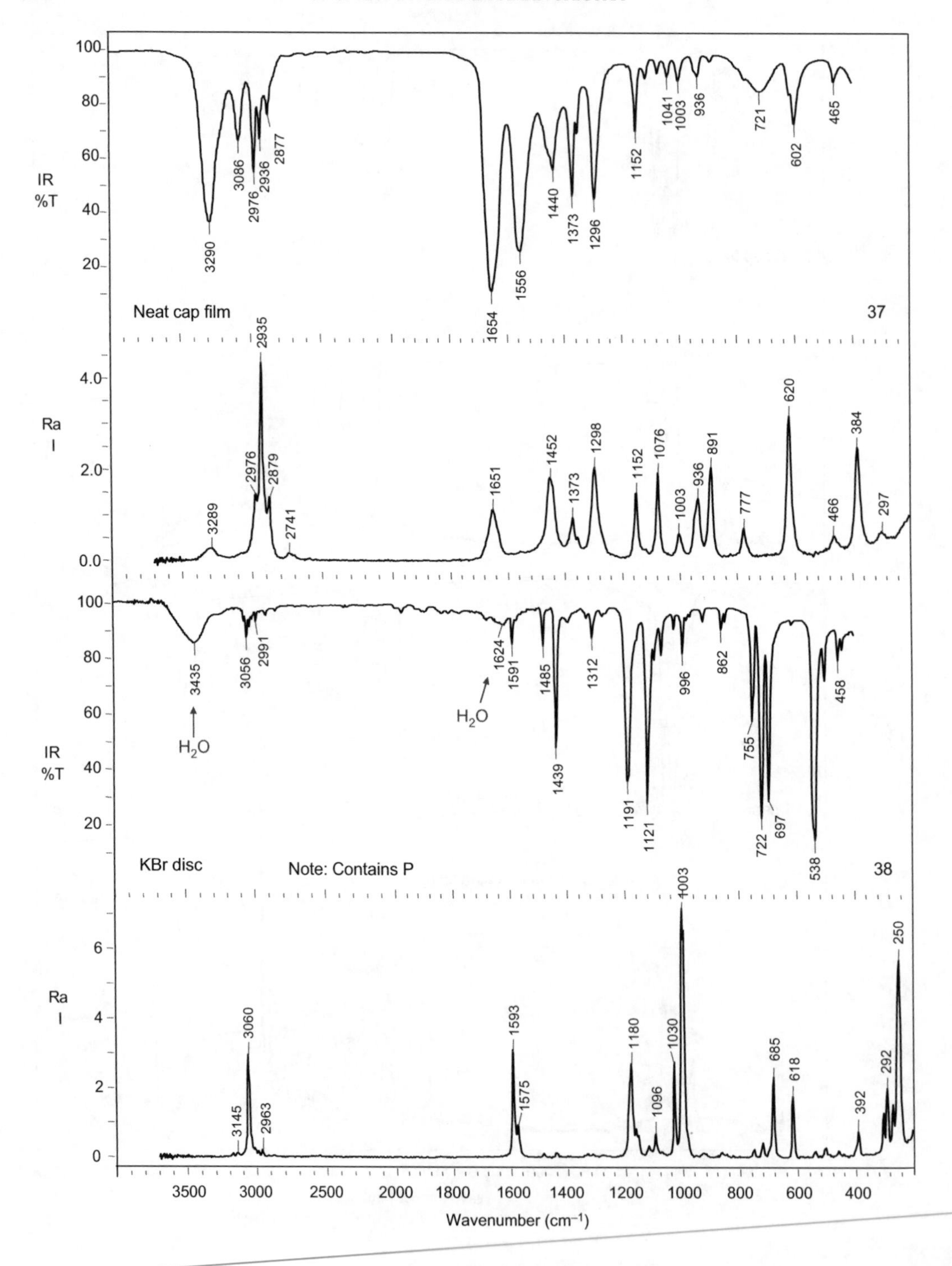

IR %T
Ra I
Neat cap film
37
KBr disc
Note: Contains P
38
H2O
Wavenumber (cm−1)

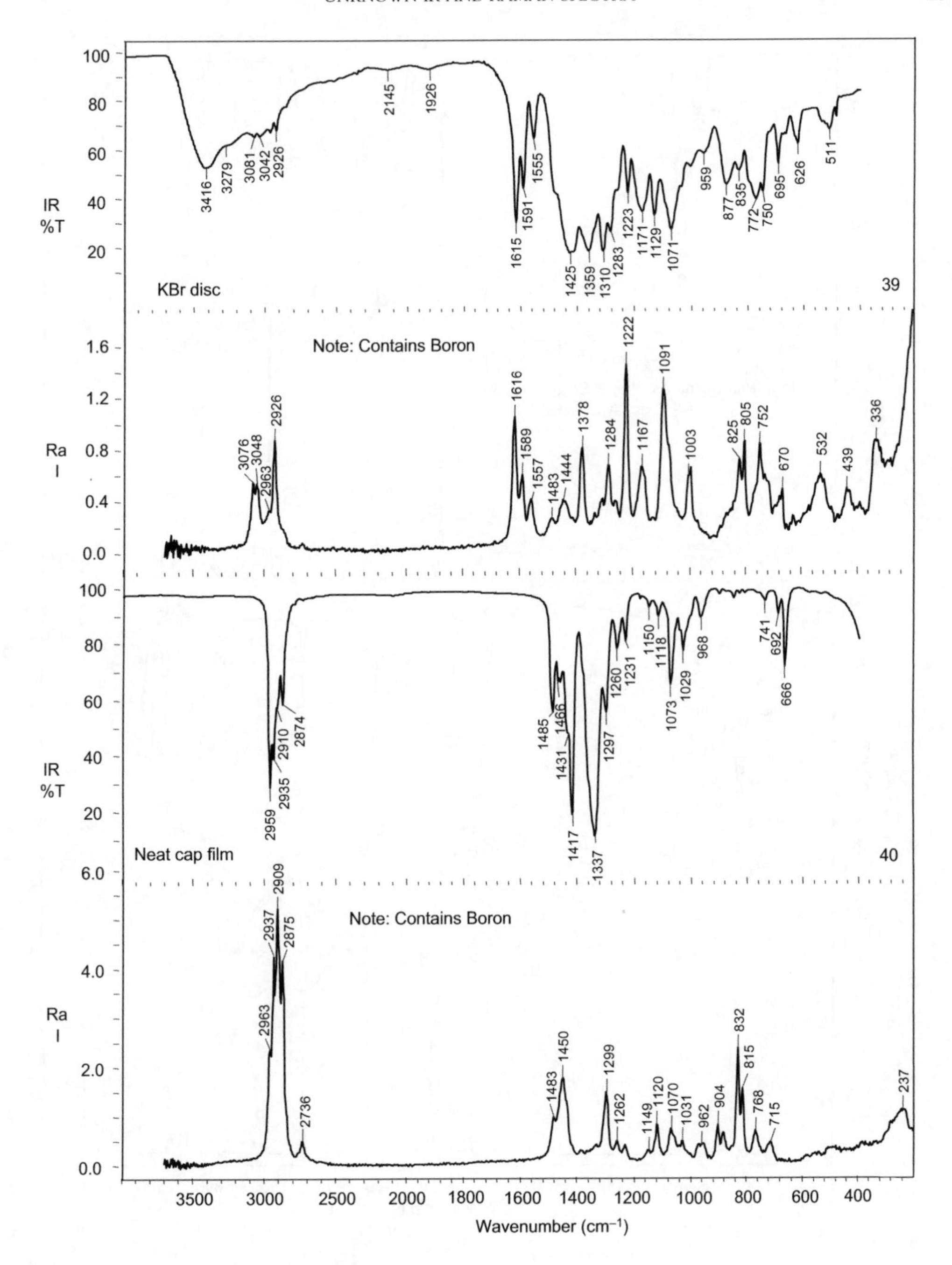
IR
%T
KBr disc
39
3416
3279
3081
3042
2926
2145
1926
1615
1591
1555
1425
1359
1310
1283
1223
1171
1129
1071
959
877
835
772
750
695
626
511
Note: Contains Boron
Ra
I
3076
3048
2963
2926
1616
1589
1557
1483
1444
1378
1284
1222
1167
1091
1003
825
805
752
670
532
439
336
IR
%T
Neat cap film
40
2959
2935
2910
2874
1485
1466
1431
1417
1337
1297
1260
1231
1150
1118
1073
1029
968
741
692
666
Note: Contains Boron
Ra
I
2963
2937
2909
2875
2736
1483
1450
1299
1262
1149
1120
1070
1031
962
904
832
815
768
715
237
Wavenumber (cm−1)

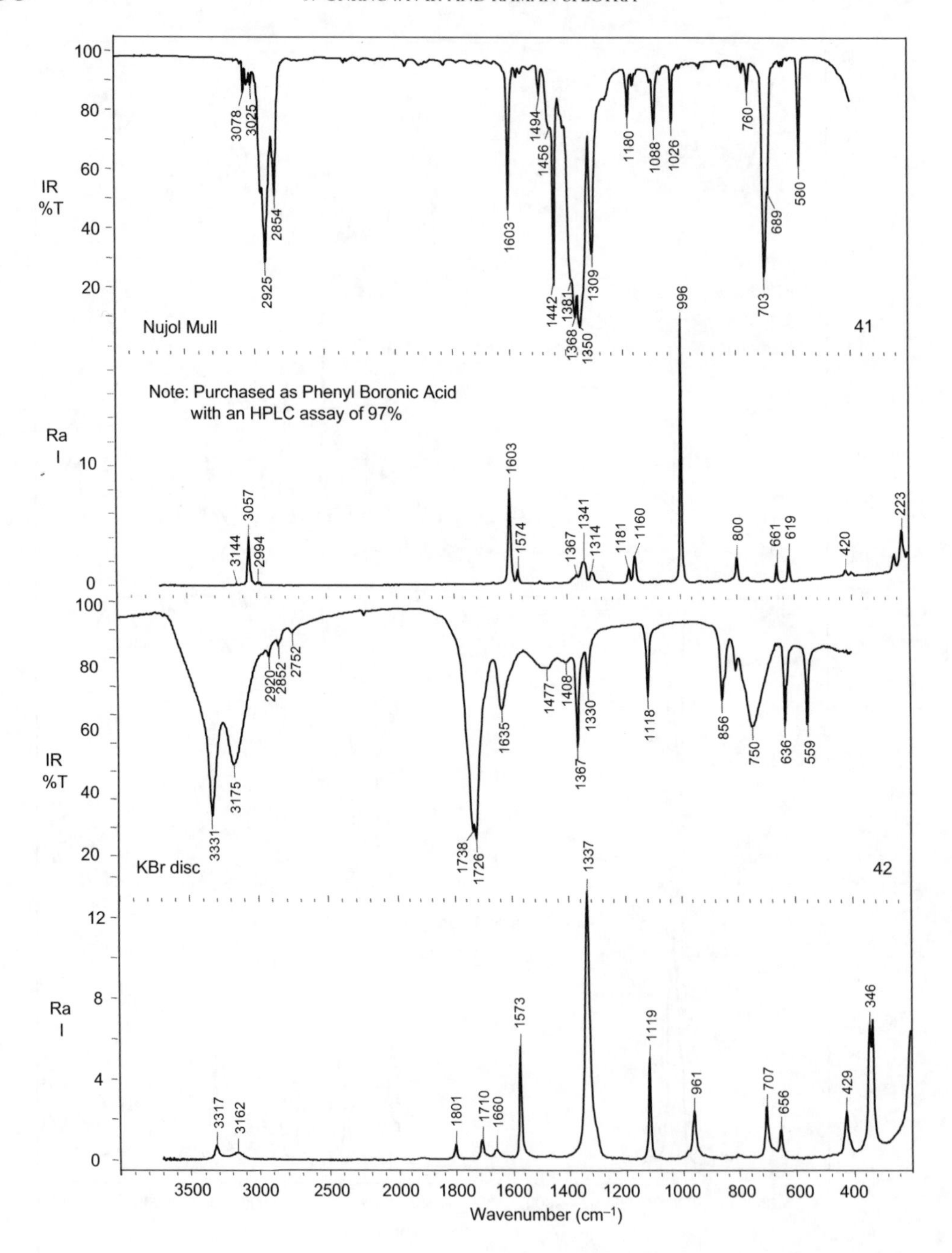

IR %T
Ra I
Nujol Mull
41
Note: Purchased as Phenyl Boronic Acid with an HPLC assay of 97%
KBr disc
42
Wavenumber (cm−1)

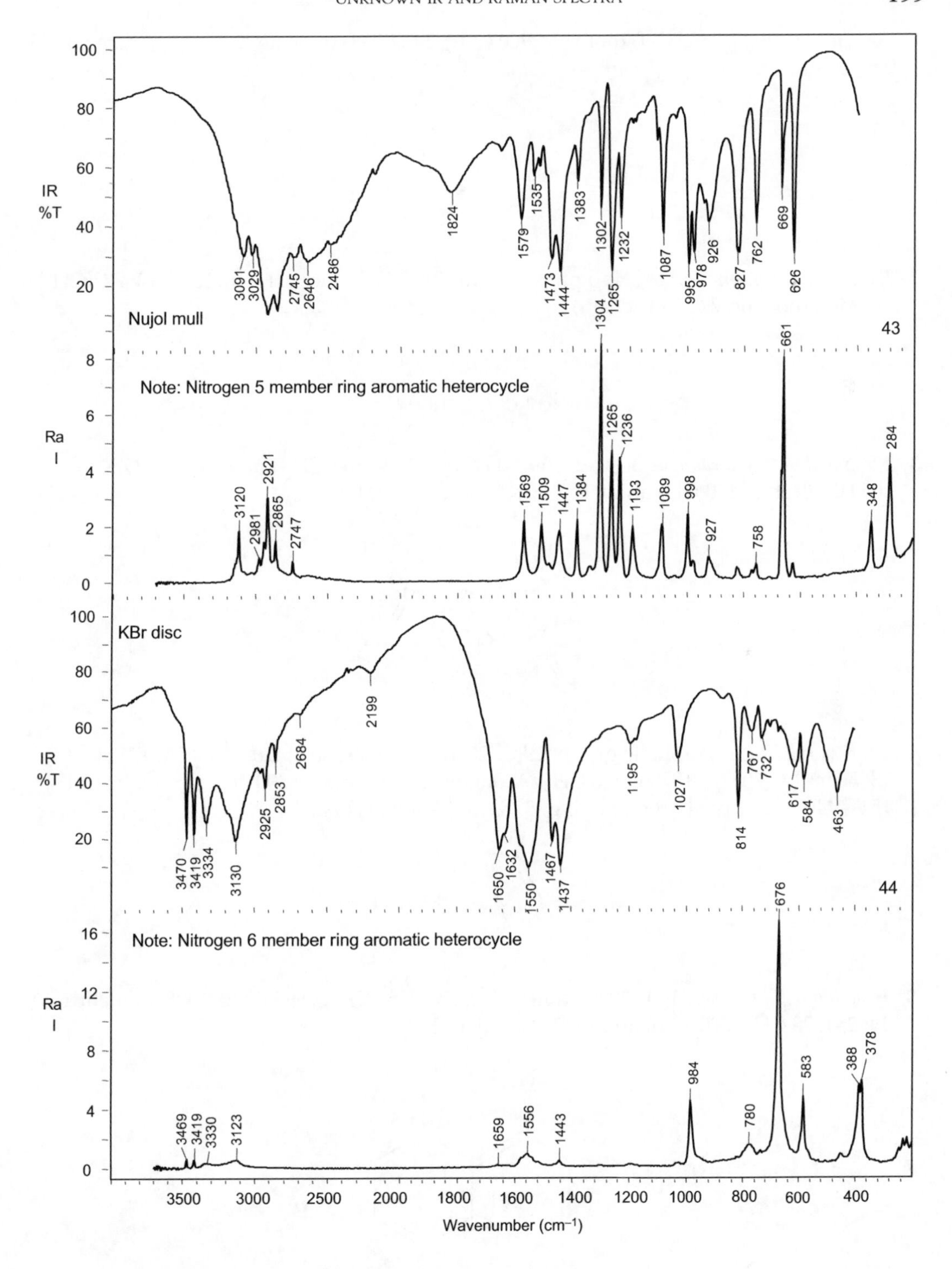
IR %T
Nujol mull
43
Ra I
Note: Nitrogen 5 member ring aromatic heterocycle
KBr disc
44
Note: Nitrogen 6 member ring aromatic heterocycle
Wavenumber (cm−1)

1. *Cyclohexane*: 2930, 2855 CH_2 (note no 2960 CH_3), 1452 CH_2 (note no 1378 CH_3), (note no 723 CH_2 chain).

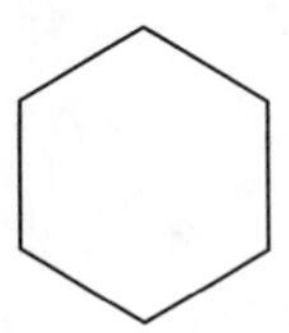

2. *2,4-dimethyl pentane*: 2965 CH_3 o.ph. str., 2940–2850 CH_3 and CH_2 str., 1387, 1369 $C(CH_3)_2$ i.ph def, (note no 723 CH_2 chain).

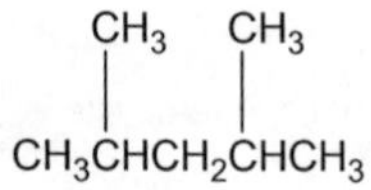

CH_3 CH_3

$CH_3CHCH_2CHCH_3$

3. *2,5-dimethyl-1,5-hexadiene*: 3080, 2980 $=CH_2$ str., 2950-2860 CH_3 and CH_2 str., 1780 2x890, 1650 C=C str., 1450 CH_2 and CH_3 def, 890 $R_2C=CH_2$ CH wag.

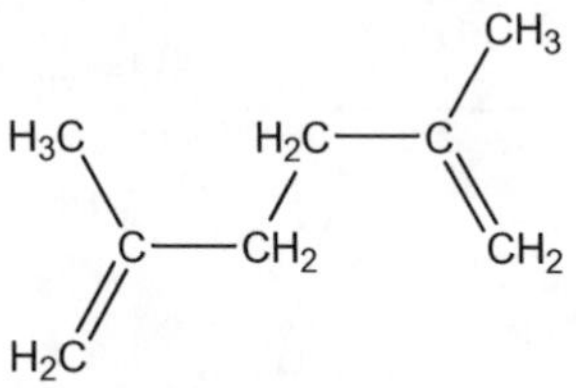

CH_3

H_3C H_2C—C

C—CH_2 CH_2

H_2C

4. *4-tert-butyltoluene*: Aliphatic group: 2965, 2876 cm^{-1} CH_3 str. 1470–1460 cm^{-1} CH_3, 1392, 1363 cm^{-1} $C(CH_3)_3$, 1375 cm^{-1} CH_3. Aromatic group: 3100–3000 cm^{-1} aromatic CH str., 1517, 1407 cm^{-1} para aromatic ring, 817 cm^{-1} 2 adj aryl H.

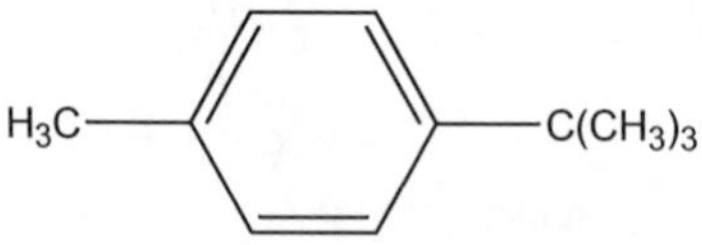

H_3C— —$C(CH_3)_3$

5. *1-methyl naphthalene*: 3100–3000 arom CH, 2975–2860 CH_3, 1597, 1510, 1398 aromatic ring, 1462, 1378 CH_3, 790 3 adj aryl H, 772 4 adj aryl H.

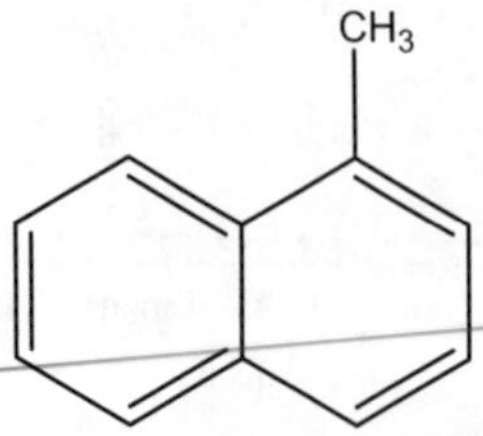

CH_3

6. *n-Butyronitrile*: 2970 CH_3, 2250 C≡N str., 1460 CH_3, CH_2 def, 1425 CH_2(–C≡N) def 1375 CH_3 i.ph. def.

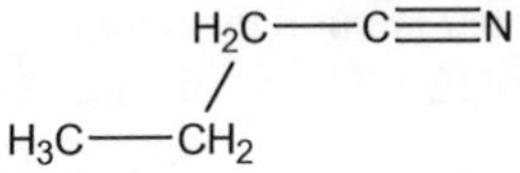

H_2C—C≡N
H_3C—CH_2

7. *2-Hexyne*: Aliphatic group 2964/2873 cm^{-1} CH_3 str., 2935, 2920 cm^{-1} CH_2 str., 1464–1450 cm^{-1} CH_3/CH_2 def, 1380 CH_3 i. ph. def., 1339 CH_2 wag. Acetylene group, C≡C str in Raman at 2301 and 2235 cm^{-1} (Fermi resonance), 375 cm^{-1} C–C≡C bend (Raman).

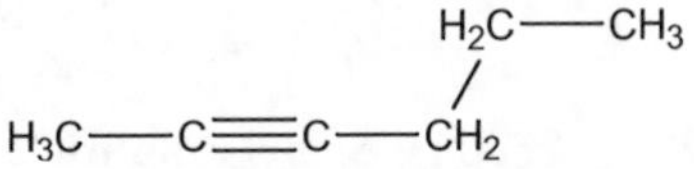

H_2C—CH_3
H_3C—C≡C—CH_2

8. *n-Butanol*: Aliphatic group: 2960, 2874 $cm^{-1}$$CH_3$ str., 2935 CH_2 str., 1470–1450 CH_3, CH_2 def, 1378 CH_3 i.ph def. Hydroxy group (primary alcohol): 3300 cm^{-1} OH str., 1420 broad OH bend, 1070–1025 involves C–O str of CH_2–OH, ~660 cm^{-1} (broad) OH wag.

HO H_2C—CH_3
H_2C—CH_2

9. *Trans-2-Hexen-1-ol*: Aliphatic group: 2959, 2873 $cm^{-1}$$CH_3$ str., 1458 CH_3, CH_2 def, 1379 CH_3 i.ph def. Olefinic group: 3004 cm^{-1} CH str., 1672 cm^{-1} C=C str., 1300 trans CH rk, 969 cm^{-1} (IR) trans CH wag. Hydroxy group (primary alcohol): 3327 cm^{-1} OH str., 1091, 1043, 1091 cm^{-1} involves C–O str., ~650 cm^{-1} (broad) OH wag.

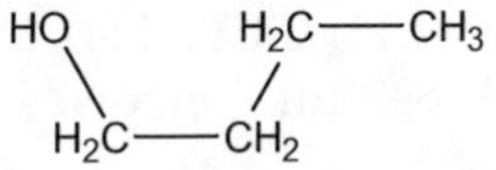

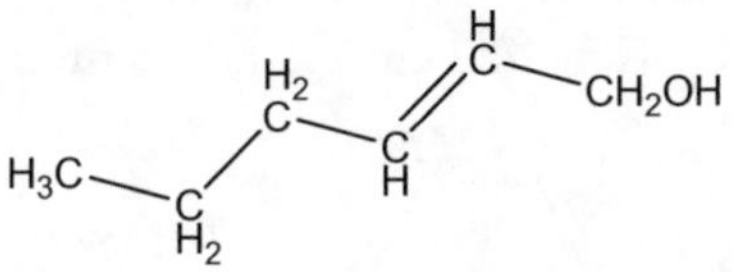

H
C
H_2 CH_2OH
C
C
H_3C H
C
H_2

10. *Benzyl alcohol*: 3330 OH, 3100–3000 arom CH, 2925, 2870 CH_2, 1600, 1493 arom ring, 1452 CH_2/arom ring, 1420 broad OH, 1015 CH_2–OH, 735 5 adj H, 696 arom ring.

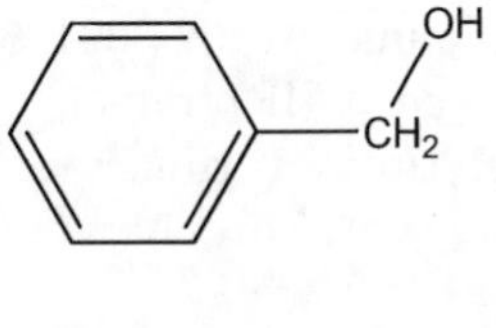

OH
CH_2

11. *Silicon dioxide*: SiO_2 1130–1000 IR strong (involves O–Si–O o. ph. str.). Note:featureless Raman spectrum.

12. *Dow Corning vacuum grease, dimethyl siloxane polymer*: 2966–2963 CH_3 o. ph. str., 2907–2906 CH_3 i. ph. str., ~1412 CH_3 o. ph. def, 1261 CH_3 i. ph. def, 1090–1020 IR involves Si–O–Si str.

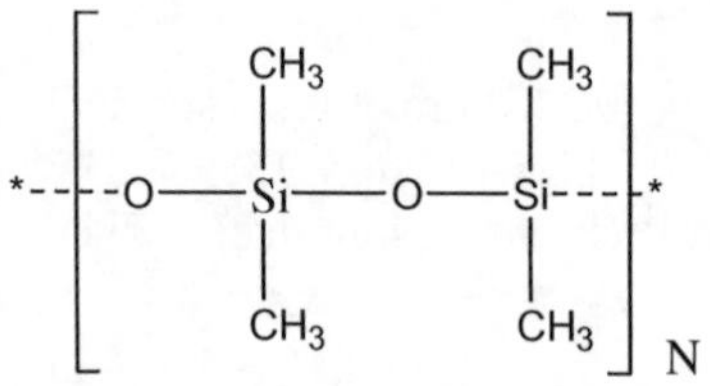

CH3 CH3
- - -O—Si—O—Si- - -
CH3 CH3
N

13. *Tetrahydrofuran*: Aliphatic 2975–2860 suggests only methylene. Two bands at 1070 and 911 cm^{-1} suggest ether (C–O–C out-of-ph. and in-ph. str., respectively).

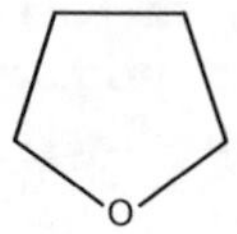

O

14. *1,4-Dioxane*: Aliphatic 2955–2850 CH_2, 1450–1442 CH_2, 1362 O CH_2 wag, 1120 cm^{-1} involves C–O–C o. ph. str. 900–830 cm^{-1} involves C–O–C in-phs. str.

O
O

15. *Anisole*: 3100–3000 aromatic CH, 3000–2838 CH_3(O), 1602, 1588, 1498 cm^{-1} aromatic ring, 1468, 1452, 1440 cm^{-1} CH_3(O) + aromatic ring, 996 cm^{-1} (Raman) 2,4,6 C Radial i.ph.str., 1247, 1040 arom-O–CH_3, 752 cm^{-1} 5 adjacent aryl H, 690 cm^{-1} aromatic ring.

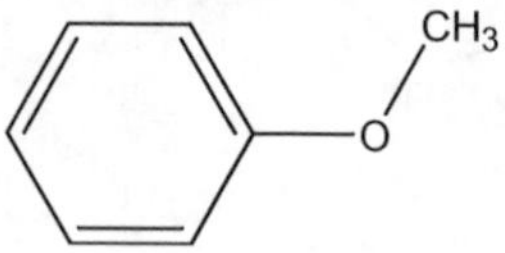

CH3
O

16. *Paper*: Mostly cellulose and carbonate, some water, with a trace of organic sizing: carbonyl and olefinic.

17. *Trans, trans-2,4-Hexadien-1-ol*: Olefinic group, 3024, 3003 cm^{-1} CH str., 1672 cm^{-1} C=C str., 1300 cm^{-1} trans CH rk, 988 cm^{-1} (IR) trans CH wag (note higher frequency due to adjacent trans, trans olefin). Hydroxy (primary alcohol) group. 3277, 3187 cm^{-1} OH str., 1097, 1071, ~1000 (shoulder) cm^{-1} involves C–O str.

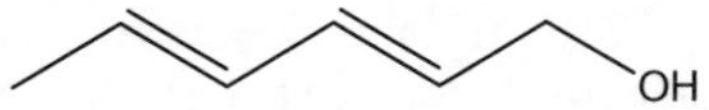

18. *1,2-Epoxypentane*: Aliphatic CH str with epoxy CH_2 o. ph. str at ca. 3050 cm^{-1} and CH_2 def. at 1483 cm^{-1}. Epoxy C–O–C in-ph str. At 1258 cm^{-1} (both IR and Raman) and C–O–C out-of-ph str at 831 cm^{-1}. Aliphatic peak intensity of methylene (2935 cm^{-1}) to methyl (2875 cm^{-1}) consistent with ca. 2 methylene to 1 methyl group.

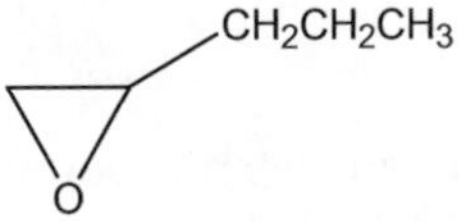

19. *Meta (3) chlorophenol*: 3360 cm^{-1} OH str., 3100–3000 aryl CH str (note no aliphatic (CH_2, CH_3) 3000–2800 cm^{-1}, 1605, 1592, 1478, 1448 cm^{-1} aryl ring, 1000 cm^{-1} (Raman) 2,4,6 C Radial i.ph.str., 1245–1190 cm^{-1} involves aryl-OH str, 852 lone H, 772 3 adj H, 680 aryl ring.

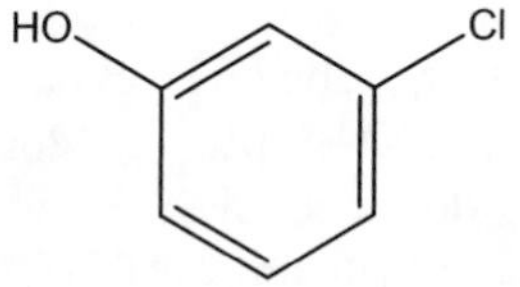

20. *Dextrin*: Polysaccaride/carbohydrate: 3421 cm^{-1} OH str., 1338 cm^{-1} OH def., 1155, 1080, 1017 cm^{-1} involves C–O str. of C–OH/C–O–C polysaccharide ring system. Axial connection similar to starch result in characteristic IR bands 930, 861, 766, and 709 involving the ring in-phase and semi-circle str. The cluster of Raman bands centered around the intense 478 cm^{-1} band also is diagnostic of the axial type connection. Presence of water is evident from bands at 2081 (combination) and 1641 cm^{-1} (def.). Biopolymeric aliphatic: CH_2 and CH str bands between 2930 and 2820 cm^{-1} and CH_2 def at 1460 cm^{-1}.

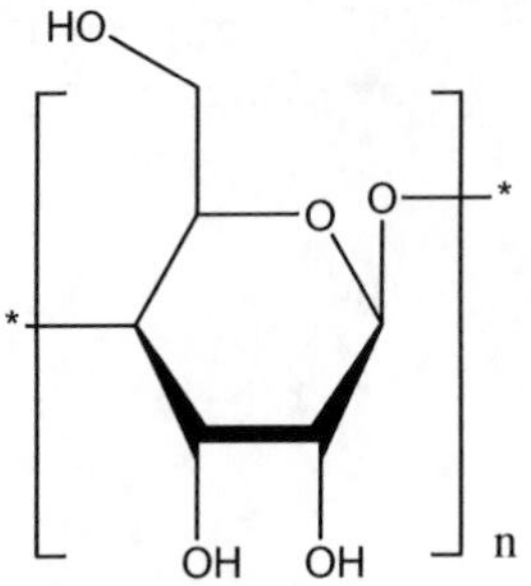

21. *2-Hexanone*: Ketone C =O str 1716 $cm^{-1}$and C =O overtone at 3417 cm^{-1}. 1412 cm^{-1} from CH_2–(C =O) and 1358–1380 cm^{-1} from CH_3 i.ph def. Aliphatic group: CH_2 and CH_3 stretches at 2935, 2875 , and 2960 cm^{-1}, respectively. Peak intensity of methylene (2935 cm^{-1}) to methyl (2875 cm^{-1}) consistent with ca. 3 methylene to 2 methyl. No evidence of branching in Raman spectra.

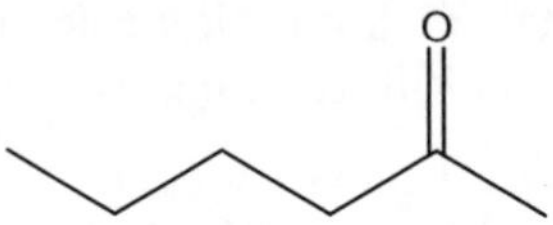

O

22. *Sodium acetate*: Simple IR and Raman spectra. Dominant IR bands at 1580 and 1444 cm^{-1} indicate a carboxylate salt (CO_2 out-of-phase and in-phase stretch, respectively).

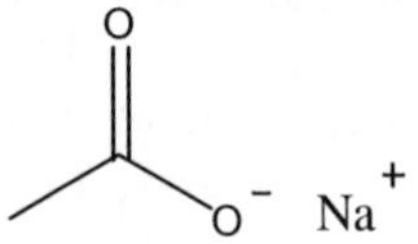

O

O^- Na^+

23. *Ethyl methyl sulfide*: Aliphatic group. The CH_2–S has bands at 2928 cm^{-1} from the CH_2 o.ph.str., 2872–875 cm^{-1} from the CH_2 i.ph.str., 1265 cm^{-1} (IR) from the CH_2 wag and 1437–1428 cm^{-1} from the CH_2 def. The CH_3–S has bands at 2969 cm^{-1} from the CH_3 o.ph.str., 2915 cm^{-1} from the CH_3 i.ph.str., the 1375 cm^{-1} is from the CH_3 i.ph.def. and bands between 1070 and 950 cm^{-1} involve the CH_3 rock + C–C str. Bands involving the C–S–C str are observed between 780 and 650 cm^{-1}.

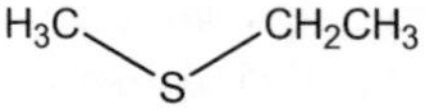

H_3C S CH_2CH_3

24. *2-Propanethiol*: SH group: SH str results in characteristic band at 2550–2580 cm^{-1}. Aliphatic group: The isopropyl group has characteristic bands at 2963 cm^{-1} (CH_3 o.ph.str), 2866 cm^{-1} (CH_3 i.ph.str), 1462–1451 cm^{-1} (CH_3 o. ph. def.), and in the IR spectrum at 1384, 1368 cm^{-1} (CH_3 i.ph. def). The CH–S group has bands at 2912–2930 cm^{-1} (CH str.) and 1247 cm^{-1} (CH wag). The isopropyl C–C/C–S str results in a strong Raman band at 631 cm^{-1} involving the in-phase $(C)_2C$–S str.

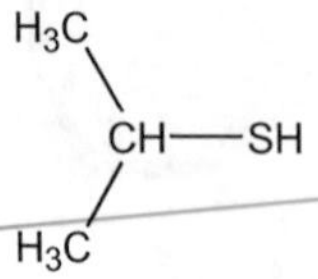

H_3C

CH—SH

H_3C

25. *3-Fluorotoluene (1-fluoro-3-methylbenzene)*: Aliphatic (CH_3) 2967–2866 cm^{-1} methyl CH_3 str., 1459 cm^{-1} CH_3 o. ph. def, 1381 cm^{-1} CH_3 i. ph. def. Aromatic group: 3100–3000 cm^{-1} aryl CH str., 1619–1590 cm^{-1} aryl ring quadrant str. (IR/Raman), 1489 cm^{-1} (IR) aryl ring semi-circle str., 1004 cm^{-1} (Raman) 2,4,6 C Radial i.ph.str., 857 cm^{-1} lone H wag, 777 cm^{-1} (IR) 3 adj H wag, 682 cm^{-1} (IR) ring pucker. Aryl-F results in IR/Raman bands at 1265, 1251 cm^{-1} involving the Aryl-F str.

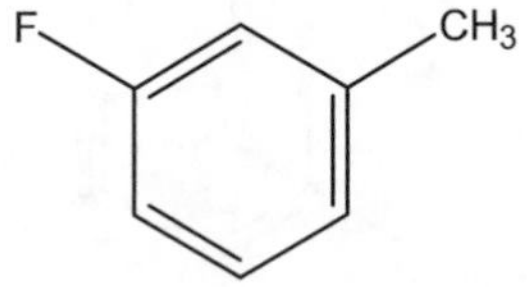

F
CH_3

26. *1,3-Bis(trifluoromethyl)benzene*: Note no aliphatic observed. Aromatic group: 3100–3000 cm^{-1} aryl CH str., 1628–1607 cm^{-1} aryl ring quadrant str. (IR/Raman), 1004 cm^{-1} (Raman) 2,4,6 C Radial i.ph.str., 911 cm^{-1} lone H wag, 810 cm^{-1} (IR) 3 adj H wag, 700 cm^{-1} (IR) ring pucker. CF_3 group: 1354, 1309, 1280 involves CF_3 o. ph. str., 1181, 1134 cm^{-1} involves CF_3 i. ph. str.

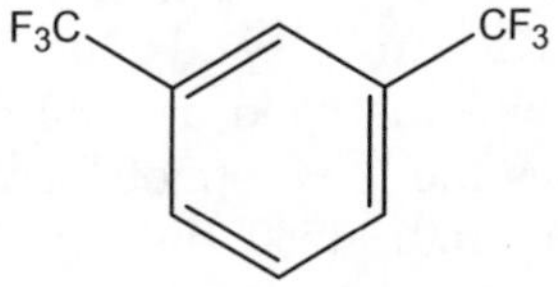

F_3C
CF_3

27. *Phenyl acetate*: Aromatic group: 3100–3000 cm^{-1} aryl CH str., 1594 cm^{-1} aryl ring quadrant str. (IR/Raman), 1493 cm^{-1} (IR) aryl ring semi-circle str 1007 cm^{-1} (Raman) 2,4,6 C Radial i.ph.str., 750 cm^{-1} (IR/Raman) 5 adjacent H wag, 692 cm^{-1} aryl ring pucker. Acetate group: 1765–1740 cm^{-1} C =O str., 1431 cm^{-1} (O)CH_3 o. ph. def., 1371 cm^{-1} (O)CH_3 i. ph. def., 1217–1163 cm^{-1} involves ϕ–O–C–C o. ph. str., 815 cm^{-1} involves ϕ–O–C–C i. ph. str.. 2939 cm^{-1} CH_3 o. ph. st.

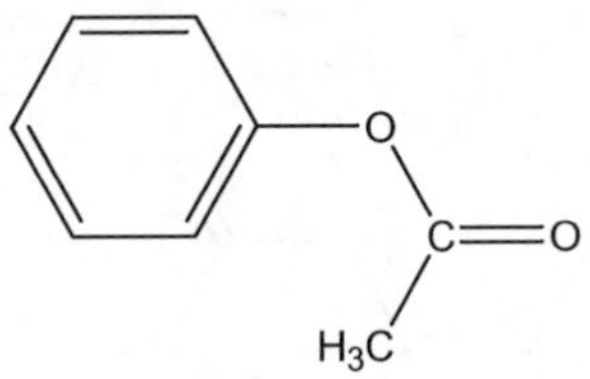

O
C=O
H_3C

28. *2,4,6-Trichlorobenzoyl chloride*: Aromatic group: 3146–3082 cm^{-1} aryl CH str., 1575–1547 cm^{-1} aryl ring quadrant str. (IR/Raman), 1434 cm^{-1} (IR) aryl ring semi-circle str. The band at 800 cm^{-1} derives from the lone H wag and the band at 720 cm^{-1} involves the aryl ring pucker. The band at 1075 cm^{-1} (IR/Raman) involves the aryl-Cl str. Aromatic acid chloride group: 1796 cm^{-1} C =O str. and the weaker band at 1767 cm^{-1} involves the overtone of the 885 cm^{-1} that is Fermi resonance enhanced with the carbonyl. The 885 cm^{-1} and 1204 cm^{-1} involves the (aryl)C–C(=O) str.

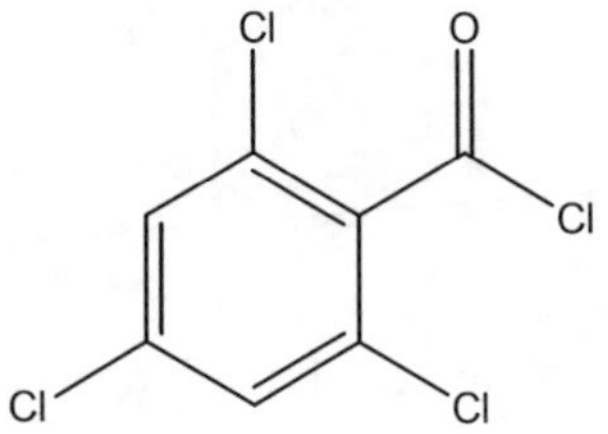

Cl O Cl Cl Cl

29. *Dioctyl phthalate*: Aliphatic group: CH_2 stretches at 2931, 2873 cm^{-1} and CH_3 stretches at 2960 and 2861 cm^{-1}, respectively. Peak intensity of methylene (2931 cm^{-1}) to methyl (2960 cm^{-1}) consistent with ca. 4 methylene to 2 methyl. Methylene and methyl deformation bands are observed at 1463 and 1381 cm^{-1}. Phthalate Carbonyl Group: C =O str (IR/Raman) at 1727 cm^{-1}, band cluster between 1290 and 1270 cm^{-1} involves C–C–O o. ph. str. and the band at 1040 cm^{-1} involves the C–C–O i. ph. str. The band at 1123 involves the O–C str of the aliphatic group. Aromatic group: 3100–3000 cm^{-1} aryl CH str., 1601–1580 cm^{-1} aryl ring quadrant str. (IR/Raman), 1488 cm^{-1} (IR) aryl ring semi-circle str. The band at 743 cm^{-1} derives from the 4 adj H wag.

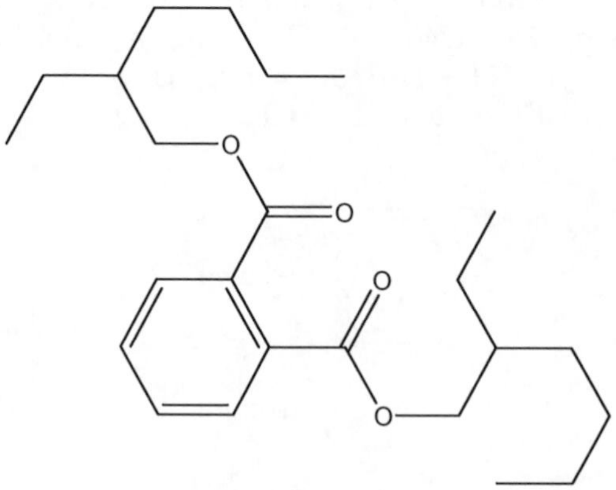

O O O O

30. *1-Hexane sulfonic acid sodium salt (sodium 1-hexanesulfonate)*: Note sample was prepared as a nujol mull which interferes with IR aliphatic bands. Raman aliphatic bands indicates both methyl and methylene groups. No aromatic observed. The water results in bands in the IR spectrum at 3544 and 3484 cm^{-1} from the OH str., 1625 cm^{-1} from the OH def., a combination bands at 2233 and 2096 cm^{-1} and the broad OH bend at ca. 622 cm^{-1}. Sulfonic acid salt group: Strong IR bands from 1213 to1187 cm^{-1} derive from the SO_3 o. ph. str. The strong IR/Raman band at ca. 1058 cm^{-1} derives from the SO_3 i. ph. str. The band at ca. 795 cm^{-1} (IR/Raman) involves the C–S str.

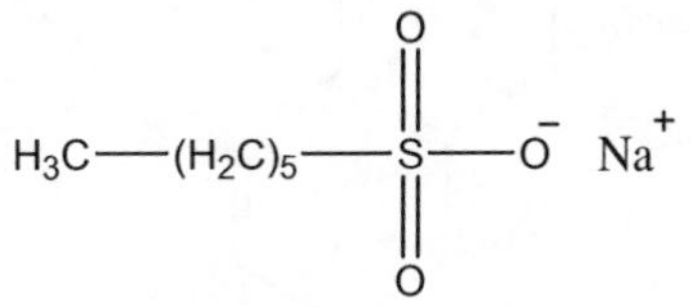

H_3C—$(H_2C)_5$—S(=O)(=O)—O^- Na^+

31. *3-Amino pentane*: Aliphatic group: CH_3 stretches at 2962, 2876 cm^{-1} and CH_2 stretch at 2922 cm^{-1}, respectively. The peak intensity of methylene (2922 cm^{-1}) to methyl (22962 cm^{-1}) is consistent with ca. 1 methylene to 1 methyl. Aliphatic bands from methyl and methylene deformation vibrations are observed at 1461, 1377 and 1356 cm^{-1}. Primary Amine group: The NH_2 stretching vibrations result in a doublet between 3380 and 3290 cm^{-1} with a shoulder at 3193 cm^{-1} which derives from an overtone (Fermi-resonance enhanced) of the 1611 cm^{-1} NH_2 deformation band. The band cluster from 921 to 769 cm^{-1} derives from the NH_2 wag and twist vibrations. Bands involving the C–N stretch occur at ~1150 and ~1045 cm^{-1}.

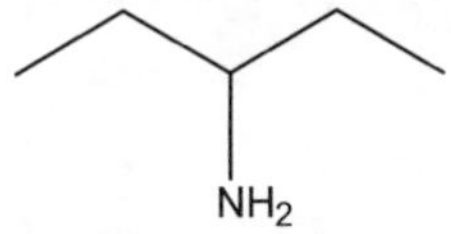

NH_2

32. *α-D-Glucose*: Water: The band at 1630 cm^{-1} indicates some water is present in the KBr disc preparation. Aliphatic group: CH_2 and CH str vibrations are observed between 2945 and 2892 cm^{-1}. The CH_2 deformation has a strong characteristic band at 160 cm^{-1}. Hydroxyl group: OH stretch at 3410, 3315 cm^{-1}. Note different hydrogen bonding environments. OH def at ca. 1345 cm^{-1}. Bands between 1150 and 996 cm^{-1} involve the C–O str of primary and secondary alcohols. Primary alcohols (–CH_2–OH) are mostly the lower frequency 1025–1000 cm^{-1} band cluster and the secondary alcohols (>CH–OH) are mostly the 1150, 1112 cm^{-1} bands. The cyclic ether has bands at ~ 1120–1110 involving the C–O–C o. ph. str. and at ~840 cm^{-1} from the C–O–C i. ph.str.

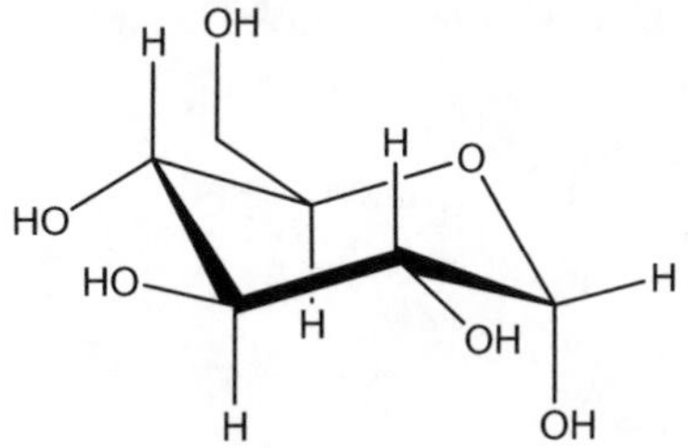

H OH H O HO HO H OH H H OH

33. *Butylated hydroxy toluene*: Note Nujol mull aliphatic interference. Aliphatic group: 2968, ~2876 cm^{-1} CH_3 str. 1472–1450 cm^{-1} CH_3 o.ph. def., *t*-butyl 1396, 1363 cm^{-1} $C(CH_3)_3$ i. ph. def., 1375 cm^{-1} CH_3 i. ph. def. Aromatic group: 3100–3000 cm^{-1} aromatic CH str., 1604–1590 cm^{-1} (IR/R) quadrant ring str., 1482, 1432 cm^{-1} (IR) semi-circle str., 866 cm^{-1} (IR) aryl lone H wag. Phenol hydroxyl group: 3626 (IR/R) OH str., 1246–1198 cm^{-1} (IR/R) involves aryl-O str.

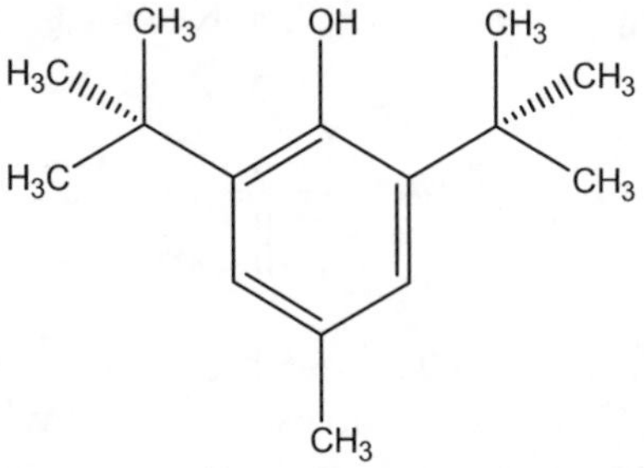

CH3 OH CH3
H3C CH3
H3C CH3
CH3

34. *Polystyrene*: Aliphatic group: 2923–2854 cm^{-1} CH_2 str. 1452 CH_2 def. Aromatic group: 3100–3000 cm^{-1} aromatic CH, 1603, 1584, 1493 cm^{-1} aromatic ring quad and semi-circle str., 1002 cm^{-1} (Raman) 2,4,6 C Radial i. ph.str., 754 cm^{-1} (IR) 5 adjacent aryl H, 695 cm^{-1} (IR) aromatic ring pucker.

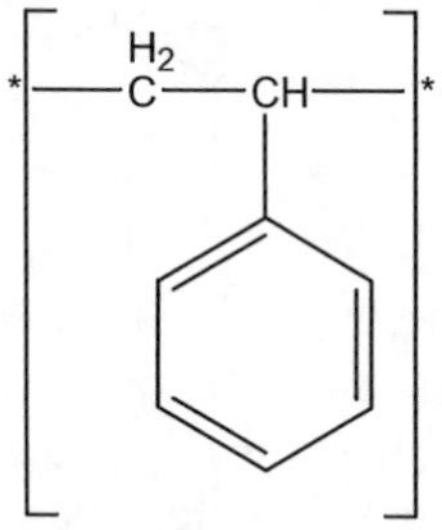

H2
* C CH *

35. *Di-n-butyl amine*: Aliphatic group: 2964/2958 cm^{-1} and 2873 cm^{-1} CH_3 str., 2928 cm^{-1} (IR) CH_2 o. ph. str., Note multiple bands in Raman from trans and gauche methylene groups. 1298 cm^{-1} (Raman) CH_2 twist of chain. 1464/1445 cm^{-1} involves methyl and methylene deformation. 1378 cm^{-1} CH_3 i. ph. def. Secondary amine group: 3287 cm^{-1} NH str., 2809 cm^{-1} CH_2 (N) i. ph. str., 1131 cm^{-1} C–N–C o. ph. str., 735 cm^{-1} NH wag.

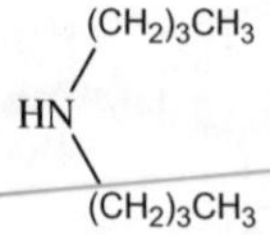

(CH2)3CH3
HN
(CH2)3CH3

36. *Benzalkonium chloride*: Aromatic group: 3100–3000 aromatic CH, ~1600, 1488 cm^{-1} aromatic ring quad and semi-circle str., 1002 cm^{-1} (Raman) 2,4,6 C Radial i. ph.str., 728 cm^{-1} (IR) 5 adjacent aryl H, 705 cm^{-1} (IR) aromatic ring pucker. Quaternary amine salt: (N^+) CH_3 has CH_3 o. ph. str. at 3002 cm^{-1}, the CH_3 i. ph. str. at 2852 cm^{-1}, the CH_3 o. ph. bend/def at 1488 cm^{-1}. The NC_4 i. ph. str. is expected between 800 and 950 cm^{-1} and may possibly be assigned to the band at 837 cm^{-1}.

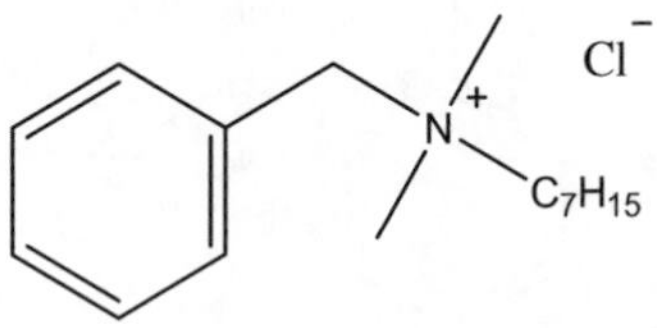

Cl^- N^+ C_7H_{15}

37. *n-Ethyl acetamide*: Aliphatic group: 2976, 2877 cm^{-1} CH_3 str., 2936 CH_2 str., 1452 cm^{-1} CH_3/CH_2 def., 1373 CH_3 i. ph. def. Amide group: 3290 cm^{-1} NH str., 1654 cm^{-1} C =O str., 1556 cm^{-1} CNH str./bend, 3086 cm^{-1} overtone of 1556 cm^{-1} band (2 × 1556), 1298 cm^{-1} CNH str./open, 721 cm^{-1} NH wag.

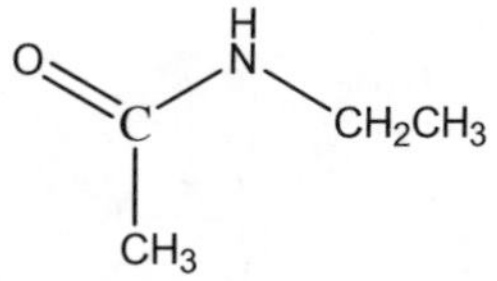

O C H N CH_2CH_3 CH_3

38. *Triphenyl phosphine oxide*: Aromatic group: 3100–3000 aromatic CH str., 1593, 1575, 1485/1439 cm^{-1} aromatic ring quad and semi-circle str., 1121 cm^{-1} involves P-aryl str., 1003 cm^{-1} (Raman) 2,4,6 C Radial i.ph.str., 755 cm^{-1} (IR) 5 adjacent aryl H, 697 cm^{-1} (IR) aromatic ring pucker. Phosphine oxide (P =O) group: 1191/1180 cm^{-1} (IR/Raman) P =O str.

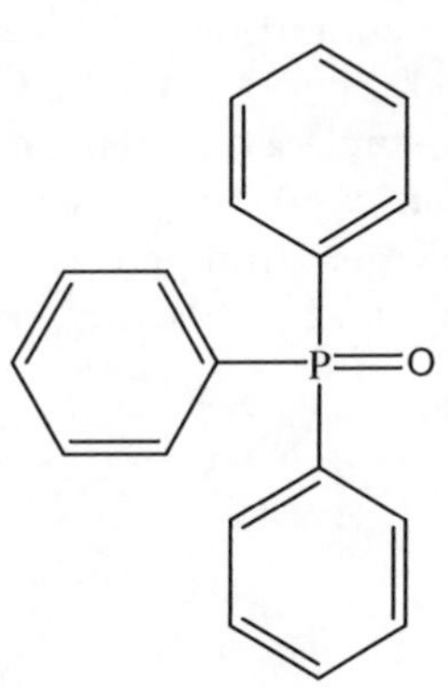

P=O

39. *4-Methyl-pyridine-3-boronic acid*: Boronic acid group: 3420–3200 cm^{-1} (broad) OH str., 1359 cm^{-1} involves B–O str., 1071 cm^{-1} B–OH def., 772 (B)OH wag. Pyridine group: 3080–3040 cm^{-1} aryl CH str., 1615, 1591, 1555 aryl quadrant str., 1480–1425 cm^{-1} involves aryl semi-circle str., 1091 cm^{-1} involves 2,4,6 C Radial i.ph.str., 1003 cm^{-1} involves the whole ring breath (str). 877 cm^{-1} lone H wag, 835 cm^{-1} 2 adj H wag.

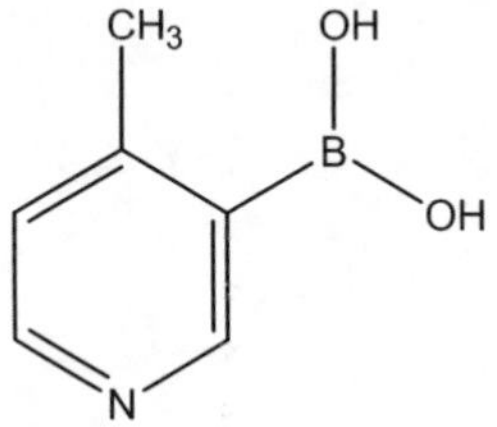

CH_3 OH B OH N

40. *Tributyl borate*: Aliphatic group: 2959 and 2874 cm^{-1} CH_3 str., 2935 CH_2 o. ph. str., 1485 cm^{-1} (O)CH_2 def., 1466–1458 cm^{-1} CH_3, CH_2 def. Borate ester group: 1337–1297 cm^{-1} involves the $B(O)_3$ o. ph. str., 832, 815 cm^{-1} (Raman) most likely involves the $B(O)_3$ i. ph. str., while the 666 cm^{-1} band (IR) is from the $B(O)_3$ out-of-plane deformation. 1407 cm^{-1} involves the OCH_2 wag + B–O str. Bands at 1073 and 1029 cm^{-1} involve mostly the C–O str.

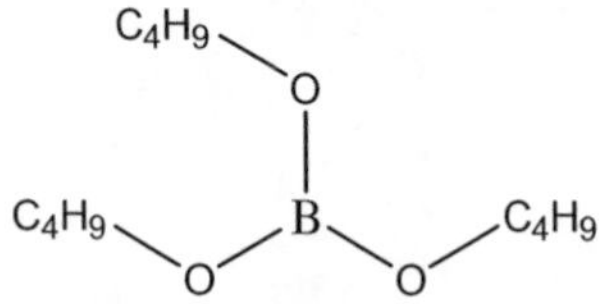

C_4H_9 O C_4H_9 B C_4H_9 O O

41. *Phenyl boroxole*: Phenyl boronic acid is a labile compound that in the solid state readily forms a Boron ether. The Boron ether when put in an acidic medium readily hydrolyzes to form the phenyl boronic acid. Raman spectrum indicates no aliphatic group is present (IR Nujol mull preparation interferes). No bands associated with the B–OH group are observed including the (B)OH str (~3250 cm^{-1}), the B–OH bend (~1000 cm^{-1}), and the B–OH wag (~800 cm^{-1}) (see interpreted IR/Raman spectra # 64 of *m*-tolylboronic acid). Aromatic group: 3100–3000 cm^{-1} aromatic CH str., 1603, 1574, 1494/1442 cm^{-1} aromatic ring quad and semi-circle str., 1180 cm^{-1} involves B-aryl str., 996 cm^{-1} (Raman) 2,4,6 C Radial i.ph.str., 760 cm^{-1} (IR) 5 adjacent aryl H, 703 cm^{-1} (IR) aromatic ring pucker. B–O linkage: Strong characteristic IR bands involving the B–O str. are observed between 1381 and 1310 cm^{-1} in both the IR (strong) and Raman (moderate).

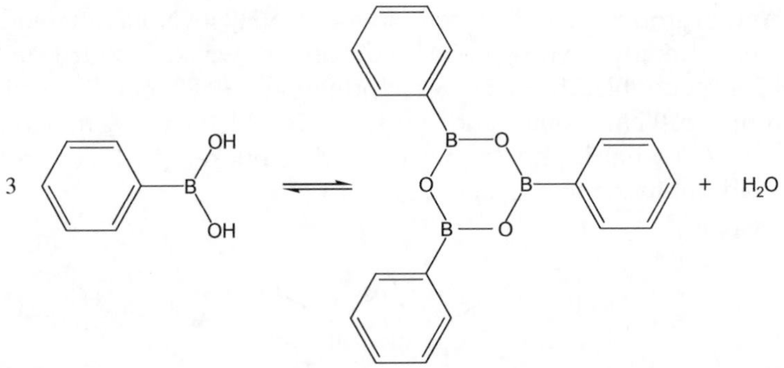

3 [phenylboronic acid] ⇌ [triphenylboroxine] + H_2O

42. *Azodicarbonamide*: NH_2 group: NH_2 o. ph. and i. ph. str at 3331 and 3175 cm^{-1}, respectively (IR and Raman). NH_2 bend/def at 1635 cm^{-1} (IR). NH_2 rk (Raman and IR) at 1118 cm^{-1} and the NH_2 wag at 750 cm^{-1} (IR) with an overtone 2 × NH_2 wag at 1477 cm^{-1}. Amide C =O group: C =O str. in IR at 1738/1726 cm^{-1}. The IR and Raman bands at 1367, 1337, and 1300 cm^{-1} involve the C–N str. Azo N =N group: Strong Raman band at 1573 cm^{-1} involves the N =N str.

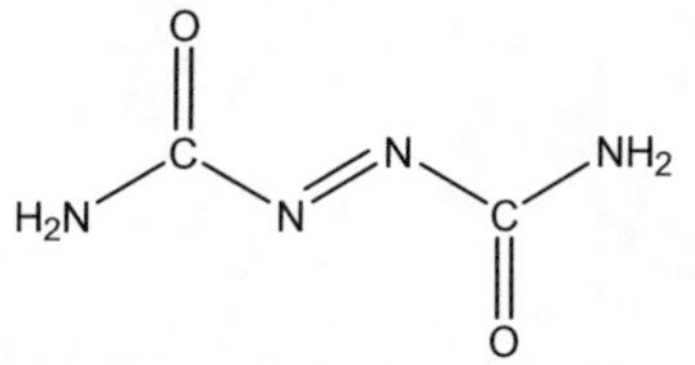

43. *4-Methyl-imidazole*: N–H group: This is characterized by complex hydrogen bonding involving N–H···N = due to π-type hydrogen bonding of the unsaturated N atom in the heterocyclic ring. Thus a broad envelope of NH str vibrations are observed from 3400 to2700 cm^{-1} in the IR spectrum. The NH wag occurs at 926 cm^{-1} and the broad band at 1824 cm^{-1} is due to the Fermi-resonance enhanced overtone of the NH wag. Imadazole ring: Ring =CH str. bands at 3120, 3091, and 3029 cm^{-1}. The out-of-phase C =C/C =N str. (Quadrant str.) results in bands between 1580 and 1509 cm^{-1} and the in-phase C =C/C=N str. (semi-circle str.) results in bands at ~1440 cm^{-1}. The band cluster at 1304, 1265 cm^{-1} may involve the imidazole ring in-phase breathing mode. Methyl group: Bands at 2921 and 2865 involve the CH_3 str. and the band at 1384 cm^{-1} involves the CH_3 i. ph. def.

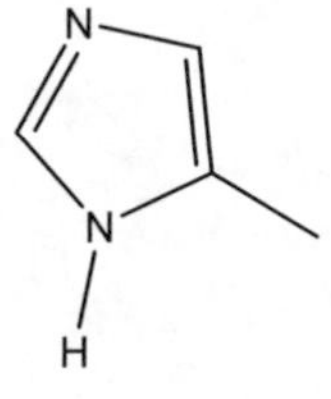

44. *Melamine*: Amino group: 3500–3100 cm^{-1} involves NH_2 str. Note differences in hydrogen bonding (i.e., strongly hydrogen-bonded versus weakly hydrogen-bonded NH). 1650–1620 cm^{-1} doublet from NH_2 deformation. ~582 cm^{-1} possibly NH_2 wag. Triazine group: 1550 cm^{-1} quadrant ring str., 1470–1430 cm^{-1} semi-circle ring str., 984 cm^{-1} (Raman) N-Radial i. ph. str., 811 cm^{-1} ring pucker (sextant out-of-plane bend), 675 cm^{-1} quadrant in-plane ring bend.

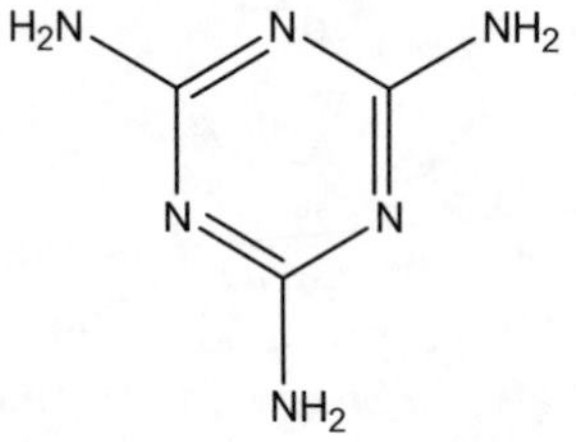

APPENDIX

IR Correlation Charts

Infrared group frequency correlation charts

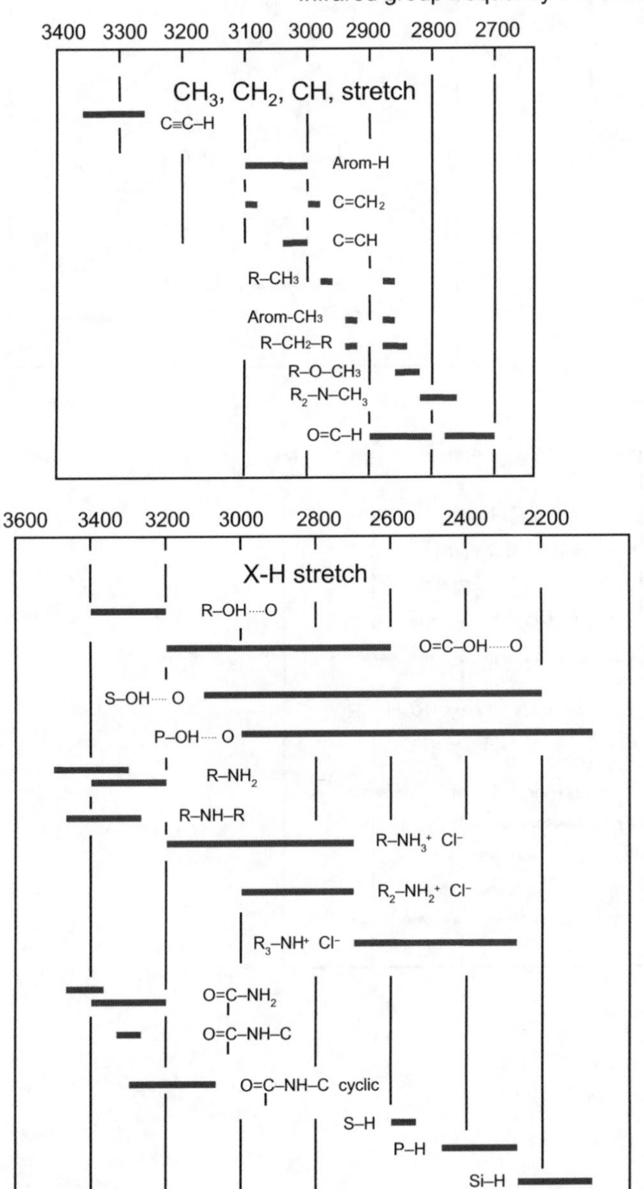

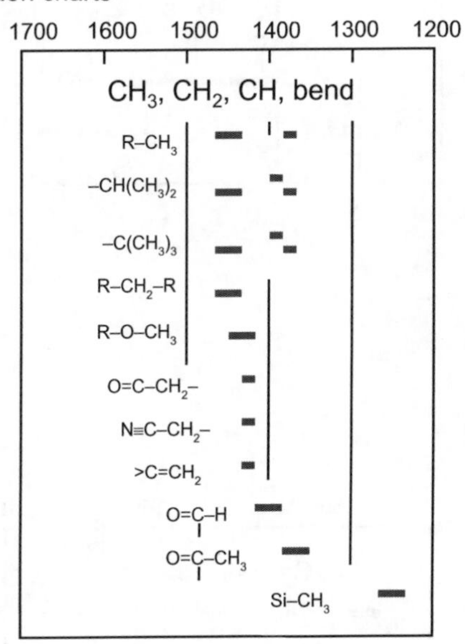

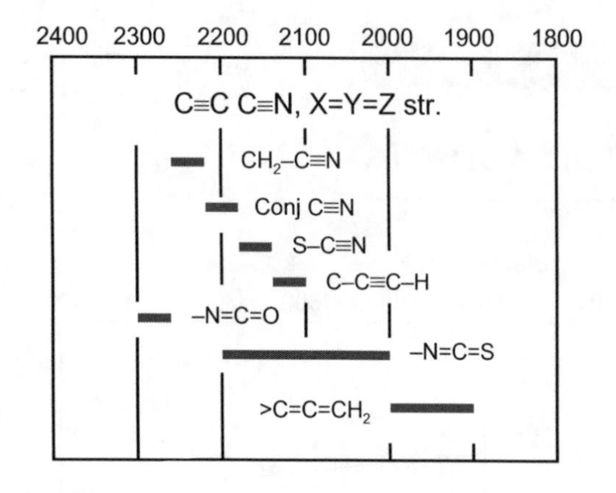
2400 2300 2200 2100 2000 1900 1800
C≡C C≡N, X=Y=Z str.
CH2–C≡N
Conj C≡N
S–C≡N
C–C≡C–H
–N=C=O
–N=C=S
>C=C=CH2

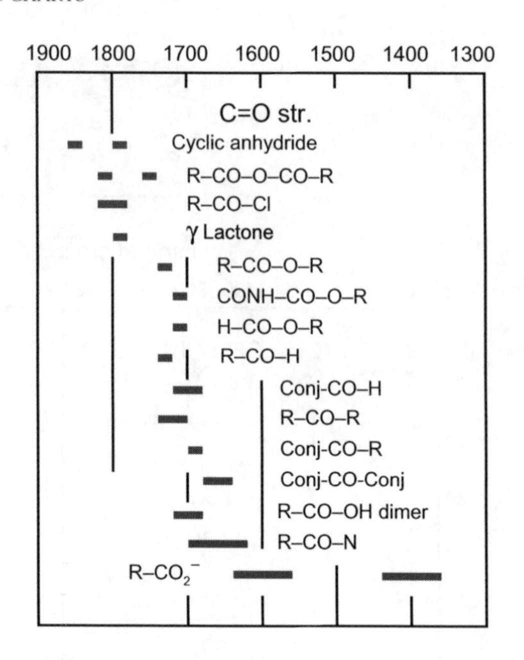
1900 1800 1700 1600 1500 1400 1300
C=O str.
Cyclic anhydride
R–CO–O–CO–R
R–CO–Cl
γ Lactone
R–CO–O–R
CONH–CO–O–R
H–CO–O–R
R–CO–H
Conj-CO–H
R–CO–R
Conj-CO–R
Conj-CO-Conj
R–CO–OH dimer
R–CO–N
R–CO2−

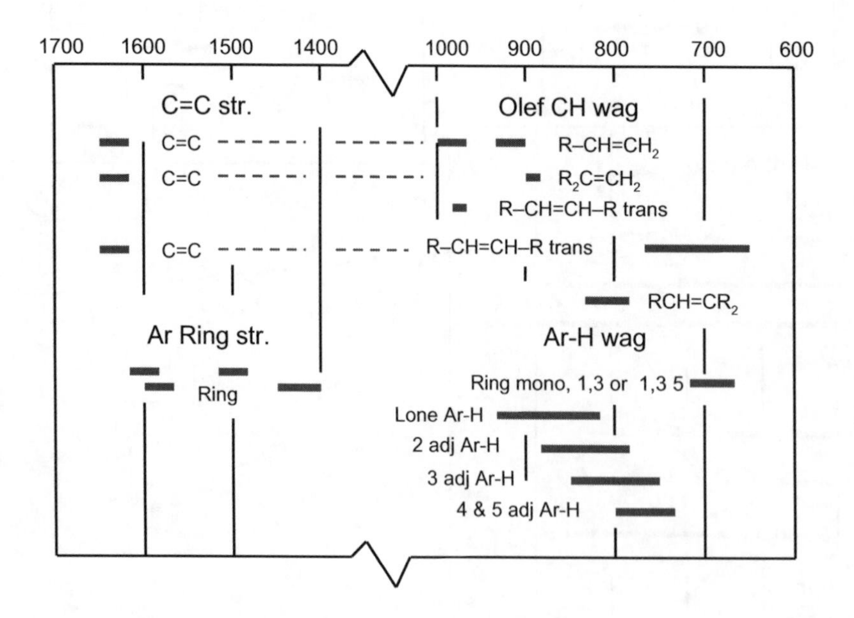
1700 1600 1500 1400 1000 900 800 700 600
C=C str.
Olef CH wag
C=C
R–CH=CH2
C=C
R2C=CH2
R–CH=CH–R trans
C=C
R–CH=CH–R trans
RCH=CR2
Ar Ring str.
Ar-H wag
Ring
Ring mono, 1,3 or 1,3 5
Lone Ar-H
2 adj Ar-H
3 adj Ar-H
4 & 5 adj Ar-H

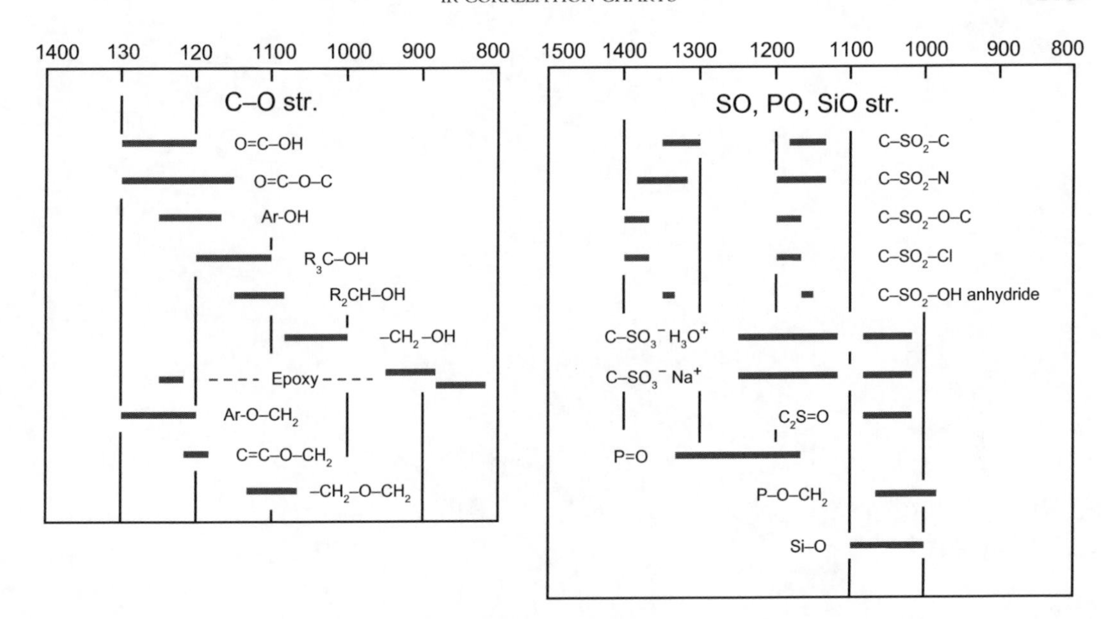
1400 130 120 1100 1000 900 800
C–O str.
O=C–OH
O=C–O–C
Ar-OH
R3C–OH
R2CH–OH
–CH2–OH
Epoxy
Ar-O–CH2
C=C–O–CH2
–CH2–O–CH2
1500 1400 1300 1200 1100 1000 900 800
SO, PO, SiO str.
C–SO2–C
C–SO2–N
C–SO2–O–C
C–SO2–Cl
C–SO2–OH anhydride
C–SO3− H3O+
C–SO3− Na+
C2S=O
P=O
P–O–CH2
Si–O

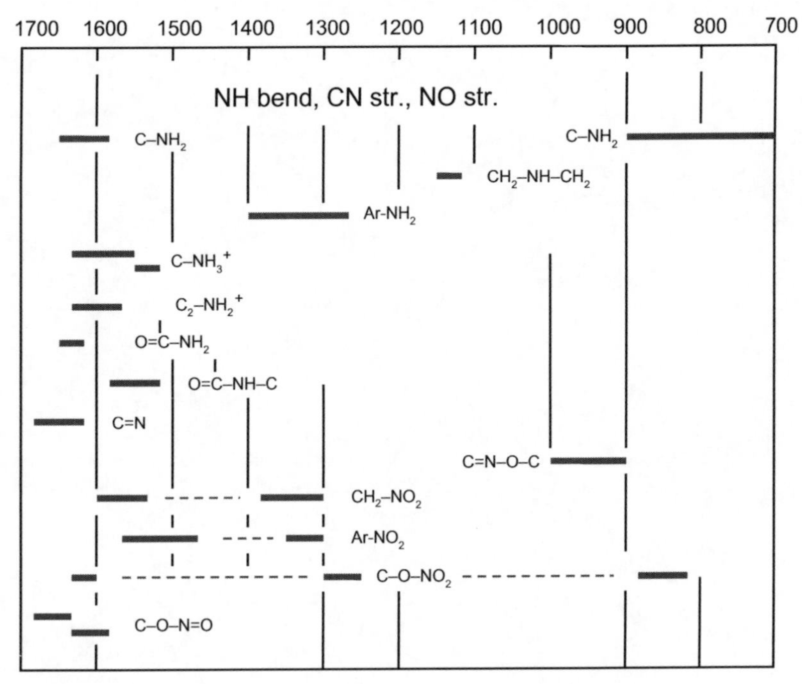
1700 1600 1500 1400 1300 1200 1100 1000 900 800 700
NH bend, CN str., NO str.
C–NH2
C–NH2
CH2–NH–CH2
Ar-NH2
C–NH3+
C2–NH2+
O=C–NH2
O=C–NH–C
C=N
C=N–O–C
CH2–NO2
Ar-NO2
C–O–NO2
C–O–N=O

Index

D

E

F

G

H

W

X

Z